T0344810

APPLICATION OF IC-MS AND IC-ICP-MS IN ENVIRONMENTAL RESEARCH

APPLICATION OF IC-MS AND IC-ICP-MS IN ENVIRONMENTAL RESEARCH

Edited by

RAJMUND MICHALSKI
Institute of Environmental Engineering of Polish Academy of Sciences, Poland

Library of Congress Cataloging-in-Publication Data:

Names: Michalski, Rajmund, editor of compilation.
Title: Application of IC-MS and IC-ICP-MS in environmental research / edited
    by Rajmund Michalski.
Description: Hoboken, New Jersey : John Wiley & Sons, 2016. | Includes
    bibliographical references and index.
Identifiers: LCCN 2016001708| ISBN 9781118862001 (cloth) | ISBN 9781119085478
    (epub)
Subjects: LCSH: Ion exchange chromatography. | Inductively coupled plasma
    mass spectrometry.
Classification: LCC QP519.9.I54 A77 2016 | DDC 543/.82–dc23
                LC record available at http://lccn.loc.gov/2016001708

Typeset in 10/12pt TimesLTStd by SPi Global, Chennai, India

Printed in the United States of America

10 9 8 7 6 5 4 3 2 1

CONTENTS

LIST OF CONTRIBUTORS

Maria Balcerzak; Department of Analytical Chemistry, Faculty of Chemistry, Warsaw University of Technology, Noakowskiego 3, 00-664 Warsaw, Poland

Klaus Fischer; Faculty VI – Regional and Environmental Sciences, Department of Analytical and Ecological Chemistry, University of Trier, Behringstr. 21, 54296 Trier, Germany

Wolfgang Frenzel; Technische Universität Berlin, Strasse des 17. Juni 135, 10623 Berlin, Germany

Jay Gandhi; Metrohm USA, 4738 Ten Sleep Lane, Friendswood, TX 77546, USA

Adam Konrad Jagielski; Department of Metabolic Regulation, Faculty of Biology, Institute of Biochemistry, University of Warsaw, Miecznikowa 1, 02-096 Warsaw, Poland

Koji Kosaka; Department of Environmental Health, National Institute of Public Health, 2-3-6 Minami, Wako, Saitama 351-0197, Japan

Jürgen Mattusch; Department of Analytical Chemistry, Helmholtz Centre for Environmental Research, Permoserstr. 15, 04318 Leipzig, Germany

Rajmund Michalski; Institute of Environmental Engineering, Polish Academy of Sciences, M. Skłodowskiej-Curie 34, 41-819 Zabrze, Poland

Michal Usarek; Department of Metabolic Regulation, Faculty of Biology, Institute of Biochemistry, University of Warsaw, Miecznikowa 1, 02-096 Warsaw, Poland

PREFACE

Environmental analytical chemistry can be regarded as the study of a series of factors that affect the distribution and interaction of elements and substances present in the environment, the ways they are transported and transferred, as well as their effects on biological systems. In recent years, the importance of monitoring and controlling environmental pollutants has become apparent in all parts of the world. As a result, analysts have intensified their efforts to identify and determine toxic substances in air, water, wastewaters, food, and other sectors of our environment. The toxicological data analyses involve constant lowering of analyte detection limits to extremely low concentration levels.

Speciation analysis, understood as research into various element forms, is gaining importance in environmental protection, biochemistry, geology, medicine, pharmacy, and food quality control. It is popular because what frequently determines the toxicological properties of a compound or element is not its total content, but in many cases, it is the presence of its various forms. Elements occurring in ionic forms are generally believed to be biologically and toxicologically interactive with living organisms. Studying low analytes concentrations, particularly in complex matrix samples, requires meticulous and sophisticated analytical methods and techniques. The latest trends embrace the hyphenated methods combining different separation and detection methods. In the range of ionic compounds, the most important separation technique is ion chromatography. Since its introduction in 1975, ion chromatography has been used in most areas of analytical chemistry and has become a versatile and powerful technique for the analysis of a vast number of inorganic and organic ions present in samples with different matrices. The main advantages of ion chromatography include the short time needed for analyses, possibility of analysis of small volume samples, high sensitivity and selectivity, and a possibility of simultaneous separation and determination of a few ions or ions of the same element at different degrees of oxidation. Mass spectrometry is the most popular detection method in speciation analysis, because it offers information on the quantitative and qualitative sample composition and helps to determine analytes structure and molar masses. The access to the structural data (necessary for the identification of the already known or newly found compounds) poses a challenge for speciation analysis as higher sensitivity of detection methods contributes to the increased number of detected element forms.

Couplings of ion chromatography with MS or ICP-MS detectors belong to the most popular and useful hyphenated methods to determine different ion forms of metals and metalloids ions (e.g., Cr(III)/Cr(VI), As(III)/As(V)), as well as others ions (e.g., bromate, perchlorate). IC-MS and IC-ICP-MS create unprecedented opportunities, and their main advantages include extremely low limits of detection and quantification, high precision, and repeatability of determinations.

The intent of this book is to introduce anyone interested in the field of ion chromatography, species analysis and hyphenated methods (IC-MS and IC-ICP-MS) the theory and practice. This book should be interesting and useful for analytical chemists engaged in environmental protection and research, with backgrounds in chemistry, biology, toxicology, and analytical chemistry in general. Moreover, employees of laboratories analyzing environmental samples and carrying out species analysis might find general procedures for sample preparation, chromatographic separation, and mass spectrometric analysis.

6 February 2016

RAJMUND MICHALSKI
Zabrze, Poland

1

PRINCIPLES AND APPLICATIONS OF ION CHROMATOGRAPHY

RAJMUND MICHALSKI

*Institute of Environmental Engineering, Polish Academy of Sciences,
M. Skłodowskiej-Curie 34, Zabrze 41-819, Poland*

1.1 PRINCIPLES OF ION CHROMATOGRAPHY

1.1.1 Introduction

The history of chromatography as a separation method began in 1903 when Mikhail Semyonovich Tsvet (a Russian biochemist working at the Department of Chemistry of the Warsaw University) separated plant dyes using adsorption in a column filled with calcium carbonate and other substances [1]. After extraction with the petroleum ether, he obtained clearly separated colorful zones. To describe this method, he used Greek words meaning *color* ($\rho\omega\mu\alpha$) and *writing* ($\gamma\rho\alpha\varphi\omega$) and coined a new word, *chromatography*, which literally meant *writing colors*. At present, chromatographic methods are among the most popular instrumental methods in the analytical chemistry as they offer quick separation and determination of substances, including complex matrix samples.

Chromatographic methods are used widely on both the preparative and analytical scales. They help to separate and determine polar and nonpolar components; acidic, neutral, and alkaline compounds; organic and inorganic substances; monomers, oligomers, and polymers. It is necessary to use an appropriate chromatography type, which depends on the physicochemical properties of the examined sample and its components. Gas chromatography (GC) and liquid chromatography (LC)

Application of IC-MS and IC-ICP-MS in Environmental Research, First Edition.
Edited by Rajmund Michalski.
© 2016 John Wiley & Sons, Inc. Published 2016 by John Wiley & Sons, Inc.

can be used to separate and determine approximately 20% and 80% of the known compounds, respectively. Ion chromatography (IC) is a part of high-performance liquid chromatography used to separate and determine anions and cations and also other substances after converting them into the ionic forms. In the literature, the term ion-exchange chromatography (I-EC) is found. It differs from ion chromatography even though both types are based on the widely known ion-exchange processes. Ion chromatography originates from ion-exchange chromatography. It uses high-performance analytical columns that are usually filled with homogenous particles with small diameters and most often conductometric detection. When compared to the classic ion-exchange chromatography, it is more efficient, faster, and more sensitive. It also offers very good repeatability of the obtained results. The ion-exchange chromatography term was used until 1975, when the first commercial ion chromatograph was available. At present, most analyses of ionic substances conducted with chromatographic techniques are performed with ion chromatography.

In the last 40 years, there were many state-of-the-art monographs that described the ion chromatography theory and applications in detail [2–5]. Some of these studies have already been republished. At present, there are three main separation methods in ion chromatography. They are based on different properties of substances used in the column phases and the resulting ion capacity. They include the following:

- Ion chromatography (IC) and can be either suppressed or nonsuppressed
- Ion exclusion chromatography (IEC)
- Ion pair chromatography (IPC).

The block diagram of an ion chromatograph (cation-exchange and anion-exchange types), together with ion-exchange reactions for the most popular suppressed ion chromatography, can be seen in Figure 1.1.

The anion separation proceeds according to the following principle: analyte ions (e.g., Cl^-) together with eluent ions pass through the analytical column in which the following ion-exchange reaction takes place:

$$\text{ion-exchange resin–}N^+HCO_3^- + Na^+ + Cl^-$$
$$\Longleftrightarrow \text{ion-exchange resin–}N^+Cl^- + Na^+ + HCO_3^-$$

The affinity of the analyte ions toward the stationary phase is diverse. Consequently, the ions are separated and leached out from the analytical column within different retention times against the background of weakly dissociated $NaHCO_3$. Afterward, they are transported into the suppressor with high-capacity sulfonic cation exchanger. The following reaction takes place:

$$\text{ion-exchange resin–}SO_3^-H^+ + Na^+ + HCO_3^-$$
$$\Longleftrightarrow \text{ion-exchange resin–}SO_3^-Na^+ + H_2CO_3$$

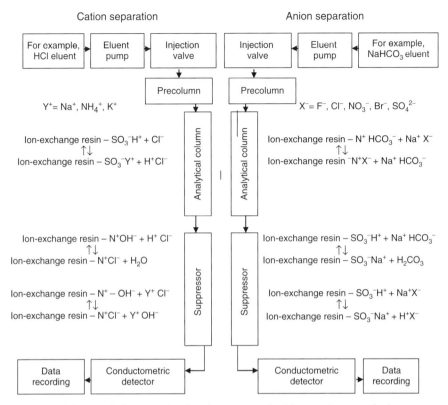

Figure 1.1 Block diagram of an ion chromatograph with a conductometric detector.

The $NaHCO_3$ eluent ions are transformed into weakly dissociated carbonic acid due to the occurring reactions. The analyte ions (e.g., Cl^-) react in accordance with the following formula:

$$\text{ion-exchange resin–}SO_3^-H^+ + Na^+ + Cl^-$$
$$\Longleftrightarrow \text{ion-exchange resin–}SO_3^-Na^+ + H^+ + Cl^-$$

Due to the reactions taking place in the analytical column and the suppressor, the analyte ions reach the detector in the form of strongly dissociated acids against the background of weakly dissociated carbonic acid. The obtained signal related to the conductivity of the analyte ions (the analyte forms a well-dissociated salt after the reactions) is high enough to use the conductometric detector to record the peaks of separated anions against the background of a weak signal related to the low eluent conductivity (forming weakly dissociated carbonic acid). Parallel reactions are observed when cations are determined. The cation-exchange column is filled with a cation exchanger with sulfonic groups. Eluent consists of water solution of, for example, hydrochloric acid. The analyte ions (e.g., Na^+) together with the eluent ions pass through the analytical column in which the following ion-exchange reaction takes place:

$$\text{ion-exchange resin--SO}_3^-\text{H}^+ + \text{Na}^+ + \text{Cl}^-$$

$$\Longleftrightarrow \text{ion-exchange resin--SO}_3^-\text{Na}^+ + \text{H}^+ + \text{Cl}^-$$

The affinity of the analyte ions toward the stationary phase is diverse. Consequently, the cations are separated and leached out from the analytical column within different retention times against the background of strongly dissociated HCl. Afterward, the ions are transported into the suppressor with high-capacity anion exchanger (e.g., with quaternary ammonium groups as functional groups). The following chemical reaction occurs:

$$\text{ion-exchange resin--N}^+\text{OH}^- + \text{H}^+ + \text{Cl}^- \Longleftrightarrow \text{ion-exchange resin--N}^+\text{Cl}^- + \text{H}_2\text{O}$$

The HCl eluent ions are transformed into water due to the reactions in the suppressor, whereas the analyte ions (Na^+) react with the exchanger in the suppression column according to the following formula:

$$\text{ion-exchange resin--N}^+\text{OH}^- + \text{Na}^+ + \text{Cl}^-$$

$$\Longleftrightarrow \text{ion-exchange resin--N}^+\text{Cl}^- + \text{Na}^+ + \text{OH}^-$$

Due to the chemical reactions in the analytical column and suppressor, the analyte ions reach the detector in the form of highly dissociated hydroxides against the water background, which allows analysis in the conductometric detector.

Ion exclusion chromatography (IEC) is a comparatively old technique, which uses the Gibbs–Donnan effect. A porous ionic-exchanger functions as a semipermeable membrane separating two water phases (mobile and stationary) contained in the exchanger pores. The membrane is only permeable for nonionized or weakly ionized substances. They are separated between two water phases, whereas their migration through the column is delayed. The ionized substances do not penetrate the inside of the pores. In I-EC cation exchanger, and occasionally anion exchanger, has generally been used for I-EC separations. They are not held in the column and leave it first. IEC is mainly used for separating weak inorganic acids, organic acids, alcohols, aldehydes, amino acids, and also for the group separation of ionic and nonionic substances [5, 6].

As an alternative to conventional ion chromatography, anions and cations can be separated on a standard reversed-phase column of the type used for HPLC. Several names have been applied for this type of separation, such as the following: ion-interaction chromatography, mobile-phase ion chromatography, and mostly IPC. In ion-pair chromatography, the X substance ions react with the lipophilic L ions (constituting the eluent component) and form the XL complex. The complex can be bound to the nonpolar surface of the stationary S phase in a reversible way. The S phase makes a reversible phase, as its polarity is lower than that of the eluent. It forms the XLS complex. The separated sample ions (XL complexes) have different retention times in the column. The retention times result from different affinities that the ions have toward the nonpolar stationary-phase surface, which causes the separation. According to the alternative model, the lipophilic eluent ions are adsorbed on the stationary-phase surface and form the LS complex. As a result, an ion exchanger forms on the nonpolar stationary-phase surface. The ions of the solved X substance react with this exchanger. The hydrophobic ions (e.g., alkyl and aryl sulfonates) may penetrate the inside of the layer formed by the LS complex. Their retention time is decided by the adsorption phenomenon. More hydrophilic ions penetrate only the external zone. Their retention time in the column is decided by the ion-exchange mechanism. This chromatography type is mainly used to determine ions such as sulfates, sulfonates, alkaloids, barbiturates, fatty acid derivatives, and selected metal ion complexes [7]. Besides these three main types, reversed-phase liquid chromatography (RPLC) [8] and hydrophilic interaction liquid chromatography (HILIC) [9] can also be used to separate selected ions.

Ion determination methods used before 1975 (gravimetric, titration, spectrometric, electrolytic, and other methods) were inexpensive and easily available; however, they were also time-consuming and required large amounts of expensive and (frequently) toxic reagents. On the other hand, the chromatographic methods used at that time were mostly applied for separation and determination of organic compounds.

The chromatographic method applications for separating metal ions were intensively investigated during WWII, when the atomic bomb was constructed (Manhattan Project). Nonetheless, the real breakthrough took place in 1971 when Small and his colleagues from Dow Physical Research Laboratory (Midland, MI) examined and proposed a chromatographic method for determination of lithium, sodium, and potassium with ion exchange and conductometric detection. Generally, ion-exchange chromatography was a preparative method because determination of ions present in the sample against the mobile-phase background caused serious difficulties. The key issues were to use proper stationary phases in the columns and to elaborate the mechanism of the eluent conductivity reduction so that the separated ions could be determined with a conductometric detector.

At the beginning of the ion chromatography development, eluents used to separate anions were alkaline (e.g., water solutions of NaOH/KOH, Na_2CO_3/$NaHCO_3$, phenolates). The eluents used to separate cations were usually sulfuric acid or methanesulfonic acid water solutions. At first, an additional column (suppression column) was used to reduce the eluent conductivity (suppression). It was placed between the analytical column and the detector. Due to the reactions occurring in it, the eluent ions formed low-conductivity products, such as H_2O (when water solutions of hydroxides were used as eluents) or H_2CO_3 (carbonate eluents). In the mid-1970s, the method was developed enough by Dow Chemicals to sell the conductometric suppression license to Durrum Chemical, which soon changed its name to Dionex for commercial reasons [10]. The company presented the first commercial ion chromatograph (Dionex, Model 10) during the American Chemical Society meeting in September 1975. The year 1975 is the official starting point of ion chromatography. The study by Small *et al.* [11] was a milestone in its development. At the turn of the 1970s and 1980s, Gjerde *et al.* [12] were the first to use the ion chromatography system without the suppression column and to apply the eluents with very low conductivity values. In this way, they created a new type of ion chromatography, that is, nonsuppressed ion chromatography. The most important events preceding the invention of ion chromatography and the stages of its development and popularization are given in Table 1.1.

The most important ion chromatography advantages are as follows:

1. Simultaneous determination of several ions in a short time (<10 min)
2. A small sample volume necessary for analyses (<0.5 ml)
3. Using different detectors
4. Simple sample preparation for analyses (liquid samples with simple matrices)
5. Simultaneous separation of anions and cations, or organic and inorganic ions
6. High separation selectivity (>1:10,000)
7. Determination of ions of the same element at different oxidation states (speciation analysis)
8. Safety and low exploitation costs (green chemistry).

The aforementioned advantages helped to elaborate a number of standardized methodologies soon after ion chromatography was invented. They mainly served to determine anions and cations in water and wastewater [19]. In the 1990s, the International Standard Organization (ISO) introduced numerous standards concerning determination of anions and cations in water and wastewater, as well as solid and gaseous samples. The list of standards is given in Table 1.2.

1.1.2 Stationary Phases

The stationary phases used in ion chromatography are also known as ion exchangers. They are macromolecular solids with cross-linked space structure. They are insoluble in water or other solvents and are able to exchange ions with the solution. The stationary phases can be divided into anion exchangers (with functional groups containing cationic counterions able to exchange anions present in the solution) and cation exchangers (with functional groups containing anionic counterions able to exchange cations present in the solution). There are also amphoteric ion exchangers that can exchange anions or cations, depending on the solution pH, and bipolar/zwitterionic ion exchangers that are able to exchange both ion types.

Ion exchangers contain functional groups whose charge is defined. Appropriate counterions, located close to the functional groups, have a counter change. Together, they constitute an electrically neutral unit. The active functional groups in cation exchangers are acidic. When they dissociate with an ion part (e.g., H^+), they can exchange this ion for other cations from the solution. The following functional groups are most often found in the active centers: sulfonic ($-SO_3H$), carboxylic ($-COOH$), iminodiacetic ($-N\{CH_2COOH\}_2$), phenolic ($-C_6H_4OH$), phosphonic ($-PO_3H_2$), and phosphinic ($-PO_2H$) ones. In the anion exchangers, the functional groups are

TABLE 1.1 Key Events Preceding the Ion Chromatography Invention and Stages of Ion Chromatography Development and Popularization

Year	Event
1903	Tsvet's discovery of chromatography as a separation method
1935	Synthesis and usage of the first sulfonated and aminated polymers (Adams' and Holmes' studies) [13]
1947	First applications of aminated resins in the form of polystyrene–divinylbenzene (PS/DVB) copolymers as ion exchangers
1952	Martin and Synge, two British scientists, were awarded the Nobel Prize in chemistry for the invention of partition chromatography, which still constitutes the theoretical basis for modern chromatographic methods
1953	Wheaton and Bauman elaborated theoretical principles for ion exclusion chromatography [14]
1971	Small elaborated a chromatographic method for separating lithium, sodium, and potassium with ion exchange and conductometric detection [15]
1975	Small et al. published their essential study on determination of anions and cations with ion chromatography with conductometric detection [11]
1975	Presentation of the first commercial ion chromatograph during the American Chemical Society meeting. Ion chromatography constituted as a new analytical method
1976–1980	Invention of ion-pair chromatography (Waters', Bidlingmeyer's, and Horvat's studies) [16]
1980	Invention of nonsuppressed ion chromatography [12]
1981	Introduction of commercial ion chromatographs in which all parts in contact with eluents were constructed without any metal parts
1981	Replacing suppression columns with membrane suppressors
1983	First studies on determination of carbohydrates and alcohols with ion chromatography with pulsed amperometric detection (PAmD)
1984	Introduction of the first computer software to operate ion chromatographs and interpret the obtained results
1984	The US Environmental Protection Agency (USEPA) and the American Society for Testing and Materials (ASTM) acknowledged ion chromatography as a reference method for ion determination in water and wastewater
1991	Introduction of electrochemical eluent generators [17]
1991	Introduction of a new suppressor generation – micromembrane suppressor (MMS)
1993	Introduction of commercial self-regenerated suppressors (SRS)
1994	Studies by Lamb et al. [18] on using macrocyclic particles in ion chromatography
1997	Invention of suppression and regeneration hybrid systems (Ion Reflux and Eluent Recycling)
1992–2015	Introduction of standards using ion chromatography for determination of ions in water and wastewater and researching gaseous air pollutants by the International Standard Organization (ISO) and other national standard organizations
Presently	Introduction of new stationary phase types, sample preparation methods, new detection modes, apparatus miniaturization, and hyphenated methods (IC-MS, IC-ICP-MS)

TABLE 1.2 ISO Standards Based on Ion Chromatography

Number	Date of Publication	Standard Title
Liquid samples		
10304-1	1992	Water Quality – Determination of dissolved fluoride, chloride, nitrite, orthophosphate, bromide, nitrate and sulfate ions using liquid chromatography of ions Part 1: Method for water with low contamination
10304-2	1997	Water Quality – Determination of dissolved anions with liquid chromatography of ions Part 2: Determination of bromide, chloride, nitrate, nitrite, orthophosphate and sulfate in wastewater
10304-3	1997	Water Quality – Determination of dissolved anions with liquid chromatography of ions Part 3: Determination of chromate, iodide, sulfite, thiocyanate and thiosulfate
10304-4	1998	Water Quality – Determination of dissolved anions with liquid chromatography of ions Part 4: Determination of chlorate, chloride and chlorite in water with low contamination
10304-1	2009	Water Quality – Determination of dissolved anions by liquid chromatography of ions Part 1: Determination of bromide, chloride, fluoride, nitrate, nitrite, phosphate and sulfate
15061	2001	Water Quality – Determination of dissolved bromate. Method by liquid chromatography of ions
11206	2011	Water quality – Determination of dissolved bromate. Method using ion chromatography (IC) and post column reaction (PCR)
14911	1998	Water Quality – Determination of dissolved Li^+, Na^+, NH_4^+, K^+, Mn^{2+}, Ca^{2+}, Mg^{2+}, Sr^{2+} and Ba^{2+} using ion chromatography method
Gaseous samples		
1911-1,2,3	1995	Air quality – Stationary source emission – Manual method of determination of HCl Part 1: Sampling and pretreatment of gaseous samples Part 2: Gaseous compounds absorption Part 3: Absorption solutions analysis and calculation

8

11632	1998	Stationary source emissions – Determination of mass concentration of sulfur dioxide – Ion chromatography method
21438-1	2007	Workplace atmospheres – Determination of inorganic acids with ion chromatography Part 1: Nonvolatile acids (sulfuric acid and phosphoric acid)
21438-2	2007	Workplace atmospheres – Determination of inorganic acids with ion chromatography Part 2: Volatile acids, except hydrofluoric acid (hydrochloric acid, hydrobromic acid, and nitric acid)
17091	2013	Workplace air – Determination of lithium hydroxide, sodium hydroxide, potassium hydroxide and calcium dihydroxide. Method of measurement of corresponding cations with suppressed ion chromatography
16740	2005	Workplace air – Determination of hexavalent chromium in airborne particulate matter Method with ion chromatography and spectrophotometric measurement using diphenyl carbazide
264125	2010	Ambient air quality – Guide for the measurement of anions and cations in PM 2.5
Solid samples		
13368-1	2001	Determination of chelating agents with ion chromatography method Part 1: EDTA, HEDTA, and DTPA
13368-2	2001	Determination of chelating agents with ion chromatography method Part 2: EDDHA and EDDHMA
14720	2013	Testing of ceramic raw and basic materials – Determination of sulfur in powders and granules of non-oxidic ceramic raw and basic materials Part 2: Inductively coupled plasma optical emission spectrometry (ICP/OES) or ion chromatography after burning in an oxygen flow

alkaline. They normally include quaternary ammonium groups ($-NR_3^+$) and tertiary and secondary protonated amines ($-NR_2H^+$, $-NRH_2^+$) or sulfonic groups ($-SR_2^+$).

A large number of functional groups in the exchanger increases the affinity of the ion-exchange resin toward the electrolyte solution, which may cause its dissolution. To prevent it, the exchanger core is formed as linear polymers that are cross-linked spatially. Such ion-exchange resins are insoluble in water, but they may expand in the solution when they absorb the solvent. The more developed the cross-linked structure is, the lower the influence is. It also depends on the ionic strength and pH of the solution, temperature, and the exchanged ion types. Most of the ion-exchanger properties are decided by the ion-exchange resin matrix (core) construction, number and type of the functional groups, and the degree to which it is cross-linked. For the separation quality, the following parameters are the most important: exchange capacity, selectivity degree, and chromatographic column performance.

Both the height of the ion-exchange resin and the column length are among the most important factors deciding on the peak resolution. If large amounts of ions should be separated and determined in a quantitative way, it is better to use a shorter column rather than increase the eluent concentration. A highly concentrated eluent can quickly overload the suppressor. Moreover, when the column is shorter, the pressure at the column outlet is lower. Consequently, a higher flow rate can be applied. For highly polarized ions that elute late in the form of broad peaks, it may be necessary to replace the used column with a shorter one. An opposite situation is encountered for the components that elute very quickly (e.g., low-molecular-weight organic acids). In such a case, good results are obtained when longer columns (sometimes, two columns connected in series) are used. Due to hydration, anions and cations with a large ionic radius can be eluted more quickly from the column filled with an ion-exchange resin cross-linked to a lower degree.

The order in which the selected anions (separated on the typically highly alkaline anion exchangers) are leached out is as follows: $OH^- < F^- < ClO_2^- < BrO_3^-$ $< HCOO^- < IO_3^- < CH_3COO^- < H_2PO_4^- < HCO_3^- < Cl^- < CN^- < NO_2^- < Br^-$ $< NO_3^- < HPO_4^{2-} < SO_3^{2-} < SO_4^{2-} < C_2O_4^{2-} < CrO_4^{2-} < MoO_4^{2-} < WO_4^{2-} < S_2O_3^{2-} < I^- < SCN^- < ClO_4^- <$ salicylates $<$ citrates. The order for the selected cations separated on the highly acidic cation exchangers is as follows: $Li^+ < H^+ < Na^+$ $< NH_4^+ < K^+ < Rb^+ < Cs^+ < Ag^+ < Tl^+ \ll UO_2^{2+} < Mg^{2+} < Zn^{2+} < Co^{2+} < Cu^{2+}$ $< Cd^{2+} < Ni^{2+} < Ca^{2+} < Sr^{2+} < Pb^{2+} < Ba^{2+} \ll Al^{3+} < Sc^{3+} < Y^{3+} < Eu^{3+} < Pr^{3+}$ $< Ce^{3+} < La^{3+} \ll Pu^{4+}$. The presented correlations help to conclude that trivalent ions are more strongly bound to the ion-exchange resin than divalent or monovalent ones. The ion-exchange resins hold ions whose valence is the same but the radius is larger in a stronger way than the ions with a small radius.

The ion-exchanger applications in the chemical analysis were overviewed in the studies by Walton and Rocklin [20] and Quershi and Varshney [21]. The stationary phases used in ion chromatography were described in the paper of Weiss and Jensen [22]. The anion exchangers can be divided into the following categories: organic polymers (including polystyrene/divinylbenzene (PS/DVB) copolymers, ethylene-divinylbenzene and divinylbenzene (EVB/DVB) copolymers, and polymethacrylate and polyvinyl polymers); exchangers in the agglomerate form; exchangers based on silica; and other types (e.g., crown ethers, cryptand phases).

Organic copolymers (i.e., polystyrene/divinylbenzene (PS/DVB) copolymers, ethylenedivinylbenzene and divinylbenzene (EVB/DVB) copolymers) and poly-methacrylate and polyvinyl copolymers are the most often used exchangers in ion chromatography. There are many commercially available anion-exchange columns filled with exchangers based on the PS/DVB copolymer with quaternary ammonium groups on the surface (as functional groups). One of the disadvantages that such anion exchangers had was the difficulty with the appropriate separation of fluoride ions against the background of the so-called water peak and the micromolecular anions of carboxylic acids. Dionex researched this problem. As a result, the company used anion exchangers with ethylenedivinylbenzene instead of polystyrene. Such anion exchangers were obtained due to the so-called grafting technology, which was applied for the first time by Schomburg [23] in the 1980s.

The agglomerate ion exchangers contain particles built of the spherical core that is surrounded by small molecules of the polymer containing functional groups. This particular type of ion exchanger is known as the latex-agglomerated anion exchanger. They are built from the PS/DVB core. On the surface, it is covered with sulfonic groups and completely aminated particles (diameters of 5–25 μm) made of polyvinyl chloride or polymethacrylate (generally known as latex). The latex particle diameters are approximately 0.1 μm. They are kept on the exchanger surface due to the electrostatic reactions and van der Waals forces. The properties of the exchanger can be modified as needed through changing the number of the latex particles bound to the basic exchanger and the degree to which they are cross-linked. In 1987, the gradient elution was introduced in ion chromatography. Since then, there have been commercially available latex exchangers adapted for the hydroxide-selective latex-based anion exchangers [24]. Latex columns with agglomerate exchangers have high mechanical and chemical resistance, high exchange capacity, and very good separation parameters. The exchanger core is completely sulfonized on the surface, which prevents the diffusion of inorganic ions into the inside of the stationary phase in the Donnan dialysis process. The diffusion processes mainly take place in the latex layer. The size and number of latex groups decide on their properties.

Methacrylate polymers are another type of exchangers used in ion chromatography. They are not as hard and firm as the PS/DVB copolymers. That is why they cannot be used in the columns working under high pressures. On the other hand, they are less hydrophobic and are better to separate large particles. The stationary phases based on silica, commonly used in high-performance liquid chromatography, are applied less often in ion chromatography as eluents with different pH are used in it. Silica itself can be used within a limited scope (pH = 2–8). When this pH range is exceeded, the silane chains are easily destroyed and the exchanger mechanical stability quickly decreases [25]. Nonetheless, the silica exchangers have better mechanical resistance than the organic ones. Similarly to silica, Al_2O_3 (common in HPLC) does not have any significant applications in ion chromatography due to the low exchange capacity and insufficient resistance to strong acids and bases. Organic and inorganic ions can be also separated on the cross-linked organic copolymers modified with the crown ethers or polycyclic cryptands. Lamb *et al.* [18] carried out the first studies on this topic. The most important advantage of the exchangers used in the gradient elution is the simultaneous elution of the polarized and nonpolarized ions.

Columns filled with traditional ion exchangers have relatively high resistances of the mobile-phase flow, which requires applying high flow rates. The maximum flow rates used in ion chromatography do not exceed $5\,ml\,min^{-1}$ and the maximum pressure is $35\,MPa$. In the monolithic columns, the obtained flow rate resistances are much lower. Their phases are made of one homogenous porous rod. The commercially available monolithic columns have diameters of a few millimeters and are filled with the PS/DVB copolymers, polymethacrylate, or silica [26]. The eluent flow rates in the monolithic columns can be up to $10\,ml\,min^{-1}$. Consequently, the analyte ions are separated within approximately 1 min. Under classic conditions, separation takes approximately 20 min. The monolithic columns are used in many ion chromatography varieties, such as low- and medium-pressure ion chromatography, ultrafast ion chromatography, capillary ion chromatography, gradient ion chromatography, or multidimensional ion chromatography [27].

Similarly to anion exchangers, the following stationary phases are used to separate organic and inorganic cations: polystyrene/divinylbenzene (PS/DVB) copolymers; ethylenedivinylbenzene and divinylbenzene (EVB/DVB) copolymers; polymethacrylate and polyvinyl polymers; cation exchangers in the agglomerate form; cation exchangers based on silica; and other cation exchangers.

Cation exchangers based on the polystyrene/divinylbenzene (PS/DVB) copolymers are commonly used in cation-exchange chromatography as the basic material to produce modified stationary phases. They are completely sulfonated on the surface, and the number of the sulfonate groups decides on their exchange capacity. The main disadvantage of the sulfonated cation exchangers based on PS/DVB is that they do not offer a possibility for the simultaneous separation of the alkaline metal and alkaline earth metal ions. The process is possible only after carboxylic groups are introduced as functional groups into cation exchangers. For nonsuppressed ion chromatography, ethylenediamine with tartaric acid (also constituting the eluent components) can be used as the complexation agent for divalent metal ions. The other option is to use a water solution of HCl and 2,3-diaminopropionic acid. In the exchangers based on EVB/DVB, a thin layer of the ion-exchange copolymer with carboxylic functional groups is bound to the copolymer particles. The main advantage is the possibility of using water acid solutions (e.g., H_2SO_4 or methanesulfonic acid – MSA) for simultaneous separation and determination of alkaline metals and alkaline earth metals.

The agglomerate cation exchangers were introduced commercially 10 years later than the latex anion exchangers, that is, in the mid-1980s. Such a delay was caused by the search for appropriate ion-exchange resins that could function as the base substrate. They contain a weakly sulfonated PS/DVB copolymer and completely aminated latex particles bound to the layer surface, on which they are held due to the electrostatic reactions and van der Waals forces. The discussed layer is covered with another completely sulfonated latex layer.

The zwitterionic ion exchangers [28] also offer interesting possibilities for ion chromatography. The presence of the negatively and positively charged groups in the exchangers results in their unique properties. When compared to traditional anion exchangers and cation exchangers, they are more mechanically stable and resistant to the expansion and drying processes. Their main advantage is the simultaneous separation of anions and cations. Even though organic exchangers are most often used in cation-exchange chromatography, silica-based exchangers are also applied. Metrohm company offers columns filled with silica modified with polybutadiene maleic acid (Metrosep Cation 1-2), whereas Alltech sells the Universal Cation columns. Such columns provide fast elution (several minutes) of the barium ions, which normally require much longer retention time. When columns contain exchangers modified with crown ethers, the order in which cations are leached out is different than in the traditional columns. Such a situation is mainly caused by the spherical reactions in them. The complexation reagent acts as a ligand, whereas the analyte cation is surrounded like a metal ion. The addition of organic solvents accelerates leaching out of some ions and improves the shape of their peaks. Importantly, columns with phases containing carboxylic functional groups cannot be used together with organic solvents. The temperature increase shortens the retention times for monovalent cations. The larger the ion radius is, the bigger the influence is. The detailed information on the ion-exchange column parameters must always be checked either in the application note or at the manufacturer's website. If the guidelines are not followed (e.g., the permissible eluent pH and addition of organic solvents), the column characteristics may change and the analytical result quality may suffer.

1.1.3 Eluents

The stationary-phase type in the analytical column and the detection method constitute the first two most important aspects affecting the separation quality. Eluents are the third factor. Parameters such as their type, concentration, pH, flow rate, and possible addition of an organic solvent must always be taken into consideration. The eluent components cannot react either with one another or with the chromatograph parts with which they are in direct contact. In ion chromatography, the eluent can be pumped through the column in either the isocratic or the gradient mode. The mobile-phase composition is stable during the whole chromatographic process in the isocratic elution. In the gradient elution, its concentration changes during separation and the strength of the elution mobile-phase increases [29]. The eluent molecules compete with the analyte molecules for the space on the stationary-phase surface. For that reason, the elution strength of a given mobile phase should be appropriately selected for a specific analysis; the stationary-phase and analyzed sample types must be taken into account.

The following principles must be remembered about when the eluent is chosen. They decide on the appropriate separation of the determined ions.

1. When the sample ion valence increases, its affinity for the functional group also rises, for example, the trivalent ions are bound in a stronger way than the divalent ions.

2. For different ions with the same valence, the bigger the ionic radius and the polarization degree are, the stronger the way in which the ion is held on the ion-exchange resin.

3. The sample ions with strong hydrophobic reactions or van der Waals forces with the matrix are leached out before the ions whose reactions are weaker.

The eluent pH also affects the separation of the polyvalent anions and cations (including organic acids and amines). For example, phosphates occur as monovalent ions in the solution with low pH; nonetheless, when pH increases, they form di- or trivalent ions. For this reason, the relative position of the phosphate ion peak in the chromatogram can change. The multifunction organic acids behave in the same way. Their retention times increase when the eluent pH is higher [30]. The eluent concentration and its flow rate in the column directly affect the separation speed. When other parameters do not change, the double increase in the flow rate shortens the retention time twice. However, it must be taken into consideration that the flow rate change influences the peak resolution and results in the pressure increase in the system.

As conductometric detectors are most often used, the eluents applied in ion chromatography can be divided into the eluents for analyses with conductometric detectors and those to be used with other detectors. They can also be classified as single- and multicomponent eluents; eluents with stable concentrations (isocratic system); eluents with changing concentrations (gradient system); and eluents modified with organic solvents.

Water solutions of sodium carbonate and sodium bicarbonate or their mixtures are the most often used eluents to separate anions with ion chromatography in the isocratic system. Water solutions of NaOH or KOH are most often applied in the gradient system. Due to the higher elution strength, water solutions of Na_2CO_3, $NaHCO_3$, or their mixtures, can be used at lower concentrations. They are safer to use than hydroxides. The carbonate/bicarbonate eluent is normally used with the molar ratio between approximately 1:1 and 3:1. The eluent component concentrations should be selected so that all the determined substances could be dissociated, that is, the eluent pH should be higher than the pK_a of the most weakly dissociating sample component.

The eluents used in nonsuppressed ion chromatography should have low conductivity. In this respect, they should differ as much as possible from the determined ions. For separation, water solutions of the organic acids are normally used (benzoic, phthalic, sulfanilic, nicotinic, citric, salicylic, tartaric, or maleic acid). In his historic article from 1975, Small *et al.* [11] stated that the eluents containing the hydronium ion (e.g., NaOH, KOH) seemed to be perfect for ion chromatography. They formed water whose conductivity was even lower than that of H_2CO_3 (the suppression product of the most often used carbonate/bicarbonate eluent) due to the suppression reaction occurring in the suppression column. Nonetheless, they had a few disadvantages that limited their usage. First of all, KOH and NaOH were weak eluents. For that reason, their concentrations had to be really high to elute all the separated ions from the column within reasonable time. As a result, the ion-exchange resins used in the suppression columns were quickly overloaded. It could also cause faster wear-off of the membrane suppressor applied later. Moreover, the hydroxide eluents could easily absorb carbon dioxide from the atmosphere, which also reduced their elution strength. Due to the progress in the new column phases and suppressor technology, water solutions of NaOH or KOH were once again used as eluents in the 1990s. They were often applied with automatic eluent generators or gradient elution.

The organic modifier addition (e.g., acetone, acetonitrile, or methanol) causes a slight reduction in retention times for monovalent ions and a significant reduction in retention times for divalent ions. The biggest influence is observed for the separation of the lipophilic ions, such as thiocyanides, perchlorate, iodides, or citrates. Not only are their retention times reduced, but also the shape of their peaks is improved.

Similarly to anions, cations can be determined with both suppressed and nonsuppressed ion chromatography. The former group encompasses mineral acid solutions that are highly diluted with water (e.g., HCl, H_2SO_4), whereas the latter one includes organic acid solutions (e.g., tartaric, dipicolinic, or oxalic acid). When cations are determined, the sample pH should be acidic; its value has a big influence on the separation quality. If the sample pH is below 2, the peaks may not be separated; when it is greater than 4 for divalent cations, the obtained results are slightly overestimated. Consequently, it is necessary to acidify the sample and repeat the analyses. Adding selected organic solvents accelerates the leaching of compounds (such as amines) and improves their peak shapes. Nevertheless, the columns with phases containing carboxylic functional groups cannot be used with organic solvents. The temperature increase results in the reduction of retention times for monovalent cations; the larger the ion radius is, the bigger is the influence. For some selected transition metal ions, the increase in the temperature reduces their retention times. An opposite situation is observed when amines are separated; their retention times decrease when the temperature rises.

Generally, ion chromatography is used for independent determinations of anions or cations. Nevertheless, their simultaneous separation is also possible [31]. To achieve that, various options are used, that is, systems containing sequentially switched anion-exchange and cation-exchange columns connected in series or in parallel [32]; separation of anions and cations in columns with mixed phases [33]; and methods consisting in the formation of complexes with the same charge sign (usually anionic ones) and their separation with an appropriate ion-exchange column [34].

The purity of water used for the sample and eluent preparation is an important factor affecting the analysis quality. It influences the noise sizes and the obtained limits of detection. Distilled water whose electric conductivity is approximately $1 \times 10^{-4}\,S\,m^{-1}$ is used in routine determinations. However, such water may still contain various ionic contaminants (up to $200\,\mu g\,l^{-1}$). That is why, it is recommended to use deionized water with electric conductivity below $5 \times 10^{-6}\,S\,m^{-1}$. The contaminants found in water used for the eluent preparation significantly affect the analysis quality and column durability. They systematically enter the column. If they are kept inside, they cause significant changes in the column characteristics, increase the noises, and deteriorate the baseline. For that reason, a precolumn should always be used as it increases the analytical column durability. The influence of the sample contaminants can also be reduced if the injected sample volume is lowered.

When the eluent is being prepared, it must be degassed. CO_2 present in the air is absorbed in water and increases its electric conductivity. Normally, helium or other neutral gas is used for degassing. Before the analysis, both the eluent and the sample should be filtered through a 0.45-μm filter to remove any solid particles. They can block the columns, which has a negative effect on the column characteristics. Modern ion chromatographs have such filters built in and the process is automatic. In general, when the eluent is selected for the ion chromatography procedure, its composition should be as chemically similar as possible to the chromatographed substances. In some cases, two eluents can be used for separation. One is optimum for the first separation phase, whereas the other is used for the final phase. If possible, the gradient elution can also be applied.

1.1.4   Suppressors

For low concentration ranges, electric conductivity constitutes the linear concentration function. Consequently, the conductometric detector is the most popular detector in ion chromatography. The greatest challenge for the ion chromatography inventors was to elaborate a method enabling them to determine separated analyte ions against the background of the eluent ions as the eluent is also an electrolyte. At first, the suppression problem was solved with the suppression column placed in the front of the analytical column. Suppression columns were replaced by suppressors in the mid-1980s. They have been playing the same role in a much more efficient way since then.

In order to reduce the solution conductivity in the suppressor, the eluent and the determined ions must be converted into compounds that differ significantly in electric conductivity. When anions are determined, the cation exchanger in the H^+ form is placed in the suppressor. When cations are determined, the anion exchanger has the OH^- form. The cation exchanger exchanges the H^+ ions for the eluent cations and converts the eluent into a weakly dissociated acid. The determined anions are converted into strongly dissociated acids. Analogous reactions take place in the suppressors used to determine cations.

The conductivity measured with a conductometric detector is the sum of the cation concentration in the sample (c^+_{sample}) multiplied by the differences in the mobility of the sample and eluent cations $(\Lambda^+_{sample} - \Lambda^+_{eluent})$ and the concentration of the sample anions (c^-_{sample}) multiplied by the difference in the mobility of the sample and eluent anions $(\Lambda^-_{sample} - \Lambda^-_{eluent})$.

$$\kappa_{measured} = c^+_{sample} \cdot |(\Lambda^+_{sample} - \Lambda^+_{eluent})| + c^-_{sample} \cdot |(\Lambda^-_{sample} - \Lambda^-_{eluent})|$$

When anions are determined with suppressed ion chromatography, the sample and eluent cations are exchanged for the H^+ ions in the suppressor. The measured conductivity is calculated as the absolute value of the product of the anion concentration in the sample and the sum of the sample anion and H^+ ion mobility.

$$\kappa_{measured} = c^-_{sample} \cdot |(\Lambda^-_{sample} + \Lambda_{H^+})|$$

A similar situation is observed for cations. The sample and eluent anions are exchanged for the OH^- ions in the suppressor. The measured conductivity is the product of the cation concentration in the sample and the sum of the sample cation and OH^- ion mobility.

$$\kappa_{measured} = c^+_{sample} \cdot |(\Lambda^+_{sample} + \Lambda_{OH-})|$$

Two biggest manufacturers of the ion chromatography apparatus and devices, Dionex (recently part of Thermo Scientific) and Metrohm (Metrohm, Herisau, Switzerland), use suppressors in systems to determine anions. There is no such approach for the cation separation. The Metrohm cation-exchange columns help to separate cations and determine them in a conductometric way without suppression. On the other hand, Dionex uses only the suppressed mode in its chromatographs for cation-exchange columns.

The suppression columns used at the very beginning of ion chromatography had to be regularly regenerated. To do that, analyses had to be stopped, which was a great difficulty. In 1981, Dionex introduced the first commercial hollow fiber suppressor, which could be regenerated in the continuous mode. The device had the same function as the suppression column, but was much more efficient [35]. When anions were determined with the eluent (water solution of sodium carbonate and bicarbonate), the eluent and analyte ions flew through the capillary fibers. Outside, the water solution of the properly concentrated sulfuric acid flew in the countercurrent. Sodium ions penetrated the cation-exchange membrane, and their place was taken by a specific number of the H^+ cations that got inside with the sulfuric acid stream. A similar mechanism was observed for cation determinations. The first suppressors of this type largely improved the analysis quality. Nevertheless, their application was limited as the number of the available regenerating reagents was limited and there was no resistance to organic solvents. Such suppressors had also other disadvantages. Their capacity to exchange Na^+ ions for H^+ ions was small as the diffusion was limited. There was also high pressure inside the suppressor. Additionally, they could be damaged by gas bubbles. The problems caused by the first capillary suppressors were largely solved when Dionex introduced the commercial micromembrane suppressors in 1985. These were multilayer membrane systems ("sandwich suppressors"), which helped to improve significantly the suppressor performance. The main advantages of the micromembrane suppressors were as follows: low dead volume, minimalizing the peak tailing effect, high capacity, possible gradient elution, and resistance to most organic solvents. Their main disadvantages included the following: necessary provision of a constant regenerating reagent flow (flow rate up to $10\,ml\,min^{-1}$) [36], relatively high price and short durability, higher noises, and service that had to be performed by highly qualified staff. The first micromembrane suppressor used to determine cations was the cation micromembrane suppressor (CMMS). To regenerate it, tetrabutylammonium hydroxide was used. The next generation of the membrane suppressors (self-regenerated suppressors, SRSs) made use of the electrode processes to produce the H^+ and OH^- ions [37].

Even though the suppression columns from the mid-1970s had many disadvantages, the researchers returned to this suppression mode 25 years later. Instead of the classic suppression columns filled with high exchange capacity of ion exchangers, modern phases in special microcolumns were applied (e.g., solid-phase chemical suppression – SPCS, Alltech Inc.) [38]. In 1996, Metrohm introduced the commercial Metrohm Suppressor Module (MSM) to determine anions. It contained three identical cartridges filled with the cation-exchanger resin. When one cartridge was regenerated with the water H_2SO_4 solution, the next one was rinsed with deionized water. The third cartridge was used for suppression in the analytical course. In the next analysis, the cartridges (suppressor elements) changed their positions. Consequently, the chromatograph worked in a continuous way. The MSM was insensitive to the backpressure. It was fully compatible with the organic solvents used in ion chromatography. As a result, the eluent composition could be modified freely. The suppressor itself was simple to use and did not need any professional service. What is more, very low noises (<0.2 nS) were introduced into the measurement system.

The suppression techniques in ion chromatography (between its invention and 2002) were described in the overview study by Haddad *et al.* [39]. When compared to the old-generation suppressors from the late 20th century, modern devices are small and their performance is high. They also offer stronger resistance to organic solvents and increased pressure [40, 41].

To sum up, the subsequent suppressor generations helped to improve the limits of detection for analytes and extend the working range. They enabled using more concentrated eluents and larger sample volumes introduced into the columns. They allowed the researchers to use the gradient elution. Finally, they helped to reduce the time in which the system balance could be restored. Selected parameters of the suppressors used in ion chromatography are given in Table 1.3.

1.1.5 Detection Methods

The most important requirements to be met by the detectors in ion chromatography are as follows: high measurement sensitivity and short response times, measurement signal that is proportional to the analyte concentration (broad linearity range), small

TABLE 1.3 Parameters of Selected Suppressors Used in Ion Chromatography

Factor	Suppression Columns	Capillary Suppressors	Micromembrane Suppressors	Auto supressors	New Generation Suppressors
Continuous regeneration	No	Yes	Yes	Yes	Yes
Noise level	High	High	Low	Low	Low
Exchange capacity	High	Low	High	High	High
Gradient elution	No	No	Yes	Yes	Yes
Limits of detection	High	Low	Low	Low	Low

TABLE 1.4   Detection Methods and Their Applications in Ion Chromatography

Detection	Measured Quantity	Most Important Applications
Conductometric	Electric conductivity	Anions and cations with pK_a or $pK_b < 7$
UV/Vis	Radiation absorption	Ions active in the UV/Vis range, metals after derivatization reactions
Amperometric	Redox reactions	Anions and cations with pK_a or $pK_b > 7$
Chemiluminescence	Light emission due to chemical reactions	Selected metals and anions
Fluorescence	Radiation absorption and emission	Amino acids after derivatization reactions
Refractometric	Changes in the light refractive index	Anions and cations in high concentrations
ICP-OES, ICP-MS, MS	Measurement of the mass-to-electric-charge ratio	Metals/metalloids, water disinfection by-products

changes in the baseline, low background noises, and low sample volume (reduction in the peak extension). The detection method selection depends on the sample type, analyte concentration, applied columns, eluents, and sample preparation for the analysis [42]. The scope of applications for selected detectors used in ion chromatography (with the measured quantity) is given in Table 1.4. The general classification of the detection methods used in ion chromatography is shown in Figure 1.2.

The electric conductivity measurement in the conductometric detector is based on the ability of the electrolyte solutions (placed in the electric field between two electrodes) to conduct the electric current. The electrodes are made of noble metals, which prevents chemical reactions on their surfaces. The following aspects affect the measured conductivity: type, equivalent conductance, and charge of ions; background (eluent) conductivity; conductometric cell constant; and temperature. According to the Kohlrausch law, the conductivity of diluted solutions is the sum of the conductivities of particular ions in the solution multiplied by their valences and concentrations:

$$\kappa = \frac{\sum \Lambda^0_i z_i c_i}{1000}$$

where

κ electric conductivity of the electrolyte ($\mu S\,cm^{-1}$)

Λ_i^0 limiting molar conductivity (ion conductivity divided by its concentration and extrapolated to an infinite dilution) ($S\,cm^2\,mol^{-1}$)

z_i ion valence

c_i molar ion concentration ($mol\,l^{-1}$)

The values of the limiting molar conductivities for most anions and cations range between 20 and 80 $S\,cm^2\,mol^{-1}$. The OH^- and H^+ ions are exceptional as their conductivities are very high (198 and $350\,S\,cm^2\,mol^{-1}$, respectively). The molar conductivity of ions is highly sensitive to the temperature changes. Such a situation requires either the detector temperature stabilization or electronic compensation. The temperature stabilization is particularly important in nonsuppressed ion chromatography. As significant changes in conductivity can prevent the determination of ions at very low concentration levels, it is crucial to thermally insulate the conductivity cell. In suppressed ion chromatography, this influence is not as important. The eluent value is approximately $2–3 \times 10^{-3}\,S\,m^{-1}$ due to its conductivity suppression effect. The range of the conductometric detector applications is very broad. The device is used in routine determinations of inorganic anions and cations and selected organic ions (e.g., carboxylic acids, amines) in surface water, underground water, drinking water, and industrial and municipal wastewater.

Other detector types are less popular in ion chromatography. Nonetheless, they have their specific uses, particularly when conductometric detection is not possible. One of such possibilities is the application of a photometric detector. It was first used to determine metal ions in 1971 [43]. Photometric detection is based on the light absorption in the visible spectrum and in the ultraviolet range. It uses the Beer–Lambert law. At the beginning, it did not have as many applications in ion chromatography as in HPLC. Generally, most inorganic ions do not contain chromophores. Consequently, they are "invisible" in the UV/Vis detector working range. At first, researchers used the UV detectors that enabled detecting the chromatographed substances at one wavelength (usually 254 nm). Later, commercial

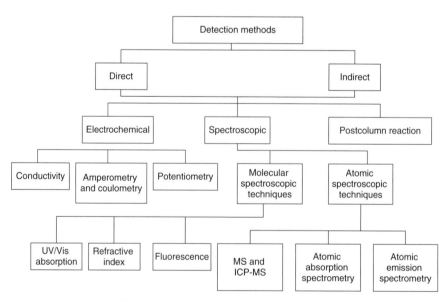

Figure 1.2   Classification of detection methods in ion chromatography.

detectors enabling detection at changing wavelengths were available. At present, researchers use detectors in which a smooth wavelength regulation within the whole ultraviolct range and part of the visible spectrum (190–900 nm) is possible. In ion chromatography, diode array detectors (DADs) and photodiode array detectors (PADs) are also used. The photometric detection can be used when ions or their complexes absorb the light within the range of such wavelengths that help to differentiate between the determined ions and the mobile-phase ions and other ions that are not analyzed. There are two possible variants of the photometric detection, that is, direct and indirect detection. In the direct method, the detector signal is measured when the analyte passes through the UV/Vis detector. Importantly, the absorption of the determined substance must exceed the eluent absorption [44].

Derivatization methods also offer good opportunities. The determined compounds that cannot be directly detected within the UV/Vis detector working range are converted into detectable forms. The reactions that lead to such couplings can be performed before the separation in the column (precolumn derivatization) or after the separation (postcolumn derivatization). The derivatization methods in ion chromatography are used to determine metal ions and inorganic by-products of water disinfection. In ion chromatography, derivatization reactions and UV/Vis detectors are used to determine, for example, Cr(VI) ions [45], transition metals [46], aluminum fluoride complexes [47], lanthanides [48], or sulfur anions [49].

The amperometric detection is based on the measurement of the current formed during the electrochemical redox reaction taking place on the working electrode surface. The working electrode potential is measured in relation to the reference electrode (e.g., Ag/AgCl), with a potential that is independent of the eluent composition. The amperometric detectors are used to detect substances that are able to oxidize or reduce themselves on the electrode surfaces. There are a few varieties of the amperometric detection, such as direct amperometric detection (DAmD), pulsed amperometric detection (PAmD), and integrated amperometric detection (IAmD). The detection potential of such detectors enables determination of the nanogram ion amounts. The measurement cell volume is only 1 µl. They can be used within a broad range of concentrations. On the other hand, they are sensitive to the changes in the eluent flow rate and pH, which negatively affects the result reproducibility. The main disadvantage of the amperometric detector is the systematic decrease in its sensitivity. Stable products of the electrochemical reactions are deposited on the working electrode surface, which reduces its active surface. In such a case, the working electrode must be exchanged or short sequential electric impulses must be used instead of the stable potential. In this amperometry variety, the detector measures the electric current only during those short time impulses, which largely reduces the risk of the working electrode surface deactivation. In the pulsed amperometry, the current is measured for a stable potential value. In the integrated amperometry, it is measured during the changes in the potential that increases in the oxidation process and decreases during the reduction. In ion chromatography, the pulsed amperometric detector was used for the first time in 1983. It helped to extend the range of the determined substances with carbohydrates, alcohols, and catecholamines [50]. In ion chromatography, the amperometric detector is used to determine, for example, sulfides [51], cyanides, sulfides, thiosulfates [52], iodides [53], biogenic amines [54], or carbohydrates [55]. The potentiometric detector operations are based on the measurement of the potential formed between the miniaturized ion-selective electrode and analyzed solution. The potential depends on the ion concentration in the solution. The application of the potentiometric detector with a copper electrode was described by Aleksander *et al.* [56]. Potentiometric detectors have shorter response times and low sensitivity. For that reason, their use in ion chromatography is limited [57].

The fluorescence detection is based on measuring the intensity of light with a determined wavelength. The light is emitted by the detected substance when it is excited with the energy whose value is higher than the emitted energy. Fluorescence detectors are very sensitive. They are mainly used in HPLC. They have much lesser importance in ion chromatography. Their use is limited to determinations preceded by the derivatization reaction before or after the analyte separation in the column. Only a few ions have a natural ability for fluorescence (e.g., UO_2^+). When ions are converted into the derivatives, the detectors can be used to determine, for example, iodides, nitrites, polyphosphates, thiosulfates, and ammonium or sodium ions. In order to improve the detectability, it is necessary to use eluents that are "invisible" for such detectors, in terms of both the exciting and emitted light wavelengths. In ion chromatography, fluorescence detectors are used to determine calcium and magnesium [58]; haloacetic acids [59]; and ascorbic acid, nitrates, sulfides, and iodides [60]. The chemiluminescence detector is one of the most sensitive detectors used in the analytical chemistry. However, its use in ion chromatography is limited due to the requirements that must be met by the analyzed substance to be detected. It is applied to detect arsenates, silicates, and phosphates [61]; chromium ions [62]; nitrogen anions [63]; and copper [64].

The refractometric detection is seldom used in ion chromatography as it is difficult to differentiate between the signals coming from the sample and eluent ions. Moreover, this detector cannot be used during the gradient elution. It is also very sensitive to any temperature changes, but it enables using concentrated eluents. In the literature, the refractometric detector applications to determine polyphosphates in water and wastewater [65] were described. There are also works related to the application of other detectors. Even though not all of them are successful, acoustic [66] or radiometric [67] detectors have been used in ion chromatography.

The detectors such as inductively coupled plasma mass spectrometry (ICP-MS) and mass spectrometry (MS) have great importance in ion chromatography. Hyphenated techniques, such as ion chromatography inductively coupled plasma mass spectrometry (IC--ICP-MS) or ion chromatography mass spectrometry (IC-MS), have many advantages and a broad use in fields such as speciation analyses of selected elements. A mass spectrometer can be directly coupled with an ion chromatograph [68] or work as the ICP-MS system [69]. Such detectors provide data on the quantitative and qualitative composition of the sample. They also help to determine the analyte structure and molar weights. The main problem is the need to maintain very low pressure in the spectrometer. On the other hand, the analyte ions leave the column under relatively high pressure. The analyte ions are separated in the spectrometer on the basis of the mass-to-charge ratio. The collected data are obtained in the spectrum form. Various ionization sources can be used in mass spectrometers, such as electrospray ionization (ESI), atmospheric pressure chemical ionization (APCI), or atmospheric pressure photochemical ionization (APPI). The MS detection can be performed in the selected ion monitoring (SIM) mode or scan mode (SM). In the first case, the information on the analyte mass is obtained. The other one provides data on mass spectra and mass distribution. There are identification difficulties for large molecules because it is possible to obtain a larger number of spectra with the same mass-to-charge ratios. The main advantages of the MS detectors are as follows: broad concentration range, high determination sensitivity, high selectivity, quick multielement analysis, and opportunity to measure isotopes. The technique has also some limitations, such as spectral and matrix interferences. It also offers lower precision than the ICP-AOS method. Moreover, the total content of the soluble salts should be $<1000 \, \text{mg} \, l^{-1}$, and the apparatus is relatively expensive.

Other mass-sensitive and generally universal detectors that can be coupled with ion chromatograph include the evaporative light-scattering detector (ELSD), which has been modified and improved for 20 years. The Corona™ charged aerosol detector (CAD) is another type of a nebulizing detector, relatively newly developed. It combines elements from MS and ELSD technologies and offers better sensitivity, reproducibility, and dynamic response than ELSD. CAD and ELSD have been mainly coupled to RPLC and HILIC separation modes. Parameters of detectors used in ion chromatography are presented in Table 1.5.

1.2 ION CHROMATOGRAPHY APPLICATIONS

Ion chromatography is one of the most important instrumental methods for determination of organic and inorganic ions in liquid, solid, and gaseous samples. It has evolved from a relatively simple method for determination of inorganic ions in water and wastewater at the milligram per liter concentration level into a sophisticated technique applied for separation and detection of organic and inorganic substances at the microgram per liter concentration level. Importantly, it is used for analyzing samples with complex matrices. The first studies describing ion chromatography were written in the 1970s. They presented its uses for analyzing drinking water, surface and underground water, condensates from steam generators, brine solutions, cooling liquids, geothermal water, special high-purity water, sea and oceanic water, and municipal and industrial wastewater [70]. Typical chromatograms illustrating separation of anions in drinking water and cations in industrial wastewater are shown in Figures 1.3 and 1.4, respectively.

Ion chromatography can also be used to determine the following: trace amounts of anions and cations in high-purity water [71]; inorganic anions in mineral acids [72]; anions and cations in mineral water [73] and sea water [74]; selected ions in alcohols [75]; ions in plant extracts [76] and natural peat bog water rich in organic compounds [77]. Additionally, ion chromatography is applied to research atmospheric precipitation [78], aerosols [79], or runoff water [80].

TABLE 1.5 Selected Parameters of Detectors Applied in Ion Chromatography

Parameter	Conductometric Detector	UV/Vis Detector (Direct Method)	UV/Vis Detector (Indirect Method)	Amperometric Detector	Potentiometric Detector	Fluorescence Detector	Chemiluminescence Detector	MS Detector
Low limits of detection	++	++	++	+++	++	+++	+++	+++
Broad linearity range	++	++	++	+	+++	++	++	++
High selectivity	−	+	−	+++	+++	++	++	+++
Unequivocal identification	−	+	−	+	−	+	+	+++
Possible application of gradient elution	++	++	−	−	+	+	+	+
Durability and stability	+++	+++	++	−	+	++	++	+
Simple use	+++	+++	++	−	+	−	−	−
Low price of purchase and service	++	++	++	++	+++	+	+	−
Universality	+++	+	++	−	−	−	−	++

+++, very good; ++, good; +, satisfactory; −, unsatisfactory.

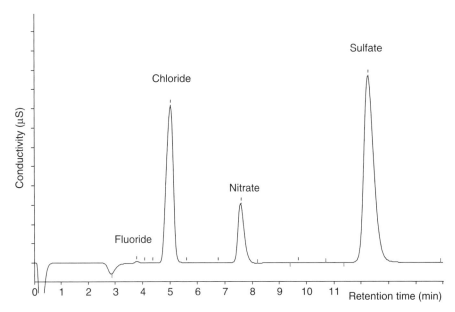

Figure 1.3 Chromatogram of anions in drinking water.

Column	Dionex IonPac AS14 (250 mm × 4 mm)
Eluent	3.5 mM Na_2CO_3 + 1.0 mM $NaHCO_3$
Eluent flow rate	0.8 ml min^{-1}
Detection	Suppressed conductometric

Ion chromatography was compared with other ion analysis methods. For instance, fluoride content analyses in the atmospheric precipitation conducted with three methods (ion chromatography, capillary electrophoresis, and potentiometric method) were described in the study [81]. The selected method parameters (precision, limits of detection, and total analysis time) were compared. The result assessment showed that the results obtained with ion-selective electrodes and capillary electrophoresis were identical. On the other hand, the fluoride concentrations obtained with ion chromatography were visibly higher. The results of the research into the snow samples from polar [82] and alpine [83] areas help to define the climatic changes taking place on the Earth over millions of years. A typical chromatogram of inorganic anions in the rain sample is presented in Figure 1.5.

Due to the matrix type, air quality analyses are a separate field in which ion chromatography is used. The first study on using ion chromatography in the gaseous sample research was published in 1979 [84]. Before the chromatographic analysis, the samples in the gaseous phase need to be converted into the solution and have ion forms. The air sampling methods used for ion chromatography analysis can be divided into three groups:

1. Passive methods (diffusion, permeation)

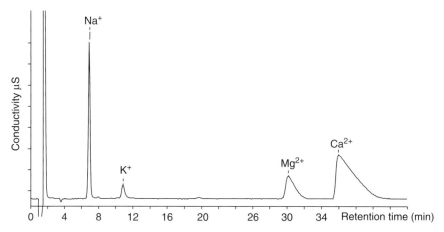

Figure 1.4   Chromatogram of cations in industrial wastewater.

Analytical column	Metrohm Metrosep C2
Eluent	4 mM tartaric acid + 0.75 mM DPA
Detection	Non-suppressed conductivity

2. Denudation methods (absorption in solution, chemisorption, adsorption, permeation)
3. Dynamic methods (freezing, absorption in solution, chemisorption).

The absorption solution composition decides on the analyte absorption efficiency. There are only a few cases in which the simultaneous quantitative absorption of a few gaseous substances is possible. For example, HF, NO_x, SO_2, and SO_3 can be absorbed in the deionized water (sometimes with a small addition of H_2O_2 to oxidize sulfides into sulfates), diluted base solutions, or a water solution of sodium carbonate. Gases, such as NH_3, are absorbed in acidic solutions (e.g., a diluted sulfuric acid solution). The SO_2 measurement methodology with ion chromatography was introduced by the USEPA in 1979. It consisted in the absorption of the atmospheric SO_2 in the 0.6% water solution of H_2O_2. After converting SO_2 into the SO_4^{2-} ions, the substance was analyzed with ion chromatography.

The optimum solution is using diffusion or permeation denuders as such determinations are automatic. The continuous measurement methods for gas pollutants and aerosols are fast and relatively inexpensive. Their main advantages include the following: simultaneous measurements of gas components and aerosol, measurement in the 24/7 mode, using the obtained results to investigate the pollutant transportation or daily changes. If it is not possible to use a system built from gas scrubbers or impregnated filters at the forced air flow, the so-called passive methods can be employed. In such a case, the samplers saturated with appropriate absorbing substances are exposed to the air influence within the specified time period (between a few hours and a few

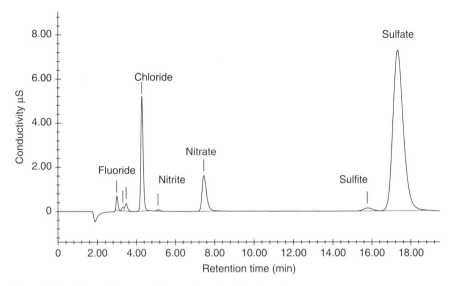

Figure 1.5 Chromatogram of inorganic anions in rain water.

Analytical column	Dionex Ion Pac AS4
Eluent	1.8 mM Na_2CO_3 + 1.7 mM $NaHCO_3$
Eluent flow rate	1.0 ml min^{-1}
Detection	Suppressed conductivity

months). After the samplers are transported into the laboratory, the obtained substances are extracted (usually with deionized water) and undergo chromatographic analyses [85].

The research into food and beverage samples is another area in which ion chromatography is used. Many analyses are not performed with the classic ion-exchange method, but with related techniques, that is, ion-exclusion chromatography, reversed-phase chromatography, chelation chromatography, or affinity chromatography. There are certain difficulties in determination of food samples with ion chromatography. Such samples have complex matrices and require appropriate preparation. Examples of ion chromatography applications in this field concern the research on [86] the following:

1. Dairy products (e.g., iodides in milk, chlorides and sodium in butter, organic acids in cheese)
2. Meat products (e.g., nitrites and nitrates)
3. Beverages (e.g., organic and inorganic anions and cations in water and beverage additives; determination of organic acids in beer, wine, and juice)
4. Canned food (e.g., chlorides, nitrites, sodium, organic acids, and transition metals in vegetables, condiments, and fish)

5. Food products for children (e.g., iodides and transition metals)
6. Breakfast cereals and bread (e.g., bromate and metal ions)
7. Fats, oils, and flour (e.g., fatty acids and carbohydrates)
8. Honey (e.g., heavy and transition metals)
9. Determination of organic compounds (e.g., carbohydrates, alcohols, amines) in sweets, canned foods, and beverages.

Introducing new high-selective phases into analytical columns and modern sample preparation methods has helped to extend the ion chromatography application range. At present, it is possible to determine virtually all ionic substances. Ion chromatography can also be used for compounds that are converted into ionic forms after appropriate precolumn or postcolumn derivatization reactions. The range of applications is becoming more and more popular and will continue to expand. It is also related to the opportunities for determination of solid samples offered by the so-called combustion ion chromatography. In this variant, the sample is combusted. Afterward, it is absorbed in a suitable solution. Finally, it is transported into the analytical column. Thus, ion chromatography is used to analyze samples from many industries, such as the following:

1. Pharmaceutical industry (e.g., analysis of purity of medicinal products, medicine production control, and determination of contents of selected substances in medicines) [87, 88]
2. Semiconductor production industry (e.g., determination of trace amounts of ions in high-purity water and substances for semiconductor production, examination of contaminants in organic solvents, analysis of trace air pollutants in production halls) [89]
3. Cement industry (e.g., analysis of contents of anions and heavy metals, such as chromium, in cement) [90]
4. Chemical industry (e.g., examining purity of chemical reagents, analysis of heavy metal contents in fertilizers, technological process control) [91]
5. Energy production industry (e.g., determination of pollutants in cooling waters containing amines used to prevent corrosion in turbines and energy installations, radioactive elements in the nuclear power plant waste) [92]
6. Petrochemical industry (e.g., determination of amines, polyphosphates, cyanides, mineral acids, and sulfur ions in the solutions for gaseous pollutant absorption) [93]
7. Metallurgical industry (e.g., analysis of compositions of steel and other alloys, researching metal ions in wastewater) [94]
8. Pulp and paper industry (e.g., investigating contents of sulfur compounds, chlorine, alkaline and alkaline earth metals, and transition metals in wastewater; technological process control) [95]

9. Electrochemical and tanning industry (e.g., determination of heavy metals, polyphosphates, and cyanides in the electrogalvanization waste; analysis of contents of chromium and its compounds in the tanning waste) [96]

1.2.1 Speciation Analysis with the Hyphenated Methods of IC-ICP-MS and IC-MS

The toxicological test results show that in many cases, particular element forms, rather than its total content, have a major influence on living organisms. For this reason, information on the various element forms in the sample is more important than knowing its total content. The word *speciation* is a borrowing from biology and derives from the Latin term *species*, which means species or species evolution. The term *speciation analysis* appeared in the literature in 1993 and was described as "the movement and transformation of element forms in the environment" [97]. The definition proposed by the International Union of Pure and Applied Chemistry (IUPAC) says that speciation is the distribution of an element in various individual chemical forms. They are defined by the isotope composition, electron structure or oxidation state, complex or molecular structure [98]. Therefore, speciation analysis is an analytical procedure that leads to identification and determination of one or more chemical element forms present in the sample. Speciation analysis is used in the following: researching biochemical cycles of selected chemical compounds, analysis of toxicity and ecotoxicity of elements, food and pharmaceutical product quality control, technological process control, and clinical analysis [99].

The toxicological data show that it is necessary to lower the limits of detection for analytes to extremely low concentration levels. The analytical methods used so far have not always met such requirements. For that reason, various separation and detection techniques are coupled as the so-called hyphenated methods. A suitable hyphenated method should be selective toward determined analytes and sensitive in a broad concentration range. It should also enable the best possible identification of the determined substances. The selection of a suitable hyphenated method should be determined by the analyte characteristics, easy coupling of different methods, required determination sensitivity, and available apparatus. Chromatographic methods are mainly used for separation, whereas spectroscopic methods serve for detection [100]. The ion chromatography applications in speciation analysis can be divided into three main groups:

1. Determination of nitrogen (NO_2^-, NO_3^-, NH_4^+), sulfur (S^{2-}, SO_3^{2-}, SO_4^{2-} $S_2O_3^{2-}$), phosphorus ($H_2PO_4^2$, HPO_3^-, HPO_4^{2-}, $P_2O_7^{4-}$), and halides (ClO_4^-, IO_3^-) ions

2. Determination of inorganic by-products of water disinfection (BrO_3^-, ClO_2^-, ClO_3^-)

3. Determination of metal and metalloid ions (e.g., Cr(III)/Cr(VI), As(III)/As(V), Fe(II)/Fe(III), Se(IV)/Se(VI)).

In the literature, there are many described ion chromatography applications for analysis of ions of nitrogen [101], sulfur [102], phosphorus [103], or halide ions [104], also in samples with complex matrices. Due to the ion chromatography development, it is used not only for routine analyses of anions and cations in water and wastewater but also for other analytes. A good example is determination of inorganic by-products in disinfected drinking water. They form because chlorine and its derivatives and ozone are used to disinfect water [105]. Bromates are the most important inorganic disinfection by-products [106]. After ozonation, their concentration in water can be between a few and several dozen micrograms per liter. The reaction course depends on the following: water pH, initial amount of bromides and organic compounds in raw water, ozone dose, and time and temperature at which the process takes place [107]. Bromate determinations with ion chromatography can be divided into three method groups [108]:

1. Direct methods (conductometric detection)
2. Indirect methods (UV/Vis detection after postcolumn derivatization)
3. Hyphenated techniques (MS, ICP-MS detection).

Direct methods are based on the selective separation of ions present in the sample, particularly chloride and bromate ions, and their detection in a conductometric detector. The International Organization for Standardization (ISO) established the standard based on ion chromatography coupled with conductometric detection [109]. It meets the requirements for the obligatory limits of quantification ($10\,\mu g\,l^{-1}$). Unfortunately, when complex matrix samples are analyzed, it is necessary to use expensive cartridges for removing chlorides, sulfates, and metal ions, which also complicates the analysis procedure.

Derivatization methods belong to the indirect group. After the analysed substance is separated in the column, it is converted into its derivatives, which can be detected with the UV/Vis detector. In derivatization reactions, the following reagents are used to determine bromate ions: fuchsine, o-dianisidine, chlorpromazine and its derivatives, potassium bromide or potassium iodide [110]. The main advantages of the derivatization methods are the limits of detection ($<1\,\mu g\,l^{-1}$) and lack of the chloride interferences [110]. In 2010, the ISO published the new standard for bromate determination, based on the indirect method with the triiodide reaction [111].

IC-ICP-MS and IC-MS constitute the final group of the bromate determination methods, that is, hyphenated techniques. Such systems provide very good detectability, precision, and limits of quantifications. Due to their high cost, they are not routinely used in laboratories. In the literature, there are many studies on the occurrence, formation, and determination of chlorate(VII) ions, whose presence in the environment is related to the pyrotechnic material production and rocket fuel use [112]. In 2014, the ISO elaborated the new method of their determination, based on ion chromatography and conductivity detection. In speciation analysis, the application of ion chromatography for determination of metal and metalloid ions seems the most prospective. The advantages of ion chromatography in cation determinations include the following: simultaneous determination of alkaline and alkaline earth metals and what is very important – ammonium ions. Interestingly, the transition metal manganese can also be analyzed together with such metal ions as alkali and alkaline earth metals, although it is generally assumed that the transition metal hydroxides formed in the suppressor system will precipitate and, thus, may not be detected by suppressed conductivity ion chromatography [113]. However, the kinetics of this reaction is not the same for all metals, especially because a weakly acidic milieu ($pH = 6$) prevails in the suppressor after the suppressor reaction, but many transition metals only precipitate at pH values above 7. Anyway, transition metal ions and eluent ions (e.g., EDTA, PDCA) form complex anion or cation forms that can be separated in a suitable ion exchanger [114, 115]. An example of the IC-ICP-MS application for the simultaneous determination of the inorganic forms of arsenic, antimony, and thallium is shown in Figure 1.6.

Even though speciation analysis with hyphenated methods has tremendously developed over the last years, it is still a relatively new field of the analytical chemistry. Its further development depends on many factors such as the development of new sample preparation, separation and detection methods, and availability of new certified reference materials. Hyphenated methods offer great possibilities and their main advantages include the following: extremely low limits of detection and quantification and very good precision and repeatability of determinations.

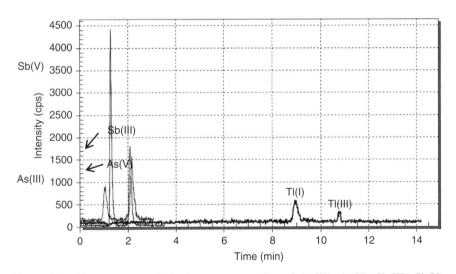

Figure 1.6 Chromatogram of simultaneous separation of As(III), As(V), Sb(III), Sb(V), Tl(I), and Tl(III) ions in the Kłodnica river water sample.

Analytical column	Dionex IonPac AS7
Eluent A	1.5 mM phthalic acid + 10 mM EDTA (1–4 min)
Eluent B	15 mM HNO_3 + 2 mM DTPA (4–12 min)
Eluent flow rate	0.8 ml min^{-1}
Detection	ICP-MS

TABLE 1.6   Examples of Metal and Metalloids Species Analyzed Using Hyphenated Methods IC-ICP-MS and IC-MS

Element	Inorganic Forms	Literature Examples
Arsenic	As(III), As(V)	[117, 118]
Selenium	Se(IV), Se(VI)	[119, 120]
Chromium	Cr(III), Cr(VI), hydroxycomplexes	[121, 122]
Tin	Sn(II), Sn(IV)	[123, 124]
Mercury	Hg(I), Hg(II)	[125, 126]
Antimony	Sb(III), Sb(V), inorganic complexes	[127, 128]
Aluminum	Al(III), fluoride, and sulfate complexes	[129, 130]
Thallium	Tl(I), Tl(III)	[131, 132]
Lead	Pb(II), Pb(IV)	[133, 134]
Other	Fe(II), Fe(III); Mn(II), Mn(III), Mn(V), Mn(VII); V(IV), V(V); Mo(V), Mo(VI); Au(I), Au(III); Pt(II), Pt(IV); Sc(I), Sc(II), Sc(III), Sc(IV), Sc(V); Ga(III), Ga(IV), Ga(V); Np(IV), Np(V), Np(VI)	[135,136, 137]
Halides	ClO_2^-, ClO_3^-, ClO_4^-, Br^-, BrO_3^-, I^-, IO_3^-, IO_4^-	[138, 139]

At the beginning of the speciation analysis development (in the 1980s and 1990s), it was most often used to determine speciation forms of arsenic, selenium, chromium, lead, mercury, and iron [116]. At present, it is also applied for the research into various forms of tin, aluminum, antimony, and halides. Typical forms of selected elements determined with hyphenated methods are enlisted in Table 1.6. Similarly to other techniques, hyphenated methods have certain limitations, which include high cost and complexity of the apparatus. As a result, they are not commonly available and used in laboratories. What is more, using hyphenated methods requires perfect understanding of the analytical methodologies and detailed knowledge of the apparatus. The systems are expensive and used for scientific research rather than routine analyses. Nonetheless, the growing number of hyphenated technique applications and studies concerning the topic prove that the importance of these methods is constantly rising [140].

1.3   SAMPLE PREPARATION FOR ION CHROMATOGRAPHY

Determination of analytes is the final part of the procedure in which sampling is one of the most crucial stages. It is enormously important as even routine processes (dilution, pH changes through sample preservation, changes in pressure and temperature) can cause irreversible changes in the primary analyte form. Sample preparation is usually the most laborious part of the analysis and constitutes the most significant source of errors. In the modern analytical chemistry, particularly in the trace analysis, sample preparation is normally more important than the determination itself [141]. Water and wastewater samples should be collected into plastic containers made of polytetrafluoroethylene (PTFE), polypropylene (PP), polystyrene (PS), or high-density polyethylene (HDPE). Glass containers are not recommended, whereas containers made of polyvinyl chloride (PVC) must not be used. Various processes can occur in the sample between sampling and analyte determination (e.g., oxidation, reduction, complexation, precipitation, biochemical and photochemical processes). They are particularly important in the speciation analysis. Generally, it is very difficult to prevent such processes. Therefore, the sample analysis should be performed as soon as possible after the sampling took place. When the sample preparation method is chosen, it is necessary to take into account the characteristics of the applied analytical column (e.g., selectivity, volume, pH range, possible use of organic solvents) [142].

In the last 40 years, ion chromatography reached a very high technical and technological level. Nonetheless, the question of sample preparation, particularly for complex matrix samples, is still essential. The most important reasons why the sample should undergo preliminary processing are [143] as follows:

1. The analyte concentration is too low in relation to the limits of detection and quantification for the method (necessity to either dilute or concentrate the sample).
2. There are significant differences between the determined ion concentrations.
3. There are ions in the sample that interfere with the separation process.
4. There are substances in the sample that can significantly change the column characteristics.
5. There are solid particles in the sample that can clog the column and cause its mechanical damage or blockage of other chromatograph elements.
6. The analysis concerns solid or gaseous samples.

Due to the sample state of matter, the sample preparation in ion chromatography can be divided into methods for [144] the following:

1. Liquid samples (filtration, dilution, pH change, standard addition, derivatization, liquid–liquid extraction, solid-phase extraction (SPE), distillation, micro-diffusion, membrane techniques)
2. Gaseous samples (absorption in solutions, absorption on solid carriers, sampling facilitated with membranes)
3. Solid samples (drying, homogenization, dissolution, extraction/leaching, digestion, combustion)

Most water samples analyzed with ion chromatography do not have complex matrices (drinking water, precipitation water, surface and underground water). Such samples do not require any additional procedures. It is sufficient to filter them through a 0.45-μm filter to remove solid particles. In many cases, wastewater requires only filtration and dilution with deionized water to obtain a suitable analyte concentration range. The materials used in the filter production differ in the pore size, filtration speed, and resistance to acids, bases, and organic solvents. For biologically active samples, it is recommended to use filters with smaller diameters (e.g., 0.2 µm) [145]. To shorten the filtration time, it is possible to use filtration under the decreased or increased pressure. The latter type is preferred as it prevents the sample from direct contact with the atmospheric air. Both filtration techniques can be used in online and off-line systems [146].

At present, the dialysis methods have an enormous importance for sample preparation. Based on the mechanism, they can be divided into passive (the so-called passive dialyses) and active (the so-called Donnan dialyses) methods, as well as methods based on electrodialysis [147]. Metrohm is the global leader in the sample preparation methodologies for ion chromatography. The company presents many related publications on its website. On the other hand, Dionex offers the accelerated solvent extractor (ASE®) system, which is very efficient and useful. It is applied for sample preparation with organic solvents extracting analytes from the sample under high pressure.

The SPE methods are also very important. Many producers offer ready-to-use systems that can be used in the manual and automatic modes. Besides cartridges, extraction disks can be used [148]. Depending on the sample and analyte types, there are many SPE cartridges filled with nonpolar materials (e.g., octadecyl C_{18}, octyl C_8, ethyl, phenyl ones), polar materials (e.g., cyanide, amine, or diol materials), or ion exchangers.

Ion chromatography can also be used for analyzing solid samples. Such samples can be prepared in many ways, but finally, they must be converted into the liquid form that can be introduced directly into the analytical column. Sample dissolution or extraction can be usually carried out at room temperature. However, the sample is heatedsometimes. The extraction method selection depends on the sample matrix and the analyte ion characteristics. Water is the preferred solvent as it prevents the appearance of additional peaks in the chromatogram. Moreover, not all the columns and suppressors can be used with organic solvents. Researchers sometimes use mixtures of water and organic solvents (e.g., methanol), water with acid or base addition, or the eluent solution. If none of these methods succeeds, it may be necessary to use the following: alkaline (with NaOH, KOH, Na_2CO_3, or K_2CO_3) or acidic (with $KHSO_4$ or $K_2S_2O_7$) fusion; combustion in the atmospheric oxygen, oxygen bomb, calorimetric bomb, or pure oxygen stream; chemical digestion in open containers; pressure digestion; or UV pyrolysis.

The combustion methods are particularly useful for analyzing organic substances containing heteroatoms (e.g., chlorine, sulfur, nitrogen ones), which can be converted into suitable ions and analyzed chromatographically. It is mainly important for biological and pharmaceutical samples, polymers, and food samples [149, 150].

1.4   SELECTED METHODOLOGICAL ASPECTS OF ION DETERMINATION WITH ION CHROMATOGRAPHY

Both in ion chromatography and other analytical methods, it is necessary to have preliminary knowledge on the analyzed sample type, expected concentration range, and matrix type. Such information is often available. If obtaining such data is not possible (e.g., when the sample is classified due to the accreditation process), it is recommended to perform a preliminary analysis with simpler methods, such as test strips. Measurements of the sample electric conductivity and pH provide preliminary information on the sample ion composition. Knowing such data prevents accidental introduction of large ion amounts into the analytical column and its overloading, which leads to the irreversible changes in the column characteristics. The selection of proper separation conditions (stationary-phase type; eluent type, concentration, and flow rate; suppression and detection types) depends on many factors, such as types of the determined ions, their stability and concentrations, and the sample pH. Some manufacturers of the ion chromatography apparatus offer help in the separation process optimization with computer simulations. The software is helpful for choosing the optimum stationary-phase composition and separation conditions, modifier concentrations, and the eluent pH and ionic strength. The computer software also facilitates selecting column types and sizes, phase particle sizes, and the mobile-phase flow rate, which helps to obtain the optimum separation conditions for given analytes. The quantitative and qualitative analyses with ion chromatography are based on the same principles that underlie other chromatographic methods. The qualitative analysis consists in comparing analyte retention times in the standard and researched samples obtained under the same chromatographic conditions. Importantly, the chromatographed substance can have slightly different retention times even under the same conditions. Such a situation might be related to the column wear-off and the analyzed sample pH and matrix. The quantitative analysis consists in measuring the height or area of the obtained peaks as they are proportional to the amount of the determined component. It must be remembered that the peak sizes for different substances whose concentrations are the same usually differ, which is caused by their different detectability in a given detector.

In suppressed ion chromatography, the so-called injection peak appears at the beginning of the chromatogram, just before the fluoride and lower carboxylic acid ion peaks (if they occur in the sample). The peak can affect the detection quality for ions whose retention times are similar (fluorides, formates, acetates). In extreme cases, it can make determinations impossible. In nonsuppressed ion chromatography, two peaks that do not come from the sample ions can appear in the chromatogram. The first one is the injection peak, whereas the other one is the system peak. Both of them depend on the eluent pH, type, and concentration; sample ion concentration; sample volume and pH; and applied detector. In ion exclusion chromatography, there are two system peaks, that is, the negative and positive ones. The first one is observed when the sample introduced into the analytical column contains fewer active ions than their concentration in the eluent. The other one occurs in the opposite situation. In the chromatographic analysis course, researchers must be certain that all the components of the previously chromatographed sample have been completely removed from the column. If such a situation is not provided, they can appear in the following chromatograms in different retention times.

Water and chemical reagents used to prepare eluents and standard and regeneration solutions should be as pure as possible. Modern ion chromatographs contain in-built filtration systems. However, preliminary sample filtration is recommended to extend the analytical column durability. All applied solutions, standard samples, regeneration solutions, water, and eluents should be free from solid particles with diameters μ of 0.45 μm. Such particles can cause the back pressure increase in the analytical column and change the column characteristics. Water for the eluent preparation should be degassed, which prevents the baseline deformation and unstable system work. To achieve that, a stream of a neutral gas (e.g., helium) is passed through the eluent solution for several minutes. It can also be done with a water pump, vacuum pump, or ultrasonic bath. Modern chromatographs have in-built degassers. It is necessary to follow strict cleanness and safety standards in the laboratory when samples are analyzed at the trace and ultratrace levels. It concerns the use of suitable gloves and glasses. The room in which such analyses are conducted must be free from inorganic pollutants. Concentrated acids or bases must not be used there. It is also important to appropriately secure the containers with samples that are placed in the automatic feeder.

When the analytical column is systematically used, its characteristics deteriorate. Consequently, the retention times of particular ions are reduced, whereas their peak shapes may be blurred and asymmetrical. The appropriate use of analytical columns requires the following: securing columns with systematically exchanged precolumns, using columns according to the manufacturer's guidelines (pH, flow rate, pressure, organic solvents), using ultrapure water for eluent preparation, avoiding bacterial growth and sediment precipitation, avoiding drying and thermal/mechanical shock, avoiding column overloading, filtering each sample, and using cartridges for sample preparation whenever it is necessary.

The bacterial and algae growth can cause serious damage (unstable baseline) and requires laborious and time-consuming cleaning of the whole system. The situation depends on the applied eluent and sample types, as well as temperature and insolation of the room in which the chromatograph works. To prevent bacterial or algae growth, it is possible to add a small amount (e.g., 2%) of methanol or acetone to the eluent. Nevertheless, organic solvents can disturb suitable operations of some suppressors and analytical columns. The main problems related to ion chromatography analyses usually concern the following: retention times, baseline instability, registered signal quantity, pressure, background conductivity, and shapes of chromatographic peaks. When it comes to the analyte retention times, the problems concern their reduction, extension, or irregular changes. Also, the problems related to the baseline noises normally concern an air-locked pump and/or an undegassed eluent. Temperature affects the noise size. That is why the chromatograph should work under stable environmental conditions and its temperature should be 5 °C higher than the ambient temperature. High noises can also appear in the system due to the contaminated analytical column or sample filter. Consequently, they should be systematically exchanged. When it comes to the registered peak size, the problems usually concern situations in which there is no signal or it is too high. Additionally, the peaks may be shortened. There are different reasons for such a situation. Some of them are fairly obvious, such as the lack of certain connections, leakages in the system, inappropriately selected eluent or detector working range. If the registered signal is too low/high, it can be caused by the insufficient volume of the introduced sample, inappropriate position of the injection valve, or incorrect parameters of the detector work. The peak shortening (the signal is too high) is caused by the analyte concentration that is too high and the resulting analytical column overloading. In such a case, it is necessary to use a lower sample volume (smaller injection loop) and dilute the sample to the appropriate concentration range. It is also required to check the settings of the measurement range and full detector working range during the chromatogram recording.

The pressure problems in the chromatographic system are usually related to the excessive pressure value, lack of pressure or its very low level, and its irregular changes (value increases and decreases). If the pressure is too high, the place in which the analytical circuit is blocked must be located. Afterward, the eluent flow must be checked throughout its whole course. The pressure increase can also be caused by using inappropriate organic solvent additions and analytical columns that are unsuitable for specific solvents. The lack of pressure or its fluctuation in the system is normally related to leakages and air-locking of the pump and/or eluent. The background conductivity problems usually concern the following: application of an unsuitable eluent, inappropriate suppressor work, CO_2 absorption in the eluent, poor quality of water and reagents, and improper parameters of the detector work. When cations are separated, the peak overlapping can be caused by the unsuitable sample pH. Importantly, cation standards for spectroscopic methods (e.g., AAS, ICP) or voltammetric methods are not suitable for ion chromatography.

1.5 ION CHROMATOGRAPHY DEVELOPMENT PERSPECTIVES

Ion chromatography was officially established in 1975. Since then, it has evolved from a simple method for separating main inorganic anions and cations in water into a sophisticated separation technique that may be coupled with modern detectors to detect trace substance amounts in various gaseous, liquid, and solid samples. Many changes introduced in the stationary phases of the ion-exchange columns have helped to extend the range of the applied eluents and detection methods. There has also been an enormous development in the sample preparation methods and suppression techniques. At present, the greatest challenges in ion chromatography are related to introducing new ion-exchange stationary phases, improving the suppressor operation efficiency, lowering the limits of detection and quantification for analyte ions, elaborating new sample preparation methods, extending the analysis range with new organic and inorganic substances, increasing the use of different ion chromatography types in the molecular biology and genetics research (genomics, proteomics, metabolomics, transcriptomics), elaborating new standard and detection methods, and apparatus miniaturization. Even though ion chromatography is in its early forties, it is still a developing and prospective instrumental analytical method that has many applications extending far beyond the determination of inorganic ions.

1.6 REFERENCES

[1] Tswett, M.S. (1906) Physikalisch-chemische studien uber das chlorophyll die adsorptionen. *Berichte der Deutschen Botanischen Gesellschaft*, **24**, 316.

[2] Small, H. (1989) *Ion Chromatography*, Plenum Press, New York.

[3] Haddad, P.R. and Jackson, P.E. (1990) *Ion chromatography: principles and applications. Journal of Chromatography Library*, **46**, 1.

[4] Weiss, J. (2004) *Handbook of Ion Chromatography*, 3rd edn, Wiley-VCH.

[5] Fritz, J.S. and Gjerde, D.T. (2009) *Ion Chromatography*, 4th edn, Wiley VCH.

[6] Fritz, J.S. (1991) Principles and applications of ion-exclusion chromatography. *Journal of Chromatography A*, **546**, 111.

[7] Cecchi, T. (2008) Ion pairing chromatography. *Critical Reviews in Analytical Chemistry*, **38**, 161.

[8] Gennaro, M.C. and Angelino, S. (1997) Separation and determination of inorganic anions by reversed-phase high-performance liquid chromatography (Review). *Journal of Chromatography A*, **789**, 181.

[9] Buszewski, B. and Noga, S. (2012) Hydrophilic interaction liquid chromatography (HILIC) – a powerful separation technique. *Analytical and Bioanalytical Chemistry*, **402**, 231.

[10] Small, H. and Bowman, B. (1998) Ion chromatography: a historical perspective. *American Laboratory*, **10**, 1.

[11] Small, H., Stevens, T.S., and Bauman, W.C. (1975) Novel ion exchange chromatographic method using conductometric detection. *Analytical Chemistry*, **47**, 1801.

[12] Gjerde, D.T., Fritz, J.S., and Schmuckler, G. (1978) Anion chromatography with low-conductivity eluents. *Journal of Chromatography A*, **186**, 509.

[13] Adams, B.A. and Holms, E.I. (1935) Adsorptive properties of synthetic resins. *Journal of the Society of Chemical Industry, London*, **54**, 234.

[14] Wheaton, R.M. and Bauman, W.C. (1953) Ion exclusion. A unit operation utilizing ion exchange materials. *Industrial and Engineering Chemistry*, **45**, 228.

[15] Small, H. (1991) Twenty years of ion chromatography. *Journal of Chromatography A*, **546**, 3.

[16] Wixom, R.L. and Gehrke, C.W. (eds) (2010) *Chromatography: A Science of Discovery*, Wiley.

[17] Dasgupta, P.K. (1992) Ion chromatography the state of the art. *Analytical Chemistry*, **64**, 775.

[18] Lamb, J.D., Smith, R.D., Anderson, R.C., and Mortensen, M.K.J. (1994) Anion separations on columns based on transition metal–macrocycle complex exchange sites. *Journal of Chromatography A*, **671**, 55.

[19] Michalski, R. (2006) Ion chromatography as a reference method for the determination of inorganic ions in water and wastewater. *Critical Reviews in Analytical Chemistry*, **36**, 107.

[20] Walton, H.F. and Rocklin, R.D. (1990) *Ion Exchange in Analytical Chemistry*, CRC Press, London.

[21] Quershi, M. and Varshney, K.G. (1991) *Inorganic Ion Exchangers in Chemical Analysis*, CRC Press, London.

[22] Weiss, J. and Jensen, D. (2003) Modern stationary phases for ion chromatography. *Analytical and Bioanalytical Chemistry*, **375**, 81.

[23] Schomburg, G. (1988) Stationary phases in high performances liquid chromatography. Chemical modification by polymer coating. *LC–GC The Magazine of Separation Science*, **6**, 36.

[24] Jackson, P.E., Weigert, C., and Pohl, C.A. (2000) Determination of inorganic anions in environmental waters with a hydroxide-selective column. *Journal of Chromatography A*, **884**, 175.

[25] Krokhin, O.V., Smolenkov, A.D., Svintsova, N.V. *et al.* (1995) Modified silica as a stationary phase for ion chromatography. *Journal of Chromatography A*, **706**, 93.

[26] Paull, B. and Nesterenko, P.N. (2005) New possibilities in ion chromatography using porous monolithic stationary-phase media. *Trends in Analytical Chemistry*, **24**, 295.

[27] Shellie, R.A., Tyrrell, E., and Pohl, C.A. (2008) Column selection for comprehensive multidimensional ion chromatography. *Journal of Separation Science*, **31**, 3287.

[28] Nesterenko, E.P., Nesterenko, P.N., and Paull, B. (2009) Zwitterionic ion-exchangers in ion chromatography: a review of recent developments. *Analytica Chimica Acta*, **652**, 3.

[29] Rocklin, R.D., Pohland, C.A., and Schibler, J.A. (1987) Gradient elution in ion chromatography. *Journal of Chromatography A*, **411**, 107.

[30] Mongay, C., Olmos, C., and Pastore, A. (1994) Prediction of inorganic and organic ion behaviour with polyvalent eluents in ion chromatography. *Journal of Chromatography A*, **683**, 335.

[31] Tarter, J.G. (1989) Eluent selection criteria for the simultaneous determination of anions and cations. *Journal of Chromatographic Science*, **27**, 462.

[32] Saari-Nordhaus, R., Nair, L., and Anderson, J.M. (1992) Dual-column techniques for the simultaneous analysis of anions and cations. *Journal of Chromatography A*, **602**, 127.

[33] Hoffmann, C.V., Reischl, R., and Maier, N.M. (2009) Stationary phase-related investigations of quinine-based zwitterionic chiral stationary phases operated in anion-, cation-, and zwitterion-exchange modes. *Journal of Chromatography A*, **1216**, 1147.

[34] Gautier, E.A., Gettar, R.T., Servant, R.E., and Batiston, D.A. (1995) Evaluation of 1,2-diaminocyclohexanetatracetic acid as eluent in the determination of inorganic anions and cations by ion chromatography. *Journal of Chromatography A*, **706**, 115.

[35] Stevens, T.S., Davis, J.C., and Small, H. (1981) Hollow fiber ion-exchange suppressor for ion chromatography. *Analytical Chemistry*, **53**, 1488.

[36] Rabin, S., Stillian, J., Baretto, K. *et al.* (1993) New membrane based electrolytic suppressor device for suppressed conductivity detection in ion chromatography. *Journal of Chromatography A*, **640**, 97.

[37] Saari-Nordhaus, R. and Anderson, J.M. (1997) Electrochemically regenerated ion suppressors, new suppressors for ion chromatography. *Journal of Chromatography A*, **782**, 75.

[38] Sato, S., Ogura, Y., Miyanaga, A. *et al.* (2002) Ion chromatographic system with a novel switching suppression device. *Journal of Chromatography A*, **965**, 53.

[39] Haddad, P.R., Jackson, P.E., and Shaw, M.J. (2003) Developments in suppressor technology for inorganic ion analysis by ion chromatography using conductivity detection. *Journal of Chromatography A*, **1000**, 725.

[40] Sedyohutomo, A., Lim, L., and Takeuchi, T. (2008) Development of packed-column suppressor system for capillary ion chromatography and its application to environmental waters. *Journal of Chromatography A*, **1203**, 239.

[41] Masunaga, H., Higo, Y., Ishii, M. *et al.* (2014) Development of a new suppressor for the ion chromatography of inorganic cations. *Analytical Sciences*, **30**, 447.

[42] Buchberger, W.W. (2001) Detection techniques in ion chromatography of inorganic ions. *Trends in Analytical Chemistry*, **20**, 296.

[43] Seymour, M.D., Sickafoose, J.P., and Fritz, J.S. (1734) Application of forced-flow liquid chromatography to the determination of iron. *Analytical Chemistry*, **43**, 1971.

[44] Connolly, D. and Paull, B. (2001) Fast separation of UV absorbing anions using ion-interaction chromatography. *Journal of Chromatography A*, **917**, 335.

[45] Poboży, E., Wojasińska, E., and Trojanowicz, M. (1996) Ion chromatographic separation of chromium with diphenylcarbazide-based spectrophotometric detection. *Journal of Chromatography A*, **736**, 141.

[46] Motellier, S. and Pitsch, H. (1996) Simultaneous analysis of some transition metals at ultra trace level by ion-exchange chromatography with on-line preconcentration. *Journal of Chromatography A*, **739**, 119.

[47] Motellier, S. and Pitsch, H. (1994) Determination of aluminium and its fluoro complexes in natural waters by ion chromatography. *Journal of Chromatography A*, **660**, 211.

[48] Kuroda, R., Wada, T., Kokubo, Y., and Oguma, K. (1993) Ion-interactive chromatography of nitrilotriacetatocomplexes at the rare earth elements with post-column reaction derivatisation. *Talanta*, **40**, 237.

[49] Miura, Y. and Tsbamoto, M. (1994) Ion chromatographic determination of sulfide, sulfite and thiosulfate in mixtures by means of their postcolumn reactions with iodine. *Analytical Sciences*, **10**, 595.

[50] Steinbach, A. and Wille, A. (2008) Determination of carbohydrates in foodstuffs by ion chromatography with pulsed amperometric detection. *LC GC Europe*, **15** (Suppl), 32.

[51] Okutani, T., Yamakawa, A., Sakuragawa, A., and Gotoh, R. (1993) Determination of a micro amount of sulfide by Ion chromatography with amperometric detection after coprecipitation with basic zinc carbonate. *Analytical Sciences*, **9**, 731.

[52] Cheng, J., Jandik, P., and Avdalovic, N. (2005) Pulsed amperometric detection of sulfide, cyanide, iodide, thiosulfate, bromide and thiocyanate with microfabricated disposable silver working electrodes in ion chromatography. *Analytica Chimica Acta*, **536**, 267.

[53] Liang, L.N., Cai, Y.Q., Mou, S.F., and Cheng, J. (2005) Comparisons of disposable and conventional silver working electrode for the determination of iodide using high-performance anion-exchange chromatography with pulsed amperometric detection. *Journal of Chromatography A*, **1085**, 37.

[54] de Borba, B.M. and Rohrer, J.S. (2007) Determination of biogenic amines in alcoholic beverages by ion chromatography with suppressed conductivity detection and integrated pulsed amperometric detection. *Journal of Chromatography A*, **1155**, 22.

[55] Li, R.Y., Liang, L.N., Mou, S.F., and Cai, Y.Q. (2009) Determination of monosaccharides and soybean oligosaccharides in white vinegar and tofu water by ion chromatography and pulsed amperometric detection. *Chinese Journal of Analytical Chemistry*, **37**, 725.

[56] Aleksander, P.W., Haddad, P.R., and Trojanowicz, M. (1985) Potentiometric detection in ion chromatography using a metallic copper indicator electrode. *Chromatographia*, **20**, 179.

[57] Sahin, M., Sahin, Y., and Ozcan, A. (2008) Ion chromatography-potentiometric detection of inorganic anions and cations using polypyrrole and overoxidized polypyrrole electrode. *Sensors and Actuators B: Chemical*, **133**, 5.

[58] Williams, T. and Bernett, N.W. (1992) Determination of magnesium and calcium by ion chromatography with post-column reaction fluorescence detection. *Analytica Chimica Acta*, **259**, 19.

[59] Simone, P.S., Anderson, G.T., and Emmert, G.L. (2006) On-line monitoring of μg/l levels of haloacetic acids using ion chromatography with post-column nicotinamide reaction and fluorescence detection. *Analytica Chimica Acta*, **570**, 259.

[60] Miura, Y., Hatakeyama, M., Hosino, T., and Haddad, P.R. (2002) Rapid ion chromatography of L-ascorbic acid, nitrite, sulfite, oxalate, iodide and thiosulfate by isocratic elution utilizing a postcolumn reaction with cerium(IV) and fluorescence detection. *Journal of Chromatography A*, **956**, 77.

[61] Fujiwara, T., Kurahashi, K., Kumamaru, T., and Sakjai, H. (1996) Luminol chemiluminescence with heteropoly acids and its application to the determination of arsenate, germanate, phosphate and silicate by ion chromatography. *Applied Organometallic Chemistry*, **10**, 675.

[62] Gammelgaard, B., Liao, X.P., and Jons, O. (1997) Improvement on simultaneous determination of chromium species in aqueous solution by ion chromatography and chemiluminescence detection. *Analytica Chimica Acta*, **354**, 107.

[63] Lucy, C.A. and Harrison, C.R. (2001) Chemiluminescence nitrogen detection in ion chromatography for determination of nitrogen containing anions. *Journal of Chromatography A*, **920**, 135.

[64] Liu, L., Zhou, G.M., and Zhang, X.S. (2002) On-line determination of trace copper(II) in water sample by low-pressure ion chromatography with chemiluminescence detection. *Chinese Journal of Analytical Chemistry*, **30**, 478.

[65] Wong, D., Jandik, P., Jones, W.R., and Haganaars, A. (1987) Ion chromatography of polyphosphates with direct refractive index detection. *Journal of Chromatography A*, **389**, 279.

[66] Yu, B., Nie, L., and Yao, S. (1997) Ion chromatography study of sodium potassium and ammonium in human body fluids with bulk acoustic wave detection. *Journal of Chromatography A*, **693**, 43.

[67] Xie, Y.T., Chen, P., and Wei, W.Z. (1999) Rapid analysis of preservatives in beverages by ion chromatography with series piezoelectric quartz crystal as detector. *Microchemical Journal*, **61**, 58.

[68] Wille, A. and Czyborra, S. (2007) IC–MS Coupling Theory*, Concepts and Applications*, Metrohm, Herisau.

[69] Michalski, R., Jablonska, M., Szopa, S., and Łyko, A. (2011) Application of ion chromatography with ICP-MS or MS detection to the determination of selected halides and metal/metalloids species. *Critical Reviews in Analytical Chemistry*, **41**, 133.

[70] Wetzel, R. (1979) Ion chromatography: further application. *Environmental Science and Technology*, **13**, 1214.

[71] Whitehead, P. (1997) Ultra-pure water for ion chromatography. *Journal of Chromatography A*, **770**, 115.

[72] Biesaga, M., Schmidt, N., and Seubert, A. (2004) Coupled ion chromatography for the determination of chloride, phosphate and sulphate in concentrated nitric acid. *Journal of Chromatography A*, **1026**, 195.

[73] Astel, A., Michalski, R., Łyko, A. *et al.* (2014) Characterization of bottled mineral waters marketed in Poland using hierarchical cluster analysis. *Journal of Geochemical Exploration*, **143**, 136.

[74] Betti, M., Giovannoni, G., Onor, M., and Papoff, P. (1991) Optimization of the procedure for the determination of alkali and alkaline-earth elements in sea water by suppressed ion chromatography. *Journal of Chromatography A*, **546**, 259.

[75] Lachenmeier, D.W., Attig, R., and Frank, W. (2003) The use of ion chromatography to detect adulteration of vodka and rum. *European Food Research and Technology*, **218**, 105.

[76] Singh, R.P., Smesko, S.A., and Abbas, N.M. (1997) Ion chromatographic characterization of toxic solutions: analysis and ion chemistry of biological liquids. *Journal of Chromatography A*, **774**, 21.

[77] Shotyk, W. (1993) Ion chromatography of organic-rich natural waters from peatlands, I, Cl^-, NO_2^-, Br^-, NO_3^-, HPO_4^{2-}, SO_4^{2-} and oxalate. *Journal of Chromatography A*, **640**, 309.

[78] Tanaka, S.K., Ohta, K., Haddad, P.R. *et al.* (2000) Simultaneous ion-exclusion/cation-exchange chromatography of anions and cations in acid rain waters on a weakly acidic cation-exchange resin by elution with sulfosalicylic acid. *Journal of Chromatography A*, **884**, 167.

[79] Dabek-Zlotorzynska, E. and Dlouhy, J. (1993) Automatic simultaneous determination of anions and cations in atmospheres' aerosols by ion chromatography. *Journal of Chromatography A*, **640**, 217.

[80] Polkowska, Z., Górecki, Z., and Namiesnik, J. (2002) Quality of roof runoff waters from an urban region (Gdańsk, Poland). *Chemosphere*, **49**, 1275.

[81] Hoop, M.A.G.T., Cleven, R.M.F., Staden, J.J., and Neele, J. (1996) Analysis of fluoride in rain water. Comparison of capillary electrophoresis with ion chromatography and ion-selective electrode potentiometry. *Journal of Chromatography A*, **739**, 241.

[82] Legrand, M., Angelis, M.D., and Maupetit, F. (1993) Field investigation of major and minor ions along summit (central greenland) ice cores by ion chromatography. *Journal of Chromatography A*, **640**, 251.

[83] Döscher, A., Schwikowski, M., and Gäggeler, H.W. (1995) Cation trace analysis of snow and firn samples from high-alpine sites by ion chromatography. *Journal of Chromatography A*, **706**, 249.

[84] Sawicki, E., Mulik, J.D., and Wittgenstein, E. (1979) *Ion Chromatographic Analysis of Environmental Pollutants*, Ann Arbor Science Publishers, Michigan, USA.

[85] Krochmal, D. and Kalina, A. (1997) A method of nitrogen dioxide and sulphur dioxide determination in ambient air by use of passive samplers and ion chromatography. *Atmospheric Environment*, **31**, 3473.

[86] Michalski, R. (2010) Food analysis: ion chromatography, in *Encyclopedia of Chromatography*, 3rd edn, vol. **2** (ed J. Cazes), Taylor & Francis, CRC Press, New York, p. 909.

[87] Bhattacharyya, L. and Rohrer, J.S. (2012) *Applications of Ion Chromatography in the Analysis of Pharmaceutical and Biological Products*, Wiley, New York, USA.

[88] Michalski, R. (2014) Application of ion chromatography in clinical studies and pharmaceutical industry. *Mini Reviews in Medicinal Chemistry*, **14** (10), 862.

[89] Ehmann, T., Mantler, C., Jensen, D., and Neufang, R. (2006) Monitoring the quality of ultra-pure water in the semiconductor industry by online ion chromatography. *Microchimica Acta*, **154**, 15.

[90] Okte, Z., Bayrak, S., Fidanci, U.R., and Sel, T. (2012) Fluoride and aluminum release from restorative materials using ion chromatography. *Journal of Applied Oral Science*, **20**, 27.

[91] Mazzei, R.A. and Scuppa, S. (2006) Ion chromatographic determination of phosphorus soluble in different extracting media in fertilizers. *Journal of AOAC International*, **89**, 1243.

[92] Jeyakumar, S., Raut, V.V., and Ramakumar, K.L. (2008) Simultaneous determination of trace amounts of borate, chloride and fluoride in nuclear fuels employing ion chromatography (IC) after their extraction by pyrohydrolysis. *Talanta*, **76**, 1246.

[93] Li, M., Yang, J., Li, H.F., and Lin, J.M. (2012) Determination of trace anions in liquefied petroleum gas using liquid absorption and electrokinetic migration for enrichment followed by ion chromatography. *Journal of Separation Science*, **35**, 1365.

[94] Al Shawi, A.W. and Dahl, R. (1996) Determination of lanthanides in magnesium alloys by ion chromatography. *Analytica Chimica Acta*, **333**, 23.

[95] Mikkanen, P., Jokiniemi, J.K., Kauppinen, E.I., and Vakkilainen, E.K. (2001) Coarse ash particle characteristics in a pulp and paper industry chemical recovery boiler. *Fuel*, **80**, 987.

[96] Zhang, X.S., Jiang, X.P., and Liu, M.H. (2001) Low pressure ion chromatography in complex chemistry. *Journal of the Society of Leather Technologists and Chemists*, **85**, 16.

[97] Forstner, U. (1993) Trace elements and their compounds in environment. *International Journal of Environmental Analytical Chemistry*, **51**, 2.

[98] Templeton, D.M., Ariese, F., Cornelis, R. *et al.* (2000) Guidelines for terms related to chemical speciation and fractionation of elements. Definitions, structural aspects, and methodological approaches (IUPAC Recommendations 2000). *Pure and Applied Chemistry*, **72**, 1453.

[99] Kot, A. and Namieśnik, J. (2000) The role of speciation in analytical chemistry. *Trends in Analytical Chemistry*, **19**, 69.

[100] Ellis, L.A. and Roberts, D.J. (1997) Chromatographic and hyphenated methods for elemental speciation analysis in environmental media. *Journal of Chromatography A*, **774**, 3.

[101] Pastore, P., Lavagnini, I., Boaretto, A., and Magno, F. (1989) Ion chromatographic determination of nitrite in the presence of a large amount of chloride. *Journal of Chromatography A*, **475**, 331.

[102] Michalski, R. and Kurzyca, I. (2006) Determination of nitrogen species (nitrate, nitrite and ammonia ions) in environmental samples by ion chromatography (Review). *Polish Journal of Environmental Studies*, **15**, 5.

[103] Divjak, B. and Goessler, W. (1999) Ion chromatography speciation of sulfur-containing inorganic anions on ICP-MS as an eluent detector. *Journal of Chromatography A*, **844**, 161.

[104] Ruiz-Calero, V. and Galceran, M.T. (2005) Ion chromatographic separations of phosphorus species: a review. *Talanta*, **66**, 376.

[105] Camel, V. and Bermond, A. (1998) The use of ozone and associated oxidation process in drinking water treatment (Review). *Water Research*, **32**, 3208.

[106] Song, R., Donohoe, C., Minear, R. *et al.* (1996) Empirical modelling of bromate formation during ozonation of bromide-containing waters. *Water Research*, **30**, 1161.

[107] 1996) *Guidelines for Drinking Water – Water Quality*, Health Criteria and Other Supporting Information, 2nd edn, vol. **2**, WHO, Geneva.

[108] Michalski, R. and Łyko, A. (2013) Bromate determination. State of the art. *Critical Reviews in Analytical Chemistry*, **43**, 100.

[109] ISO 15061:2001 – Water Quality – Determination of Dissolved Bromate. Method by Liquid Chromatography of Ions.

[110] Michalski, R. and Łyko, A. (2010) Determination of bromate in water samples using post column derivatization method with triiodide. *Journal of Environmental Science and Health, Part A*, **45**, 1275.

[111] ISO 11206:2011 – Water Quality – Determination of Dissolved Bromate. Method Using Liquid Chromatography of Ions and Post Column Reaction (PCR).

[112] Urbansky, E.T. (2000) Quantitation of perchlorate ion: practices and advances applied to the analysis of common matrices. *Critical Reviews in Analytical Chemistry*, **30**, 311.

[113] Michalski, R. (2009) Application of ion chromatography for the determination of inorganic cations. *Critical Reviews in Analytical Chemistry*, **39**, 230.

[114] Shaw, M.J. and Haddad, P.R. (2004) The determination of trace metal pollutants in environmental matrices using ion chromatography. *Environmental International*, **30**, 403.

[115] Jones, P. and Nesterenko, P.N. (1997) High-performance chelation ion chromatography a new dimension in the separation and determination of trace metals. *Journal of Chromatography A*, **789**, 413.

[116] Das, A.K., Guardia, M., and Cervera, M.L. (2001) Literature survey of on-line elemental speciation in aqueous solutions (Review). *Talanta*, **55**, 1.

[117] Jabłońska-Czapla, M., Szopa, S., Grygoyć, K. *et al.* (2014) Development and validation of HPLC-ICP-MS method for the determination inorganic Cr, As and Sb speciation forms and its application for Pławniowice reservoir (Poland) water and bottom sediments variability study. *Talanta*, **120**, 475.

[118] Terasahde, P., Pantsar-Kallio, M., and Maninen, P.K.G. (1996) Simultaneous determination of arsenic species by ion chromatography ICP-MS. *Journal of Chromatography A*, **750**, 83.

[119] Li, F., Goessler, W., and Irlogic, K.J. (1999) Determination of trimethylselenium iodide, selenomethiamine, selenious acid and selenic acid using high performance liquid chromatography with on-line detection by ICP-MS or FAAS. *Journal of Chromatography A*, **830**, 337.

[120] Jackson, B.P. and Miller, W.P. (1999) Soluble arsenic and selenium species in fly ash/organic waste amended soils using ion chromatography inductively coupled plasma mass spectrometry. *Environmental Science and Technology*, **33**, 270.

[121] Seby, F., Gagman, M., Garraud, H. *et al.* (2003) Development of analytical procedures for determination of total chromium by quadrupole ICP-MS and high-resolution ICP-MS, and hexavalent chromium by HPLC-ICP-MS, in different materials used in the automotive industry. *Analytical and Bioanalytical Chemistry*, **377**, 685.

[122] Meng, Q., Fan, Z.T., Buckley, B. *et al.* (2011) Development and evaluation of a method for hexavalent chromium in ambient air using IC–ICP-MS. *Atmospheric Environment*, **45**, 2021.

[123] Kumar, U.T., Dorsey, J.G., Caruso, J.A., and Evans, E.H. (1993) Speciation of inorganic and organotin compounds in biological samples by liquid chromatography with ICP-MS detection. *Journal of Chromatography A*, **654**, 261.

[124] Yu, Z.H., Jing, M., Wang, X.R. *et al.* (2009) Simultaneous determination of multi-organotin compounds in seawater by liquid–liquid extraction-high performance liquid chromatography-inductively coupled plasma mass spectrometry. *Spectroscopy and Spectral Analysis*, **29**, 2855.

[125] Shum, S.C.K., Pang, H., and Houk, R.S. (1992) Speciation of mercury and lead compounds by microbore column liquid chromatography/ICP-MS with direct injection nebulisation. *Analytical Chemistry*, **64**, 2444.

[126] Hong, Y.S., Rifkin, E., and Bouwer, E.J. (2011) Combination of diffusive gradient in a thin film probe and IC–ICP-MS for the simultaneous determination of CH_3Hg^+ and Hg^{2+} in oxic water. *Environmental Science and Technology*, **45**, 6429.

[127] Ulrich, N. (1998) Speciation of antimony(III), antimony(V) and trimethylstiboxide by ion chromatography with inductively coupled plasma atomic emission spectrometric and mass spectrometric detection. *Analytica Chimica Acta*, **359**, 245.

[128] Canepari, S., Marconi, E., Astolfi, M.L., and Perrino, C. (2010) Relevance of Sb(III), Sb(V), and Sb-containing nano-particles in urban atmospheric particulate matter. *Analytical and Bioanalytical Chemistry*, **397**, 2533.

[129] Happel, O. and Seubert, A. (2006) Characterization of stable aluminium-citrate species as reference substances for aluminium speciation by ion chromatography. *Journal of Chromatography A*, **1108**, 68.

[130] Frankowski, M. (2014) Aluminium and its complexes in teas and fruity brew samples, speciation and ions determination by ion chromatography and high-performance liquid chromatography-fluorescence analytical methods. *Food Analytical Methods*, **7**, 1109.

[131] Krasnodebska-Ostrega, B., Sadowska, M., and Ostrowska, S. (2012) Thallium speciation in plant tissues-Tl(III) found in *Sinapis alba* L. grown in soil polluted with tailing sediment containing thallium minerals. *Talanta*, **93**, 326.

[132] Chu, Y., Wang, R., and Jiang, S. (2011) Speciation analysis of thallium by reversed-phase liquid chromatography – inductively coupled plasma mass spectrometry. *Journal of the Chinese Chemical Society*, **58**, 1.

[133] Barałkiewicz, D., Kózka, M., Piechalak, A. *et al.* (2009) Determination of cadmium and lead species and phytochelatins in pea (*Pisum sativum*) by HPLC–ICP-MS and HPLC–ESI-MS[n]. *Talanta*, **79**, 493.

[134] Ebdon, L., Hill, S.J., and Rivas, C. (1998) Lead speciation in rainwater by isotope dilution-high performance liquid chromatography inductively coupled plasma mass spectrometry. *Spectrochimica Acta*, **53**, 289.

[135] Coetzee, P.P., Fischer, J.L., and Hu, M. (2003) Simultaneous separation and determination of Tl(I) and Tl(III) by IC–ICP-OES and IC–ICP-MS. *Water SA*, **29**, 17.

[136] Wolle, M.M., Fahrenholz, T., Rahman, G.M. *et al.* (2014) Method development for the redox speciation analysis of iron by ion chromatography-inductively coupled plasma mass spectrometry and carryover assessment using isotopically labeled analyte analogues. *Journal of Chromatography A*, **1347**, 96.

[137] Arias-Borrego, A., Garcia-Barrera, T., and Gomez-Ariza, J.L. (2008) Speciation of manganese binding to biomolecules in pine nuts (Pinus pines) by two-dimensional liquid chromatography coupled to ultraviolet and inductively coupled plasma mass spectrometry detectors followed by identification by electrospray ionization. *Rapid Communications in Mass Spectrometry*, **22**, 3053.

[138] Moreda-Pineiro, J., Alonso-Rodriguez, E., Moreda-Pineiro, A. *et al.* (2012) Use of pressurized hot water extraction and high performance liquid chromatography-inductively coupled plasma-mass spectrometry for water soluble halides speciation in atmospheric particulate matter. *Talanta*, **101**, 283.

[139] Barron, L. and Paull, B. (2006) Simultaneous determination of trace oxyhalides and haloacetic acids using suppressed ion chromatography–electrospray mass spectrometry. *Talanta*, **69**, 621.

[140] Michalski, R. (2012) Application of IC–MS and IC–ICP-MS in environmental research. *Current Trends in Mass Spectrometry*, (LC-GC North America Suppl. S), 30/2.

[141] Smith, R. (2003) Before the injection – modern methods of sample preparation for separation techniques. *Journal of Chromatography A*, **1000**, 3.

[142] Slingby, R. and Kaiser, R. (2001) Sample treatment techniques and methodologies for ion chromatography. *Trends in Analytical Chemistry*, **20**, 288.

[143] Saubert, A., Frenzel, W., Schafer, H. *et al.* (2004) *Sample Preparation Techniques for Ion Chromatography*, Metrohm, Herisau, Switzerland.

[144] Haddad, P.R., Doble, P., and Macka, M. (1999) Developments in sample preparation and separation techniques for the determination of inorganic ions by ion chromatography and capillary electrophoresis. *Journal of Chromatography A*, **856**, 145.

[145] Michalski, R. (2010) Sample preparation for ion chromatography, in *Encyclopedia of Chromatography*, 3rd edn, vol. **III** (ed J. Cazes), Taylor & Francis, CRC Press, New York, p. 2106.

[146] Montgomery, R.M., Saari-Nordhaus, R., Nair, L.M., and Anderson, J.M. (1998) On-line sample preparation techniques for ion chromatography. *Journal of Chromatography A*, **804**, 55.

[147] Steiner, F. and Bohn, P. (2003) Electrodialysis as a Sample Preparation Tool for Capillary Electrophoresis of Small Anions – A Proof of Principle. *Chromatographia*, 58/1, 59–66

[148] Saari-Nordhaus, R., Nair, L.M., and Anderson, J.M. (1994) Elimination of matrix interferences in ion chromatography by the use of solid phase extraction discs. *Journal of Chromatography A*, **671**, 159.

[149] Erkelens, C., Billiet, H.A.H., Galan, L.D., and Leer, E.W.B. (1987) Origin of system peaks in single-column ion chromatography of inorganic anions using high pH borate-gluconate buffers and conductivity detection. *Journal of Chromatography A*, **404**, 67.

[150] Morita, Y., Okazaki, Y., and Kawakita, H. (2012) Determination of halogens in organic materials for RoHS directives by air combustion-ion chromatography. *Bunseki Kagaku*, **61**, 367.

[151] Pantsar-Kallio, M. and Manninen, P.K.G. (1998) Speciation of halogenides and oxyhalogens by ion chromatography–inductively coupled plasma mass spectrometry. *Analytica Chimica Acta*, **360**, 161.

[152] Echigo, S., Minear, R.A., Yamada, H., and Jackson, P.E. (2001) Comparison of three post-column reaction methods for the analysis of bromate and nitrite in drinking water. *Journal of Chromatography A*, **920**, 205.

2

MASS SPECTROMETRIC DETECTORS FOR ENVIRONMENTAL STUDIES

MARIA BALCERZAK

Department of Analytical Chemistry, Faculty of Chemistry, Warsaw University of Technology, Noakowskiego 3, Warsaw 00-664, Poland

2.1 INTRODUCTION

Analytical control of the environment for identification and quantification of chemical substances naturally occurring or released into it from various sources is strongly limited by the possibilities of analytical techniques and procedures applied. Sensitivity and selectivity of a detector exhibiting potential for the registration of the analyte signals occurring in the examined sample are critical parameters, in particular, in trace and ultratrace analysis of environmental materials of highly complex matrices, for example, soils, sediments, airborne particles, and wastes. Low concentrations of the analytes of interest and large excess of the interfering sample components make the choice of the detection method a fundamental problem. Sample preparation procedures providing an effective isolation of the analytes from the matrix and their preconcentration can broaden the applicability of the available detection techniques [1]. The development of instrumental techniques and sample preparation methods resulting in the improvement in sensitivities and selectivities of the detection provides the possibility of the determination of lower amounts of substances in more complex materials. This can result in revealing of new contaminants of potential environmental and human health impact not earlier recognized. Better understanding of environmental processes occurred is a direct consequence of such information.

Application of IC-MS and IC-ICP-MS in Environmental Research, First Edition.
Edited by Rajmund Michalski.
© 2016 John Wiley & Sons, Inc. Published 2016 by John Wiley & Sons, Inc.

Characterization of the chemical nature of substances occurring in the environment, their sources, transport, ecological and human health effects is a fundamental task of environmental analytical research. Identification and quantification of a large number and diverse groups of chemicals occurring in a large variety of complex matrices are a challenge in such studies. Personal care products, pharmaceuticals, endocrine-disrupting compounds, and hormones occurring in various environmental compartments provide the examples of a large number of substances of quite different chemical characters require monitoring [2–8]. Pesticides, perfluorinated compounds (PFCs), textile dyes, drinking water disinfection by-products, polychlorinated biphenyls (PCBs), polycyclic aromatic hydrocarbons (PAHs), flame retardants, sunscreens/UV filters, dioxins, nutrients, artificial sweeteners, microorganisms, heavy metals, and other potentially toxic inorganics are chemicals and contaminants of special environmental importance [9–12]. Metals, in particular heavy metals (e.g., Hg, Cd, Pb) of potential negative health effects, occurring in environment at elevated amounts as compared with natural levels as a result of their wide applications, present a traditional and extensive field of the environmental interest [13]. Platinum group metals (PGMs) discharged into the environment, in particular, from autocatalysts and from hospital wastes, provide a special example of relatively recent challenges in elemental and speciation analysis of a large variety of environmental materials for ultratraces of the metals [14–19]. The interest in the assessment of safety and environmental impact of nanomaterials released into the environment from various sources is growing [20, 21].

Monitoring of air, environmental waters, wastes, plants, soils, sediments, and geological materials for a large variety of chemicals occurring at different levels is a fundamental goal of analytical environmental studies [22–26]. The evaluation of potential health hazard of environmental pollutants requires analytical procedures also applicable to the examination of biological materials such as food and clinical samples (physiological fluids and tissues). A large variety of analytical approaches has been described for the investigation of the areas of focus. Procedures allowing the determination of the analytes at low, ppb, and sub-ppb levels, in complex matrices are of special interest. A number of analytical instrumental techniques have found applications in such investigations. Atomic absorption (both flame and electrothermal) spectrometry, optical emission spectrometry employing different excitation modes, atomic fluorescence spectrometry, inductively coupled plasma mass spectrometry, electroanalytical techniques, and neutron activation analysis are most often used for the evaluation of elemental composition of the examined materials. Inductively coupled plasma mass spectrometry (ICP-MS) owing to high sensitivity and selectivity is particularly attractive for such studies [27–32]. Multielement capabilities, wide linear working concentration ranges, unique possibility of element isotopic analysis [33, 34], and the possibility of high sample throughput are attractive properties of ICP-MS technique. The combinations of the technique with various separation methods, for example, chromatographic (GC and LC) [35–45] and CE [41, 46, 47], provide very effective procedures for speciation analysis of metals and metalloids in environmental and biological samples and are widely used in the examination of various metal species in biochemistry, nutrition, and medicine.

Mass spectrometric detectors coupled with different separation techniques and employing various ionization modes are commonly applied for the investigation of a large variety of organic compounds and their biotransformation processes in environmental and biological materials [48–62]. The power for specific detection and quantification of heteroatoms such as chlorine, bromine, iodine, sulfur, and phosphorous provides possibilities of the determination of various environmentally important organic compounds by ICP-MS.

Ion chromatography is particularly suitable for environmental monitoring of a large variety of ionic, inorganic, and small organic species occurring or transformed into them upon analytical procedure used, in various kind of samples: waters, wastes, plants, soils, and air [36, 63]. The technique is well established for routine monitoring of common inorganic ions in various kinds of waters and wastes [64, 65]. The development of the technique, in particular, in terms of stationary phases, eluent methodologies, detection modes, as well as sample preparation procedures, provides possibilities of rapid and relatively fast methods for a large variety of the analytes: elements and their species with inorganic and organic complexing agents, organic acids, amines, amino acids, surfactants, carbohydrates, and so on [66–73]. The determination of inorganic anions presents the largest area of applications of IC technique in environmental analysis.

Hyphenation of IC with mass spectrometric detectors results in substantial broadening of the possibilities of the application of the technique in the environmental analysis as compared with commonly used conductivity and the other detection systems [73–76]. Superior sensitivities and selectivities offered for the examination of various analyte species in more complex matrices are advantageous properties of the IC systems combining mass spectrometric, both elemental (ICP-MS) and molecular (MS with soft ionization methods), detectors [76, 77]. ICP-MS technique is one most versatile tool for elemental quantification in a large variety of environmental samples.

2.2 MASS SPECTROMETRIC DETECTORS

The detection of the analytes (atoms or molecules) by mass spectrometry requires their conversion into ions (positively or negatively charged) existing in the gaseous form, separation according to their mass-to-charge ratio (m/z), and recording. The identification of m/z ratio of the ion of interest in the mass spectrum obtained and scanning its intensity make the basis of qualitative and quantitative analysis.

Positively charged element ions make the basis of inorganic analysis. Chemical digestion of the examined sample or physical processes, for example, evaporation of volatile species, and their dissociation are preliminary steps of the elemental analysis. Ionized organic compounds, molecules or the derived molecule fragments, make the basis of their detection directly in the samples of simple matrices or after preliminary isolation from more complex examined materials. The ions (ionized atoms, molecular ions, ionized adducts, or molecule fragments) generated in the ion source are transformed into mass analyzer, which sorts them according to their m/z ratios. The signals of particular ions are recorded by the detector employed, and data obtained (*m/z* ratio and intensity) are used for the identification of the analytes occurring in the sample examined and calculation of their amounts. The investigation of the fragmentation process by identifying the nature of molecule fragments generated in the ion source is a key task for the evaluation of the structure of the molecule occurring in the examined samples and identifying of unknown compounds. The possibility of the determination of the isotopes of the elements is a unique property of MS technique substantially broadening its applicability in the analysis of complex materials, for example, biological, archeological, and environmental origin.

Technologies and characteristics of MS detectors (ion sources and mass analyzers) most often used in analytical laboratories for the examination of environmental samples are shortly described next. More detailed information is available in numerous books dealing with MS technique [29, 30, 78–88].

2.2.1 Ionization Methods

The process of ionization is fundamental, limiting the applicability of MS technique in the analysis of the examined samples for the occurrence of the elements or organic compounds. The choice of the ionization mode depends on the nature of the sample and the type of information required. Potentials of various ionization methods (electron ionization, chemical ionization, field ionization/field desorption, fast atom bombardment, thermospray ionization, electrospray ionization, atmospheric pressure chemical ionization, inductively coupled plasma and laser desorption) of the widest applications in the analysis of environmental materials are discussed next.

2.2.1.1 Electron Ionization   Electron ionization (EI), also called electron impact ionization, is the oldest and best characterized ionization method. The process of ionization occurs under collisions of a beam of high energetic electrons (mostly of 70 eV) generated by a filament (rhenium or tungsten wire) with neutral analyte molecules in an ionization chamber under vacuum conditions. The technique is particularly suitable for the analysis of gaseous samples for the occurrence of volatile organic compounds as well as element vapors. Both molecular and fragmented ions being the products of the collisions can be used for the elucidation of the molecular mass and structure of the organic analytes. Extensive fragmentation caused by an excess of energy imparted on the analyte molecules is characteristic for this ion source. The technique is considered a "hard" ionization one. Complete fragmentation of the molecular ions before they leave the ion source provides difficulties with the determination of molecular mass of the compound of the interest (no signals of the parent molecular ions in the obtained mass spectrum). Decomposition of thermally labile compounds prior the ionization due to temperature required for vaporization is a drawback of the method.

The examinations of PCBs and their metabolites isolated from environmental samples [89], polychlorinated *n*-alkanes in biota [90], 170 organic contaminants of different chemical characteristics (PAHs, PCBs, polybrominated diphenyl ethers (PBDEs), a number of pesticides and relevant metabolites) in environmental waters [91], brominated flame retardants in fish muscle tissue [92], organometals in biota, waters, and sediments [93–95] and nitrosamines in water samples [96] provide the examples of the applications of GC–EI-MS detectors in the analysis of various environmental samples for important contaminants. EI-MS can be successfully coupled with nanoscale LC. A number of various organic contaminants occurring in the environment, for example, pesticides, hormones, nitro-PAH, and endocrine-disrupting compounds were examined by direct coupling of EI-MS with nano-LC [97–99]. Simplicity of operation, the absence of matrix effects, the tolerance for salts and non-volatile buffers are strong features of LC–EI-MS systems.

2.2.1.2 Chemical Ionization Chemical ionization (CI) is based on the interaction of the ionized reactant gas (most often methane, isobutane, or ammonia) species present in a large excess in the ion source with the neutral analyte molecules resulting in their ionization. Reactant gas molecules are preliminarily ionized under the collision with electrons. Substantially higher (10^6) reactant gas pressure in the ionization chamber limits direct interaction of electrons with the analyte molecules. The ionization of the analyte is usually achieved by proton transfer from generated reactant gas ionic species. Quasimolecular ions $[M+H]^+$ are the main products of such ionization modes. The technique is classified as a soft ionization one. Substantially lower internal energy imparted on the analyte molecules as compared with EI results in less fragmentation and higher abundance of quasimolecular ions. This facilitates the interpretation of the mass spectrum obtained and, in particular, identification of the analyte signal differing by one mass unit from the signal of the molecular ion often not occurring in the EI spectrum. Acidity of the ionic reactant gas species and proton affinity of the analyte molecule affect the amount of energy exchanged as well as fragmentation process. So, the use of various reactant gases can result in differences in the mass spectra obtained by CI method. Negatively charged analyte ions being the products of the reaction with electrons or proton elimination $[M−H]^-$, as well as the adducts, for example, $[M+NH_4]^+$ in case of the use of ammonia as a reagent gas, can be formed in the CI source. Negative chemical ionization is widely used in the investigation of the analytes of high electronegativity, for example, halogens. The technique can be particularly useful in the analysis of environmental and biological samples for the occurrence of halogen-containing contaminants, for example, PCBs and pesticides.

Chemical ionization requires relatively volatile analytes. It is particularly suitable for coupling with GC. The determination of nitrosamines in various kinds of waters (surface water, groundwater treatment plants, and water from swimming pools) [100], endocrine-disrupting pesticide residues in apple matrix [101], organochloride pesticides in plasma of wild birds [102], pyrethroids and mirex insecticides in soil [103, 104], permethrin and cyfluthrin in water and sediment samples [105], fungicides captan and folpet in khaki (persimmon; flesh and peel) and cauliflower [106], organophosphate and phthalate esters in indoor air and dust [107] provide the examples of the use of CI-GC-MS for the determination of ultratrace amounts of pollutants in environmental samples. Better selectivity and sensitivity in ultratrace analysis of endocrine-disrupting pesticides were reported for CI method as compared with EI mode [101]. Solid-phase extraction is often used for separation of the analytes from the samples prior to the determination by GC-MS technique, for example, PBDEs from water samples [108]. Chemical ionization-MS technique has become a powerful tool for *in situ* analysis of traces of various air gases, for example, reactive nitrogen species (HNO_3, HO_2NO_2, PAN, and NH_3) [109].

2.2.1.3 *Field Ionization/Field Desorption* Field ionization (FI) and field desorption (FD) are closely related methods using intense electrostatic field (10^8 V cm^{-1}) for extraction of the gas-phase analyte molecules from the examined samples and producing ions for MS detection [110]. The electrostatic field is generated by applying a potential of 10 kV across the electrodes, anode and cathode.

The anode (an emitter) in FI source is in the form of a sharp tip or a tungsten wire, the surface of which can be increased by the formation of carbon whiskers or dendrites on it. The sample is introduced into the ion source in the form of volatile species generated under heating. The analyte molecules passing near the emitter are ionized by electron tunneling, and the ions formed are accelerated toward the cathode and then into the mass-analyzing system. The technique is applicable to the examination of volatile, thermally stable compounds. FI is classified as a very soft ionization technique because it produces molecular radical ions with little or no fragmentation. Lower as compared, for example, with EI, sensitivity of the detection is a disadvantage of the FI technique.

Field desorption ionization technique allows the examination of nonvolatile or thermally unstable compounds, for example, biological molecules or synthetic polymers. In FD, the sample is deposited as a film, small crystals, or solution directly on the emitter (tungsten or rhenium filament covered with carbon microneedles or whiskers similar to those used in FI) surface. Ionization and desorption of the analyte molecules occur under increasing emitter heating when the electric field is applied across the electrodes. Simple mass spectra, containing molecular and, in some cases, quasimolecular ions [M+H]$^+$ or [M+alkali metal cation]$^+$ depending on the acidity of the sample solution and the presence of alkali metal ions, are obtained. No or little fragmentation is the reason of classifying the FD as one of extremely mild ionization techniques. It is particularly suited for the ionization of high-molecular-mass and nonpolar compounds, for example, hydrocarbons. The applications of FI-MS and FD-MS in biochemical, medical, and environmental studies were discussed [111]. Continuous-flow sample introduction allows changing a pulsed ion source to continuous source, which can result in improvement of resolution, mass accuracy, and dynamic range [112]. Successful use of the FD technique to the identification of organophosphorous pesticides can serve as the example of environmental applications [113]. FD interfaced with Fourier transform ion cyclotron resonance (FT-ICR) MS provides high resolution of nonpolar species in complex mixtures [114–116].

2.2.1.4 *Fast Atom Bombardment* In the fast atom bombardment (FAB) technique, ionization is achieved through the impact of fast nonionized heavy atoms (usually Ar or Xe) targeting the sample [117]. The sample is preliminarily dispersed in a nonvolatile liquid matrix (such as glycerol, thioglycerol, *m*-nitrobenzylic alcohol, or diethanolamine) and introduced into an ionization chamber after deposition onto a

metal plate. High kinetic energy of the gas atoms transferred into sample molecules results in the ionization. The ions are generated at room temperature directly in the examined sample, volatilization is not required. Molecular, quasimolecular, and fragment positive or negative ions are created. Long lifetime of the sample and stability of beam currents (>20 min) are attractive features of FAB technique. Simultaneous ionization of the matrix can lead to fairly large background noise, which can render the investigation of small mass ions. Bombarding the sample with fast ions, Cs^+ or Ar^+, provides a complementary to FAB technique, called secondary ion mass spectrometry (SIMS). The technique is claimed to provide better sensitivity and enhanced fragmentation.

FAB technique is particularly suitable for the examination of polar and non-volatile compounds, biomolecules ($>10^4$ Da), and thermally unstable compounds [118, 119]. Analysis of quaternary amine pesticides [120] and surfactants [121] provides examples of the application of the FAB technique in the examination of environmentally important compounds.

2.2.1.5 *Thermospray Ionization*

Thermospray ionization (TI) technique uses a controlled vaporization of ions from liquid phase containing a volatile electrolyte and flowing ($1-2$ ml min^{-1}) through electrothermally heated stainless steel capillary. The technique was developed as particularly suitable for coupling with LC. The electric current supplied onto capillary is of power to vaporize the liquid sample. The aerosol formed under rapid expanding of sample vapors consists of the analytes and the liquid matrix. Droplet evaporation into the region of hot vapors, their charging, and ion ejection under the electric field are the main processes occurring under the TI conditions. The ejected ions are directly guided into the mass spectrometer. The vaporizer temperature requires optimization in the analysis of particular group of compounds to prevent their evaporation before reaching the outlet of the capillary.

Positive ($[M+H]^+$) and negative ($[M-H]^-$) analyte ions are generated dependently on the acidity or basicity of the sample solution. They may be accompanied by some degree of fragmentation ions. The ions produced are apparently identical to those present in solutions. The reaction of primary ions formed with neutral gas-phase molecules can result in formation of new product ions. The technique is best suited for the examination of low-molecular-mass (<1000 Da) compounds. It has found applications in the analysis of nonvolatile environmental contaminants, for example, pesticides, and other environmentally important compounds [122–124].

2.2.1.6 *Electrospray Ionization*

Electrospray ionization (ESI) is related to introduction of the examined sample solution under the atmospheric pressure into a strong electric field generated by applying a high voltage ($3-6$ kV) to the spraying capillary with a counter electrode located $0.3-2$ cm away. Charge accumulation on the surface of fine droplets formed leads to their disintegration, solvent evaporation, and ion desorption [125, 126]. Desolvation of the droplets is assisted by a flow of the warm nitrogen gas introduced between the spraying capillary tip and the sampling orifice located on the counter electrode for ions transportation into a mass spectrometric analyzer. The ions released from the sample examined are mostly molecular, multiply charged, protonated $[M+zH]^{z+}$ or deprotonated $[M-zH]^{z-}$ in the positive and negative modes, respectively, dependently on the conditions used. The ionization process is enhanced in the presence of, for example, formic or acetic acid (in the positive mode) and ammonium hydroxide (in the negative mode) introduced into the sample solution. Adducts with sodium, potassium, or ammonium cations allow the examination of compounds not possessing active sites being able to interact with protons.

The generation of multicharged ions provides important feature of such ionization method, allowing broadening of the mass-to-charge range that can be detected, in particular, when using mass analyzers of restricted upper m/z values, for example, quadrupole instruments, owing to reduced m/z ratios as compared with molecules that are not intact. The electrospray spectrum directly reflects the number of basic sites in the examined molecules. Deconvolution of the peaks identified in electrospray spectrum into the peak of single charge provides possibility of the evaluation of the molecular mass of the examined compound.

The ESI technique is valuable for the examination of highly ionic and exceptionally labile molecules. It is of great importance for the investigation of large biologically important compounds (e.g., proteins, peptides, aminoacids), organometallic compounds, polymers, and so on [127]. The detection of molecular mass up to at least 100,000 Da and at ppb levels is possible. ESI is particularly suited for combination with liquid chromatography. The LC-ESI-MS systems employing various chromatographic procedures and mass analyzers are widely used for the investigation of environmentally important compounds, for example, pesticides [128–132] and pharmaceuticals [133, 134] in various materials. Mobile phase containing easily sprayed solvents suitable for the further analytical steps is advantageous. Methanol–H_2O and ACN–H_2O are commonly used. The effect of the eluent composition on the ionization efficiency of ESI and atmospheric pressure chemical ionization (characterized next), also widely used for coupling with LC, was discussed [135]. Electrospray mass spectrometric detectors coupled with ion chromatography (IC) provide advantages in selectivity and sensitivity compared to widely used conductometric detection. The problem of sensitivity and selectivity is particularly important when examining complex environmental samples of extremely low concentrations of the analytes. ESI-MS is valuable for confirmation of the nature of chromatographic peaks, for example, bromate, perchlorate, haloacetic acids, and selenium species isolated from environmental and drinking waters by ion chromatography and identified only by the use of retention times [77, 136]. IC–ESI-MS technique was successfully used for the determination of oxyhalides and haloacetic acids in both soil extracts and drinking water samples [137]. Electrospray ionization-MS is a powerful tool for coupling with CE for analyzing a large variety of samples, in particular those of biological matrices [138–142].

2.2.1.7 Atmospheric Pressure Chemical Ionization   Atmospheric pressure chemical ionization (APCI) is particularly suited for the examination of compounds in solutions. It is most widely used for coupling with liquid chromatography. The mobile phase containing the eluent and the analytes is introduced into a pneumatic nebulizer with a high-speed air or nitrogen beam and droplets formed undergo desolvation followed by vaporization of the sample components under heating (300–500 °C) in the desolvation chamber. The ionization of the compounds of interest occurs under the collision with the reactant gas-phase ion at atmospheric pressure. Primary reagent gas ions are generated from the solvent vapor by corona discharges or β-particle emitters being the electron sources. The ionization can be carried out in positive or negative mode by proton transfer, charge exchange, and adduct formation. Proton transfer yields pseudomolecular and multicharged ions of the analyte molecules (in the positive mode). Adduct formation or proton subtraction provides negatively charged analyte species. The technique is best suited for relatively stable and small analyte molecules of molecular mass less than 2000 Da. High efficiency of the ionization due to numerous gas ions–analyte molecules collision is characteristic of APCI technique. Generation of molecular species with slight fragmentation classifies the technique as a soft ionization one.

The APCI was found suitable for LC-MS pesticides analysis and, for some group of such compounds, was more attractive in terms of sensitivity and limited effects of the conditions used as compared with ESI [143, 144]. Fragmentation voltage applied affects the sensitivity of the detection by APCI-MS and for some compounds, for example, in the investigation of herbicides and their transformation products in water samples [145], can provide difficulties with identification of all compounds of the interest. The APCI was found attractive for coupling with LC-MS for the determination of 42 priority pesticides and 33 priority organic pollutants from European Directive EC/464 in river water [146], caffeine in surface water [147] aminopropanone in wastewater [148], and PFCs in water, sludge, and sediment [149]. The technique may complementary act to ESI for the detection of different groups of organic compounds, for example, cytotoxic compounds in wastewaters [150]. Trace chlorophenols in waste and river water samples [151] and clam tissue [152] samples were determined by ion chromatography with APCI-MS. The APCI was successfully coupled with GC for the determination of 142 pesticide residues in fruits and vegetables [153] and 42 pesticides and 33 other organic pollutants in river water [146]. The GC–APCI system is advantageous to EI for coupling with tandem mass spectrometry owing to higher abundance of parent molecular ions [154]. Direct sampling of ambient air into the APCI source allowed continuous real-time trace analysis for the occurrence of benzene, toluene, ethylbenzene, and xylenes [155].

2.2.1.8 *Inductively Coupled Plasma*
Inductively Coupled Plasma   Inductively coupled plasma (ICP) ionization source utilizes a high-temperature (6000–10,000 K) plasma for vaporization, dissociation (atomization), excitation, and final ionization of the sample components introduced into it. The ions generated under plasma conditions are principally atomic and singly charged (ca. 1% fraction of multiply charged), which makes such source an ideal for MS detection.

The ICP source consists of ionized, electrically conductive gas (mostly argon) of high concentration of ions and electrons. The plasma is generated by the application of a high voltage spark to an argon stream flowing through one of the concentric quartz tubes of plasma torch. Electrons introduced into the argon interact with radio-frequency (RF) energy applied on a RF coil surrounding the plasma torch, which results in a chain reaction of collisions inducing ionization and ICP discharge.

Plasma is supported by continuous heating of argon by RF electric current. The examined samples (gases, liquids, or solids) can be introduced into plasma directly (gas) or after generation of volatile species (e.g., hydrides) [156, 157] of the compounds of the interest, by classical pneumatic or ultrasonic nebulization, electrothermal vaporization (ETV) [158], thermospray nebulization [159], and laser ablation (LA) [160–163]. In the plasma, the sample constituents are decomposed into atoms and ionized at a high (>90%) degree of ionization for the most elements. The generated ions are transported into the mass analyzer and the detector. A sensitivity of the ICP-MS detection strongly depends on the transmission efficiency of the ions from the atmospheric-pressure plasma to the low-pressure mass spectrometer.

The ICP-MS technique provides the most versatile element-specific detector of wide applications in the analysis of a large variety of materials: environmental and geological samples, food and agricultural products, clinical samples, pharmaceutical and numerous industrial products [28, 61, 164–167]. Virtually, all elements, metals, metalloids, and some of nonmetals can be determined by ICP-MS technique. Organic compounds undergo destruction under plasma conditions. They can be examined through the identification and quantification of specific elements, for example, heteroatoms, of their molecules.

High-speed, excellent DLs (sub-nanogram per milliliter level in most cases), wide dynamic range (up to nine orders in a single acquisition), the possibility of accurate multielemental analysis, and the unique capability of measuring isotopic ratios make ICP-MS technique the most promising for the determination of ultratraces of the elements in complex matrices. Hyphenation of ICP-MS with separation techniques, in particular with GC, LC, and CE, provides possibility of characterizing various element species occurring in the individual fractions from separation system used [37, 168–173]. Modern mass spectrometers allow simultaneous element detection by ICP-MS and organic species by, for example, ESI-MS, using a split flow from a chromatographic system.

Direct use of ICP-MS detection in the analysis of complex environmental materials can be problematic in case of extremely low concentrations determined and the occurrence of matrix and spectral interferences. High salt concentrations in the examined sample can lead to suppression or enhancement of the ICP-MS signals of the analytes. Spectral interferences, isobaric from different element ions of the same mass (e.g., ^{114}Sn and ^{114}Cd, ^{136}Ba^{2+} and ^{68}Zn^{+}) and from polyatomic groups of the same m/z ratio as the analyte isotope used for the measurement (e.g., ^{40}Ca^{16}O and ^{40}Ar^{16}O (^{56}Fe detection) and ^{40}Ar^{38}Ar (^{78}Se detection)) can occur in case of a limited resolution of the analyzer and arise with the increase in the concentrations of the interfering species. Mathematical correction, alternative sample introduction techniques, instrumental modifications, or chemical separation procedures are applied for the elimination of spectral interferences. Mathematical correction involving quantification of the interferent effect and subtraction of the analyte signal by the use of a suitable equation is the first common approach. Ease of application, no need for additional sample manipulation and speed are the advantages of such method. The use of mathematical correction can be limited by the concentration of the interferent in the examined samples, for example, 50:1 (Hf:Pt) concentration ratio in case of the determination of platinum in the presence of hafnium widely occurring in environmental samples [174]. Modification of a classical pneumatic nebulization sample introduction system to reduce the amount of water introduced into the ionization source provides the possibility of the elimination or minimizing the intensity of O^- and OH^- containing molecular ions in case of their interference with the detection. ETV, thermospray nebulization, and membrane desolvation of sample aerosol are often applied for such purposes. Direct sample introduction in the form of slurry [175], laser ablation [160, 161], and glow discharge [176] techniques can allow overcoming of some problems with selective detection of the analyte signals in the analysis of samples of complex matrices. Heterogeneity of the examined samples and low concentrations to be determined can limit the use of direct sample introduction into the plasma. In case of simple matrices, reduction of plasma power to cool conditions (600–900 W) allows elimination of interferences, for example, from argon species. Mass spectrometric instruments employing collision/dynamic reaction cells capable of the conversion of the interferent into nonactive species are widely used for improving (two orders of magnitude is possible) the selectivity of MS detection [177–180]. Substantial improvement in resolving power can be achieved by the use of double-focusing sector field mass spectrometer (SF-ICP-MS) offering substantially better resolution than commonly used quadrupole instruments [181, 182]. Preliminary chemical isolation of the analytes from matrix elements is the most effective method for overcoming spectral interferences in ICP-MS detection [183]. Apart from effective separation from the interferents, chemical procedures allow preconcentration of the analytes prior to the detection.

2.2.1.9 *Laser Desorption*

Laser desorption (LD) ionization reflects the evaporation of the compounds of the interest from the examined sample (usually in solids) under the irradiation of the sample surface by a laser pulses (of a few nanoseconds and 10^6–10^{10} W cm^{-2}). The desorbed molecules and charged particles react in vapor-phase region near the sample surface producing analyte ions guided via electric field into the inlet of a mass analyzer compatible with pulsed ionization methods. Modification of the technique by the use of a matrix capable of absorbing the laser energy provided a valuable soft ionization technique, matrix-assisted laser desorption ionization (MALDI), of substantially lower fragmentation of the examined molecules as compared with LD [184–186]. MALDI is a nondestructive vaporization and ionization technique. The sample is preliminary dissolved in a suitable solvent (typically water, methanol, acetone) and mixed with a large excess of a matrix compound (mostly weak organic acids, e.g., 2,5-dihydroxybenzoic, cinnamic, and sinapinic acids, at the ratio of 1000:1 (matrix to the analyte)) capable of absorbing laser energy, vaporizing, carrying simultaneously desorbed the analyte molecules, and ionizing them. The mixture is deposited on the MALDI sample holder and dried prior to the laser irradiation. The cocrystallization of the sample with the matrix occurs. The ions are desorbed from the solid samples through bombardment by a laser beam, most often a nitrogen laser at a wavelength of 337 nm and 4 ns pulse width. Protonation of the analyte molecules by proton and cation transfer from the excited matrix molecules can occur under the conditions used. The matrix can also serve as a proton acceptor; thus, the ionization of the analytes can be carried out in the positive or negative ion mode. Singly and multiply charged ions are the products of the ionization process [184]. Far fewer charged ions as compared with ESI are produced by MALDI. Ionization mechanism in MALDI is complex and relates a kind of the sample examined, a kind of the matrix used, the interaction of the analyte and the matrix, the laser intensity, and other conditions [184, 187]. Recent use of carbon nanotubes as a matrix allows almost elimination of background ion interference in the examination of low-molecular-mass compounds [188]. MALDI is most often combined with TOFMS analyzers allowing the examination of unlimited mass range. Coupling with ultrahigh resolution mass analyzers such as the FT-ICR improves the accuracy of the exact mass measurements [189].

The nondestructive vaporization and ionization of both large and small molecules and high sensitivity (picomolar level) make MALDI a powerful technique for the examination of high-molecular-mass (typically up to 500,000 Da) fragile, and nonvolatile biomolecules (peptides, proteins, oligonucleotides, and sugars), polymers, natural products, organic and inorganic compounds [184, 185, 189, 190]. The tendency to produce mainly singly charged species provides possibility of rapid molecular mass evaluation. The technique is of great importance for proteomic research. It has also been found applicable for the examination of organometallic and inorganic coordination complexes [185, 191]. MALDI is a valuable technique for monitoring of environmental contaminants, for example, PFCs in environmental waters [192], PAHs and estrogen [193], arsenic species in traditional medicines and diphenylolpropane in water samples [188], agrochemicals in plants [194, 195], environmental bacteria strains [196], and humic substances [197]. The applicability of UV-MALDI technique in the analysis of halogenated phenyl pesticides has recently been described [198]. The use of graphene as the matrix provides higher desorption/ionization efficiencies for PAHs and no fragmentation as compared with conventional matrix [193].

2.2.2 Mass Analyzers

The sensitivity and selectivity of MS detection substantially depend on a kind of mass analyzer capable of the separation of ions of different m/z and transmission of a sufficient number of ions to the detector yielding measurable ion currents. Mass analyzer is located between the ion source and the detector. Its potential for distinguishing between m/z differences (two peaks) is characterized by the resolution (R), which is defined as $R = m/\Delta m$ (m – the mass to be measured, Δm – the difference in masses to be identified). In general, two peaks are considered resolved if the height of the overlapping portion between them is $\leq 10\%$ of the peak height. The resolution is a characteristic property of various mass analyzers differing in the design, construction, and mode of ion separation. Quadrupole (Q), quadrupole ion trap (QIT), time of flight (TOF), and Fourier transform (FT) are the main commercially available mass analyzers [80–83, 186, 199–201]. They differ in the power of the resolution, mass range to be detected, and accuracy of the mass measurement. The choice of the mass analyzer greatly depends on the analytical problem being solved, the complexity of the examined sample, and the occurrence of interferences from the sample components, requirements in terms of selectivity and accuracy of the results.

2.2.2.1 Quadrupole Analyzers

Quadrupole (Q) analyzers are the most commonly used today. They work with ranges up to 3000–4000 m/z and readily resolve ions differing in mass by one unit (1 Da). Quadrupole analyzer consists of two electrically connected parallel pairs of cylindrical metal roads, which serve as the electrodes of a mass filter. A combination of two voltage components, a direct current (DC) and alternating radio frequency (RF), applied to the pairs of the roads makes a field within the quadrupole influencing the trajectory of ions passing through the channel between the roads. A band of ions of limited m/z value is transmitted at the adjusted combination of DC and RF potentials while the other ions collide with the roads due to unstable trajectories. Continuous rapid changes of the applied potentials allow scanning of a range of m/z masses.

The resolution achievable by using quadrupole analyzers does not ensure the separation of interferences from the same nominal mass, for example, $^{56}Fe^+$ and $^{40}Ar^{16}O^+$, $^{75}As^+$ and $^{40}Ar^{35}Cl^+$, $^{80}Se^+$ and $^{40}Ar_2^+$. Substantial improvements in the separation of most of polyatomic and isobaric interferences offer sector-field (high-resolution) mass analyzers based on double-focusing magnetic- and electric-sector field. High-resolution mass analyzer is able to focus both the energy and the m/z ratio. Each high-resolution analyzer consists of electrostatic (energy filter) and magnetic (mass filter) sectors. Electric field reduces the spread of kinetic energy of ions of m/z ratio, which results in substantially better separation of the ions of m/z ratios as compared with single separation mode. Resolving power at the 10^5 level is possible. High cost of such instruments limits their wide applications.

2.2.2.2 Tandem Mass Analyzers

Higher selectivity (a resolution of 0.1 Da) as compared with unit-mass resolution offer quadrupole analyzers arranged in linear series. They make the system of tandem mass spectrometry employing mostly triple quadrupole mass analyzers with two of them (first and third) working as mass analyzers and the second as collision/dissociation cell. The ion scan can be carried out in two modes: multiple-reaction monitoring (MRM) or neutral loss (NL). In MRM, the first quadrupole acts as a mass filter isolating precursor ions, the second as a reaction region where a selected ion is broken into fragments, and the third scanning the resulting fragments (fragment-ion scan or daughter scan). In neutral mode, the third analyzer scans a constant mass M_n being a fragmented part leaving the collision cell, providing the possibility of the elucidation of the fragmentation that leads to such loss. Neutral scan mode is used for group-specific detection, for example, CO_2 ($m/z = 44$) when detecting aromatic acids, a propylene group ($m/z = 42$) in the examination of triazine herbicides. Tandem systems with sequential combination of more analyzers (MS^n) are possible. In practice, their number is limited to three or four due to low transmission of ions in each step and loss in sensitivity. Higher ion transmission is performed in tandem systems employing the other kind of the analyzers: ion trap, time of flight, and FT-ICR [202–204].

Tandem mass spectrometers are of great potential for the analysis of a large variety of environmental pollutants, for example, bioactive pharmaceutical compounds in environmental waters and sewage sludge [205–212]; pesticide residues in environmental waters, fruits, and vegetables [213–218]; flame retardants in marine sediments and biota [219, 220]; X-ray contrast and artificial sweeteners in environmental waters [221]; polycyclic aromatic hydrocarbons in soils [222]; PFCs in water, sludge, and sediment samples [149, 223]; tributyltin in marine sediment and waters [224]; phosphate metabolites in plants [225]; and organic UV-filters and their transformation products in the aquatic environment [226]. LC–MS and LC–MS/MS systems of different mass spectrometers (single quadrupole, ion trap, and triple quadrupole) applied for the determination of steroids sex hormones, drugs, and alkylphenolic surfactants in the aquatic environment were reviewed [227]. A comprehensive coverage of theory, instrumentation, and applications of tandem mass spectrometry in the examination of biological samples (drug metabolites, proteins, and complex lipids) is available [228].

2.2.2.3 Ion Trap Analyzers   Ion trap (IT) analyzer is a device that can store the ions for an extended period of time prior to the detection [229–232]. It consists of three hyperbolically shaped electrodes, a central ring electrode, and a pair of end-cap electrodes. Both end-cap electrodes have orifices for ion injection and ion extraction. The top end-cap allows ions to be introduced into the trap. Trapping or storage of externally produced ions is achieved by applying variable RF voltage to the ring electrode, while the two end caps are kept grounded. Gaseous ions introduced through an orifice in the upper end cap circulate in a stable orbit within the ring cavity. Helium gas at a pressure of about 1 mTorr is used to reduce kinetic energy of injected ions through elastic collisions with helium atoms making ion trajectories collapse toward the center of the ion trap after several milliseconds. Depending on the value of RF voltage, ions of different m/z are trapped inside the ion trap and stored. By increasing the applied voltage, the paths of the ions of successive m/z ratios are rendered unstable and exit the trap via the bottom end-cap toward an external detector.

High sensitivities, small size, a relatively low cost as compared to the other mass spectrometers, and ease coupling to various ion sources are important attributes of ion trap analyzers. Enhanced sensitivity is a result of ion confinement in the center of the ion trap, which reduces ion losses. Ion trap analyzer can operate as stand-alone or in combination with other mass analyzers in hybrid instruments [230–233]. Possibilities of the application of IT analyzers in monitoring of organic contaminants (pharmaceuticals, pesticides) in environmental waters and wastewaters were examined [233–235].

2.2.2.4 Time-of-Flight Mass Analyzers   *Quasi*-simultaneous detection of all ions extracted from the ion source, high ion transmission, and practically unlimited mass range offer time-of-flight mass analyzers (TOF-MS) [236–238]. The separation of ions of different m/z ratios in TOF analyzers is accomplished on the basis of their different times of passing a fixed path (a flight tube, typically 2×0.5 m length, kept at high vacuum) in field free region of the instrument. The ions are introduced into the flight tube after the acceleration to the same kinetic energy that reflects in different velocities acquired by the ions of different masses. Extremely low time differences (nanoseconds) between the adjacent masses allow *quasi*-simultaneous detection of all ions introduced into the field free separation region. Detail discussions of coupling of TOF-MS analyzers with various ionization sources were published [238–241].

TOF-MS analyzers are particularly suitable for the detection of fast and variable signals. TOF ion separation mode has become the method of choice for real-time scanning of chromatographic [242] and MALDI [187] signals. Direct coupling with LC or as a part of high-resolution systems provides potential for the determination of pesticides [243–245] and pharmaceuticals [246, 247] in environmental waters and food. The potential of TOF-MS coupled with GC for the determination of various organic pollutants in environmental samples was discussed [248, 249].

2.2.2.5 *Fourier Transform Mass Analyzers*

A significant increase in mass resolution and accuracy achievable even for small numbers of ions provide FT-ICR and Orbitrap analyzers. The use of Fourier transform mathematical operations for generating mass spectra in both configurations results in their classification as Fourier transform mass spectrometric (FTMS) devices [250]. Fourier transform mass spectrometers contain an ion trap within ions are allowed to circulate in defined orbits over extended period of time [80]. The Fourier transform operation converts the time domain signal into a complex mass spectrum. Cyclotron frequency of ions of m/z ratio in a fixed magnetic field makes the basis of the detection in FT-ICR. The Orbitrap operates by radially trapping ions about a central spindle electrode. An outer barrel-like electrode is coaxial with the inner spindle-like electrode and mass/charge values are measured from the frequency of harmonic ion oscillations, along the axis of the electric field, undergone by the orbitally trapped ions [251]. Orbitrap differs from FT-ICR by the absence of magnetic field and hence significantly slower decrease in resolving power with increasing m/z.

In FTMS, the ions are observed over a period of time (time-domain signal) potentially extending up to several seconds. The related signal is generated by image currents induced on detection electrodes. It contains all characteristic frequencies of the detected ions with intensities corresponding to their amount. The frequencies contained in this signal can be separated by Fourier transformation and converted into a mass spectrum. The high mass accuracy in FTMS is due to the possibility of very accurate measurement of frequencies, which represent masses. The descriptions of theory and principles of both FT-ICR and Orbitrap mass analyzers are available [250–257].

FT-ICR mass spectrometry has become a powerful tool for the investigation of biological macromolecules [253, 254, 257]. Recent applications of high-resolution Orbitrap mass spectrometry coupled with LC for identification of pharmaceutical metabolites in reclaimed water [258], isoflavones in wastewater effluent [259], pesticide residues in fruits and vegetables [129, 260, 261], pesticides, veterinary drugs, and mycotoxins in food products [262], and acidic contaminants in wastewater effluents

[204] provide examples of the potential of the technique in the examination of environmental samples.

ACKNOWLEDGMENTS

Financial support of the work by the Warsaw University of Technology is kindly acknowledged.

2.3 REFERENCES

[1] Dean, J.R. (2003) *Methods for Environmental Trace Analysis*, John Wiley & Sons Ltd, Chichester.

[2] Kudłak, B. and Namieśnik, J. (2008) Environmental fate of endocrine disrupting compounds – analytical problems and challenges. *Critical Reviews in Analytical Chemistry*, **38**, 242.

[3] Caliman, F.A. and Gavrilescu, M. (2009) Pharmaceuticals, personal care products and endocrine disrupting agents in the environment – a review. *Clean: Soil, Air, Water*, **37**, 277.

[4] Li, Z.H. and Randak, T. (2009) Residual pharmaceutically active compounds (PhACs) in aquatic environment – status, toxicity and kinetics: a review. *Veterinarni Medicina*, **52**, 295.

[5] Corcoran, J., Winter, M.J., and Tyler, C.R. (2010) Pharmaceuticals in the aquatic environment: a critical review of the evidence for health effects in fish. *Critical Reviews in Toxicology*, **40**, 287.

[6] Kosjek, T. and Heath, E. (2011) Occurrence, fate and determination of cytostatic pharmaceuticals in the environment. *Trends in Analytical Chemistry*, **30**, 1065.

[7] Zwiener, C. (2012) Analytical challenges in environmental and geosciences. *Analytical and Bioanalytical Chemistry*, **403**, 2469.

[8] Besse, J., Latour, J.F., and Garric, J. (2012) Anticancer drugs in surface waters. What can we say about the occurrence and environmental significance of cytotoxic, cytostatic and endocrine therapy drugs? *Environment International*, **39**, 73.

[9] Richardson, S.D. (2000) Environmental mass spectrometry. *Analytical Chemistry*, **72**, 4477.

[10] Richardson, S.D. (2004) Environmental mass spectrometry: emerging contaminants and current issues. *Analytical Chemistry*, **76**, 3337.

[11] Richardson, S.D. (2006) Environmental mass spectrometry: emerging contaminants and current issues. *Analytical Chemistry*, **78**, 4021.

[12] Lange, F.T., Scheurer, M., and Brauch, H.J. (2012) Artificial sweeteners – a recently recognised class of emerging environmental contaminants: a review. *Analytical and Bioanalytical Chemistry*, **403**, 2503.

[13] Apostoli, P. (2002) Elements in environmental and occupational medicine. *Journal of Chromatography B*, **778**, 63.

[14] Rosner, G., König, H.P., and Coenen-Stass, D. (1991) *WHO. Platinum*, Environmental Health Criteria Series, No. 125. International Programme on Chemical Safety, WHO, Geneva.

[15] Zereini, F. and Alt, F. (eds) (2000) *Anthropogenic Platinum-Group Element Emissions: Their Impact on Man and Environment*, Springer-Verlag, Berlin.

[16] Melber, C., Keller, D., and Mangelsdorf, I. (2002) *WIIO. Palladium*, Environmental Health Criteria Series, No. 226. International Program on Chemical Safety, WHO, Geneva.

[17] Ravindra, K., Bencs, L., and Van Grieken, R. (2004) Platinum group elements in the environment and their health risk. *The Science of the Total Environment*, **318**, 1.

[18] Gagnon, Z.E., Newkirk, C.E., and Hicks, S. (2006) Impact of platinum group metals on the environment: a toxicological, genotoxic and analytical chemistry study. *Journal of Environmental Science and Health Part A*, **41**, 397.

[19] Reith, F., Campbell, S.G., Ball, A.S. *et al.* (2014) Platinum in Earth surface environments. *Earth-Science Reviews*, **131**, 1.

[20] Farré, M., Sanchís, J., and Barceló, D. (2011) Analysis and assessment of the occurrence, the fate and the behavior of nanomaterials in the environment. *Trends in Analytical Chemistry*, **30**, 517.

[21] López-Serrano, A., Muñoz Olivas, R., Sanz Landaluze, J., and Cámara, C. (2014) Nanoparticles: a global vision. Characterization, separation, and quantification methods. Potential environmental and health impact. *Analytical Methods*, **6**, 38.

[22] Cave, M.R., Butler, O., Cook, J.M. *et al.* (2000) Environmental analysis. *Journal of Analytical Atomic Spectrometry*, **15**, 181.

[23] Hill, S.J., Arowolo, T.A., Butler, O.T. *et al.* (2002) Atomic spectrometry update. Environmental analysis. *Journal of Analytical Atomic Spectrometry*, **17**, 284.

[24] Koester, C.J. and Moulik, A. (2005) Trends in environmental analysis. *Analytical Chemistry*, **77**, 3737.

[25] Sobrova, P., Zehnalek, J., Vojtech, A. *et al.* (2012) The effects on soil/water/plant/animal systems by platinum group elements. *Central European Journal of Chemistry*, **10**, 1369.

[26] Butler, O., Cairns, W.R.L., Cook, J.M., and Davidson, C.M. (2014) 2013 Atomic spectrometry update – a review of advances in environmental analysis. *Journal of Analytical Atomic Spectrometry*, **29**, 17.

[27] Begerow, J., Turfeld, M., and Dunemann, L. (2000) New horizons in human biomonitoring of environmentally and occupationally relevant metals – sector field ICP MS versus electrothermal AAS. *Journal of Analytical Atomic Spectrometry*, **15**, 347.

[28] Ammann, A.A. (2007) Inductively coupled plasma mass spectrometry (ICP MS): a versatile tool. *Journal of Mass Spectrometry*, **42**, 419.

[29] Becker, S. (2007) *Inorganic Mass Spectrometry: Principles and Applications*, Wiley & Sons.

[30] Hill, S.J. (ed) (2007) *Inductively Coupled Plasma Spectrometry and Its Application*, Blackwell Publishing Ltd.

[31] Beauchemin, D. (2010) Inductively coupled plasma mass spectrometry. *Analytical Chemistry*, **82**, 4786.

[32] Solà-Vázquez, A., Costa-Fernández, J.M., Pereiro, R., and Sanz-Medel, A. (2011) Plasma-based mass spectrometry for simultaneous acquisition of elemental and molecular information. *Analyst*, **136**, 246.

[33] Rodrígues-Gonzáles, P., Epov, V.N., Pecheyran, C., Amouroux, D., and Donard, O.F.X. (2012) Species-specific stable isotope analysis by the hyphenation of chromatographic techniques with MC-ICPMS. *Mass Spectrometry Reviews*, **31**, 504.

[34] Vanhaecke, F. and Degryse, P. (eds) (2012) *Isotopic Analysis. Fundamentals and Applications using ICP-MS*, Wiley VCH.

[35] Dopp, E., Hartmann, L.M., Florea, A.M., Rettenmeier, A.W., and Hirner, A.V. (2004) Environmental distribution, analysis, and toxicity of organometal(loid) compounds. *Critical Reviews in Toxicology*, **34**, 301.

[36] Jackson, P.E. (2006) Ion chromatography in environmental analysis, in *Encyclopedia of Analytical Chemistry* (ed R.A. Meyers), John Wiley & Sons Ltd, Chichester.

[37] Wang, T.B. (2007) Liquid chromatography–inductively coupled mass spectrometry (LC–ICP-MS). *Journal of Liquid Chromatography and Related Technologies*, **30**, 807.

[38] Wang, R.Y., Hsu, Y.L., Chang, L.F., and Jiang, S.J. (2007) Speciation analysis of arsenic and selenium compounds in environmental and biological samples by ion chromatography–inductively coupled plasma dynamic reaction cell mass spectrometer. *Analytica Chimica Acta*, **590**, 239.

[39] Popp, M., Hann, S., and Koellensperger, G. (2010) Environmental application of elemental speciation analysis based on liquid or gas chromatography hyphenated to inductively coupled plasma mass spectrometry – a review. *Analytica Chimica Acta*, **668**, 114.

[40] Michalski, R., Jabłońska, M., Szopa, S., and Łyko, A. (2011) Application of ion chromatography with ICP-MS or MS detection to the determination of selected halides and metal/metalloids species. *Critical Reviews in Analytical Chemistry*, **41**, 133.

[41] Pröfrock, D. and Prange, A. (2012) Inductively coupled plasma-mass spectrometry (ICP-MS) for quantitative analysis in environmental and life sciences: a review of challenges, solutions, and trends. *Applied Spectroscopy*, **66**, 843.

[42] Bouchet, S. and Bjorn, E. (2014) Analytical developments for the determination of monomethylmercury complexes with low molecular mass thiols by reverse phase liquid chromatography hyphenated to inductively coupled plasma mass spectrometry. *Journal of Chromatography A*, **1339**, 50.

[43] Cheng, H.Y., Wu, C.L., Shen, L.H. *et al.* (2014) Online anion exchange column preconcentration and high performance liquid chromatographic separation with inductively coupled plasma mass spectrometry detection for mercury speciation analysis. *Analytica Chimica Acta*, **828**, 9.

[44] Clough, R., Harrington, C.F., Hill, S.J. *et al.* (2014) Atomic spectrometry update. Review of advances in elemental speciation. *Journal of Analytical Atomic Spectrometry*, **29**, 1158.

[45] Grotti, M., Terol, A., and Todoli, J.L. (2014) Speciation analysis by small-bore HPLC coupled to ICP-MS. *Trends in Analytical Chemistry*, **61**, 92.

[46] Timerbaev, A.R., Küng, A., and Keppler, B.K. (2002) Capillary electrophoresis of platinum-group elements. Analytical, speciation and biochemical studies. *Journal of Chromatography A*, **945**, 25.

[47] Muse, J., Tripodi, V., and Lucangioli, S. (2014) An overview of capillary electrophoresis in element speciation analysis of the environment. *Current Analytical Chemistry*, **10**, 225.

[48] Ferrer, I. and Thurman, E.M. (2003) Liquid chromatography/time-of-flight/mass spectrometry (LC/TOF/MS) for the analysis of emerging contaminants. *Trends in Analytical Chemistry*, **22**, 750.

[49] Kosjek, T., Heath, E., Petrović, M., and Barceló, D. (2007) Mass spectrometry for identifying pharmaceutical biotransformation products in the environment. *Trends in Analytical Chemistry*, **26**, 1076.

[50] Moriwaki, H. (2007) Liquid chromatography–mass spectrometry for the analysis of environmental mutagens. *Current Analytical Chemistry*, **3**, 69.

[51] Barceló, D. and Petrović, M. (2008) LC–MS^2 extends to more analytes in food and the environment. *Trends in Analytical Chemistry*, **27**, 191.

[52] Wang, J. (2009) Analysis of macrolide antibiotics, using liquid chromatography–mass spectrometry, in food, biological and environmental matrices. *Mass Spectrometry Reviews*, **28**, 50.

[53] Barceló, D. and Petrović, M. (2010) Tandem MS for environmental and food analysis. *Analytical and Bioanalytical Chemistry*, **398**, 1143.

[54] Krauss, M., Singer, H., and Hollender, J. (2010) LC-high resolution MS in environmental analysis: from target screening to the identification of unknowns. *Analytical and Bioanalytical Chemistry*, **397**, 943.

[55] Petrović, M., Farré, M., Lopez de Alda, M. *et al.* (2010) Recent trends in the liquid chromatography–mass spectrometry analysis of organic contaminants in environmental samples. *Journal of Chromatography A*, **1217**, 4004.

[56] Evans, E.H., Pisonero, J., Smith, C.M.M., and Taylor, R.N. (2014) Atomic spectrometry updates: review of advances in atomic spectrometry and related techniques. *Journal of Analytical Atomic Spectrometry*, **29**, 773.

[57] Gavina, J.M.A., Yao, C.H., and Feng, Y.L. (2014) Recent developments in DNA adduct analysis by mass spectrometry: a tool for exposure biomonitoring and identification of hazard for environmental pollutants. *Talanta*, **130**, 475.

[58] Gika, H.G., Theodoridis, G.A., Plumb, R.S., and Wilson, I.D. (2014) Current practice of liquid chromatography–mass spectrometry in metabolomics and metabonomics. *Journal of Pharmaceutical and Biomedical Analysis*, **87**, 12.

[59] Hirayama, A., Wakayama, M., and Soga, T. (2014) Metabolome analysis based on capillary electrophoresis-mass spectrometry. *Trends in Analytical Chemistry*, **61**, 215.

[60] Jakimska, A., Kot-Wasik, A., and Namieśnik, J. (2014) The current state-of-the-art in the determination of pharmaceutical residues in environmental matrices using hyphenated techniques. *Critical Reviews in Analytical Chemistry*, **44**, 277.

[61] Vogiatzis, C.G. and Zachariadis, G.A. (2014) Tandem mass spectrometry in metallomics and the involving role of ICP-MS detection: a review. *Analytica Chimica Acta*, **819**, 1.

[62] Barón, E., Eljarrat, E., and Barceló, D. (2014) Gas chromatography/tandem mass spectrometry method for the simultaneous analysis of 19 brominated compounds in environmental and biological samples. *Analytical and Bioanalytical Chemistry*, **406**, 7667.

[63] Michalski, R. (2010) Environmental research: ion chromatography, in *Encyclopedia of Chromatography*, 3rd edn, vol. **1** (ed J. Cazes), Taylor & Francis, CRC Press, p. 802.

[64] Michalski, R. (2006) Ion chromatography as a reference method for determination of inorganic ions in water and wastewater. *Critical Reviews in Analytical Chemistry*, **36**, 107.

[65] Michalski, R. (2010) Ion chromatography: water and waste water analysis, in *Encyclopedia of Chromatography*, 3rd edn, vol. **1** (ed J. Cazes), Taylor & Francis, CRC Press, p. 1251.

[66] Weiss, J. and Jansen, D. (2003) Modern stationary phases for ion chromatography. *Analytical and Bioanalytical Chemistry*, **375**, 81.

[67] Shaw, M.J. and Haddad, P.R. (2004) The determination of trace metal pollutants in environmental matrices using ion chromatography. *Environment International*, **30**, 403.

[68] Sarzanini, C. and Bruzzoniti, M.C. (2005) New materials: analytical and environmental applications in ion chromatography. *Analytica Chimica Acta*, **540**, 45.

[69] Bruzzoniti, M.C., De Carlo, R.M., and Sarzanini, C. (2011) The challenging role of chromatography in environmental problems. *Chromatographia*, **73** (Suppl. 1), S15.

[70] Karu, N., Hutchinson, J.P., Dicinoski, G. *et al.* (2012) Determination of pharmaceutically related compounds by suppressed ion chromatography: IV. Interfacing ion chromatography with universal detectors. *Journal of Chromatography A*, **1253**, 44.

[71] Olkowska, E., Polkowska, Z., and Namieśnik, J. (2013) A solid phase extraction-ion chromatography with conductivity detection procedure for determining cationic surfactants in surface water samples. *Talanta*, **116**, 210.

[72] Xue, Y., Hang, Y.P., and Luan, G.C. (2014) Ion chromatography for rapid and sensitive determination of three alkylamines in wastewater after headspace single-drop microextraction. *Analytical Letters*, **47**, 25.

[73] Brent, L.C., Reiner, J.L., Dickerson, R.R., and Sander, L.C. (2014) Method for characterization of low molecular weight organic acids in atmospheric aerosols using ion chromatography mass spectrometry. *Analytical Chemistry*, **86**, 7328.

[74] Buchberger, W.W. (2001) Detection techniques in ion chromatography of inorganic ions. *Trends in Analytical Chemistry*, **20**, 296.

[75] Seubert, A. (2001) Online coupling of ion chromatography with ICP AES and ICP MS. *Trends in Analytical Chemistry*, **20**, 274.

[76] Haddad, P.R. (2004) Ion chromatography. *Analytical and Bioanalytical Chemistry*, **379**, 341.

[77] Roehl, R., Slingsby, R., Avdalovic, N., and Jackson, P.E. (2002) Applications of ion chromatography with electrospray mass spectrometric detection to the determination of environmental contaminants in water. *Journal of Chromatography A*, **956**, 245.

[78] Ashcroft, A.E. (1997) *Ionization Methods in Organic Mass Spectrometry*, RSC Publishing.

[79] Barshick, C.M., Duckworth, D.C., and Smith, D.H. (eds) (2000) *Inorganic Mass Spectrometry, Fundamentals and Applications*, Marcel Dekker, Inc., New York.

[80] Barker, J. (1999) *Mass Spectrometry*, John Wiley & Sons Ltd, Chichester.

[81] Herbert, C.G. and Johnstone, R.A. (2002) *Mass Spectrometry Basics*, CRC Press, Boca Raton.

[82] Hoffmann, E. and Stroobant, V. (2007) *Mass Spectrometry: Principles and Applications*, John Wiley & Sons Ltd, Chichester.

[83] Dass, C. (2007) *Fundamentals of Contemporary Mass Spectrometry*, John Wiley & Sons, Hoboken, New Jersey.

[84] Boyd, T.K., Basic, C., and Bethem, R.A. (2008) *Trace Quantitative Analysis by Mass Spectrometry*, John Wiley & Sons Ltd, Chichester.

[85] Ekman, R., Silberring, J., Westman-Brinkmalm, A., and Kraj, A. (eds) (2009) *Mass Spectrometry: Instrumentation, Interpretation, and Applications*, John Wiley & Sons Ltd, United Kingdom.

[86] Cole, R.B. (ed) (2010) *Electrospray and MALDI Mass Spectrometry: Fundamentals, Instrumentation, Practicalities, and Biological Applications,* John Wiley & Sons, Hoboken, New Jersey.

[87] Gross, J. (2011) *Mass Spectrometry,* Springer-Verlag, Berlin.

[88] Lebedev, A.T. (ed) (2012) *Comprehensive Environmental Mass Spectrometry,* ILM Publications, UK.

[89] Li, X., Robertson, L.W., and Lehmler, H.L. (2009) Electron ionization mass spectral fragmentation study of sulfation derivatives of polychlorinated biphenyls. *Chemistry Central Journal,* **3**, 5.

[90] Zencak, Z., Reth, M., and Oehme, M. (2004) Determination of total polychlorinated *n*-alkane concentration in biota by electron ionization-MS/MS. *Analytical Chemistry,* **76**, 1957.

[91] Portolés, T., Mol, J.G.J., Sancho, J.V., and Hernández, F. (2014) Use of electron ionization and atmospheric pressure chemical ionization in gas chromatography coupled to time-of-flight mass spectrometry for screening and identification of organic pollutants in waters. *Journal of Chromatography A,* **1339**, 145.

[92] Kalachova, K., Cajka, T., Sandy, C. *et al.* (2013) High throughput sample preparation in combination with gas chromatography coupled to triple quadrupole tandem mass spectrometry (GC-MS/MS): A smart procedure for (ultra)trace analysis of brominated flame retardants in fish. *Talanta,* **105**, 109.

[93] Moreno, M.J., Pacheco-Arjona, J., Rodríguez-González, P. *et al.* (2006) Simultaneous determination of monomethylmercury, monobutyltin, dibutyltin and tributyltin in environmental samples by multi-elemental species-specific isotope dilution analysis using electron ionisation GC–MS. *Journal of Mass Spectrometry,* **41**, 1491.

[94] Centineo, G., Rodríguez-González, P., Blanco González, E. *et al.* (2006) Isotope dilution GC–MS routine method for the determination of butyltin compounds in water. *Analytical and Bioanalytical Chemistry,* **384**, 908.

[95] D'Ulivo, L., Yang, L., Feng, Y.L. *et al.* (2014) Speciation of organometals using a synchronizing GC–EIMS and GC–ICPMS system for simultaneous detection. *Journal of Analytical Atomic Spectrometry,* **29**, 1132.

[96] McDonald, J.A., Harden, N.B., Nghiem, L.D., and Khan, S.J. (2012) Analysis of *N*-nitrosamines in water by isotope dilution gas chromatography–electron ionisation tandem mass spectrometry. *Talanta,* **99**, 146.

[97] Cappiello, A., Famiglini, G., Palma, P., and Siviero, A. (2005) Liquid chromatography–electron ionization mass spectrometry: fields of application and evaluation of the performance of a direct-EI interface. *Mass Spectrometry Reviews,* **24**, 978.

[98] Famiglini, G., Palma, P., Pierini, E. *et al.* (2008) Organochlorine pesticides by LC–MS. *Analytical Chemistry,* **80**, 3445.

[99] Palma, P., Famiglini, G., Trufelli, H. *et al.* (2011) Electron ionization in LC–MS: recent developments and applications of the direct-EI LC–MS interface. *Analytical and Bioanalytical Chemistry,* **399**, 2683.

[100] Pozzi, R., Bocchini, P., Pinelli, F., and Galletti, G.C. (2011) Determination of nitrosamines in water by gas chromatography/chemical ionization/selective ion trapping mass spectrometry. *Journal of Chromatography A,* **1218**, 1808.

[101] Húšková, R., Matisová, E., Švorc, L. *et al.* (2009) Comparison of negative chemical ionization and electron impact ionization in gas chromatography–mass

spectrometry of endocrine disrupting pesticides. *Journal of Chromatography A*, **1216**, 4927.

[102] Rivera-Rodríguez, L.B., Rodríguez-Estrella, R., Ellington, J.J., and Evans, J.J. (2007) Quantification of low levels of organochlorine pesticides using small volumes (≤100 µl) of plasma of wild birds through gas chromatography negative chemical ionization mass spectrometry. *Environmental Pollution*, **148**, 654.

[103] Ali, M.A. and Baugh, P.J. (2003) Sorption-desorption studies of six pyrethroids and mirex on soils using GC/MS-NICI. *International Journal of Environmental Analytical Chemistry*, **83**, 923.

[104] Yasin, M., Baugh, P.J., Bonwick, G.A. *et al.* (1996) Analytical method development for the determination of synthetic pyrethroid insecticides in soil by gas chromatography–mass spectrometry operated in negative-ion chemical-ionization mode. *Journal of Chromatography A*, **754**, 235.

[105] Bonwick, G.A., Sun, C., Abdul-Latif, P. *et al.* (1995) Determination of permethrin and cyfluthrin in water and sediment by gas chromatography–mass spectrometry operated in the negative chemical ionization mode. *Journal of Chromatography A*, **707**, 293.

[106] Barreda, M., López, F.J., Villarroya, M. *et al.* (2006) Residue determination of captan and folpet in vegetable samples by gas chromatography/negative chemical ionization-mass spectrometry. *Journal of AOAC International*, **89**, 1080.

[107] Bergh, C., Torgrip, R., and Östman, C. (2010) Simultaneous selective detection of organophosphate and phthalate esters using gas chromatography with positive ion chemical ionization tandem mass spectrometry and its application to indoor air and dust. *Rapid Communications in Mass Spectrometry*, **24**, 2859.

[108] Chen, X.F., Cheng, C.G., Wang, X., and Zhao, R.S. (2012) Sensitive determination of polybrominated diphenyl ethers in environmental water samples with etched stainless steel wire based on solid-phase microextraction prior to gas chromatography–mass spectrometry. *Analytical Methods*, **4**, 2908.

[109] Huey, L.G. (2007) Measurement of trace atmospheric species by chemical ionization mass spectrometry: speciation of reactive nitrogen and future directions. *Mass Spectrometry Reviews*, **26**, 166.

[110] Lattimer, R.P. and Schulten, H.R. (1989) Field-ionization and field-desorption mass spectrometry: past, present, and future. *Analytical Chemistry*, **61**, 1201A.

[111] Schulten, H.R. (1979) Biochemical, medical and environmental applications of field-ionization and field-desorption mass spectrometry. *International Journal of Mass Spectrometry and Ion Physics*, **32**, 97.

[112] Schaub, T.M., Linden, H.B., Hendrickson, C.L., and Marshall, A.G. (2004) Continuous-flow sample introduction for field desorption/ionization mass spectrometry. *Rapid Communications in Mass Spectrometry*, **18**, 1641.

[113] Schulten, H.R. and Sun, S.E. (1981) Field desorption mass spectrometry of standard organophosphorus pesticides and their identification in waste water. *International Journal of Environmental Analytical Chemistry*, **10**, 247.

[114] Schaub, T.M., Hendrickson, C.L., Qian, K.N. *et al.* (2003) High-resolution field desorption/ionization Fourier transform ion cyclotron resonance mass analysis of nonpolar molecules. *Analytical Chemistry*, **75**, 2172.

[115] Schaub, T.M., Hendrickson, C.L., Quinn, J.P. *et al.* (2005) Instrumentation and method for ultrahigh resolution field desorption ionization Fourier transform ion cyclotron resonance mass spectrometry of nonpolar species. *Analytical Chemistry*, **77**, 1317.

[116] Smith, D.F., Schaub, T.M., Rodgers, R.P. *et al.* (2008) Automated liquid injection field desorption/ionization for Fourier transform ion cyclotron resonance mass spectrometry. *Analytical Chemistry*, **80**, 7379.

[117] Barber, M., Bordoli, R.S., Elliott, G.J. *et al.* (1982) Fast atom bombardment mass spectrometry. *Analytical Chemistry*, **54**, 645A.

[118] Tomer, K.B. (1989) Development of fast-atom bombardment combined with tandem mass spectrometry for the determination of biomolecules. *Mass Spectrometry Reviews*, **8**, 445.

[119] Capasso, R. (1999) A review on the electron ionization and fast atom bombardment mass spectrometry of polyphenols naturally occurring in olive wastes and some of their synthetic derivatives. *Phytochemical Analysis*, **10**, 299.

[120] Tondeur, Y., Sovocool, G.W., Mitchum, R.K. *et al.* (1987) Use of FAB MS/MS for analysis of quaternary amine pesticide standards. *Biomedical and Environmental Mass Spectrometry*, **14**, 733.

[121] Borgerding, A.J. and Hites, R.A. (1992) Quantitative analysis of alkylbenzenesulfonate surfactants using continuous-flow fast-atom-bombardment spectrometry. *Analytical Chemistry*, **64**, 1449.

[122] Bellar, T.A. and Budde, W.L. (1988) Determination of non-volatile organic compounds in aqueous environmental samples using liquid chromatography/mass spectrometry. *Analytical Chemistry*, **60**, 2076.

[123] Barceló, D. (1988) Application of thermospray liquid chromatography/mass spectrometry for determination organophosphorous pesticides and trialkyl and triaryl phosphates. *Biomedical Mass Spectrometry*, **17**, 363.

[124] Fischer, J.B. and Michael, J.L. (1995) Thermospray ionization liquid chromatography–mass spectrometry and chemical ionization gas chromatography–mass spectrometry of hexazinone metabolites in soil and vegetation extracts. *Journal of Chromatography A*, **704**, 131.

[125] Fenn, J.B. and Rutan, S.C. (eds) (2000) *Mechanism of Electrospray Ionization*, Elsevier, Amsterdam.

[126] Crotti, S., Seraglia, R., and Traldi, P. (2011) Review: Some thoughts on electrospray ionization mechanisms. *European Journal of Mass Spectrometry*, **17**, 85.

[127] Pramanik, B.N., Ganguly, A.K., and Gross, M.L. (eds) (2002) *Applied Electrospray Mass Spectrometry*, Marcel Dekker, New York, Basel.

[128] Wang, J., Leung, D., and Chow, W. (2010) Applications of LC/ESI-MS/MS and UHPLC QqTOF MS for the determination of 148 pesticides in berries. *Journal of Agricultural and Food Chemistry*, **58**, 5904.

[129] Wang, J., Chow, W., Leung, D., and Chang, J. (2012) Application of ultra high-performance liquid chromatography and electrospray ionization quadrupole orbitrap high-resolution mass spectrometry for determination of 166 pesticides in fruits and vegetables. *Journal of Agricultural and Food Chemistry*, **60**, 12088.

[130] Zhang, K., Wong, J.W., Yang, P. *et al.* (2012) Protocol for an electrospray ionization tandem mass spectral product ion library: development and application for identification of 240 pesticides in foods. *Analytical Chemistry*, **84**, 5677.

[131] Köck-Schulmeyer, M., Olmos, M., de Alda, M.L., and Barceló, D. (2013) Development of a multiresidue method for analysis of pesticides in sediments based on isotope dilution and liquid chromatography–electrospray-tandem mass spectrometry. *Journal of Chromatography A*, **1305**, 176.

[132] Wang, J., Chow, W., Chang, J., and Wong, J.W. (2014) Ultrahigh-performance liquid chromatography electrospray ionization Q-Orbitrap mass spectrometry for the analysis of 451 pesticide residues in fruits and vegetables: method development and validation. *Journal of Agricultural and Food Chemistry*, **62**, 10375.

[133] López-Serna, R., Pérez, S., Ginebreda, A. *et al.* (2010) Fully automated determination of 74 pharmaceuticals in environmental and waste waters by online solid phase extraction–liquid chromatography–electrospray-tandem mass spectrometry. *Talanta*, **83**, 410.

[134] García-Galán, M.J., Díaz-Cruz, S., and Barceló, D. (2013) Multiresidue trace analysis of sulfonamide antibiotics and their metabolites in soils and sewage sludge by pressurized liquid extraction followed by liquid chromatography–electrospray-quadrupole linear ion trap mass spectrometry. *Journal of Chromatography A*, **1275**, 32.

[135] Kostiainen, C. and Kauppila, T.J. (2009) Effect of eluent on the ionization process in liquid chromatography–mass spectrometry. *Journal of Chromatography A*, **1216**, 685.

[136] Urbansky, E.T., Gu, B., Magnuson, M.L. *et al.* (2000) Survey of bottled waters for perchlorate by electrospray ionization mass spectrometry (ESI-MS) and ion chromatography (IC). *Journal of the Science of Food and Agriculture*, **80**, 1798.

[137] Barron, L. and Paull, B. (2006) Simultaneous determination of trace oxyhalides and haloacetic acids using suppressed ion chromatography–electrospray mass spectrometry. *Talanta*, **69**, 621.

[138] Maxwell, E.J. and Chen, D.D.Y. (2008) Twenty years of interface development for capillary electrophoresis-electrospray ionization-mass spectrometry. *Analytica Chimica Acta*, **627**, 25.

[139] Klampfl, C.W. (2009) CE with MS detection: a rapidly developing hyphenated technique. *Electrophoresis*, **30**, S83.

[140] Timerbaev, A.R. (2009) Capillary electrophoresis coupled to mass spectrometry for biospeciation analysis: critical evaluation. *Trends in Analytical Chemistry*, **28**, 416.

[141] Bonvin, G., Schappler, J., and Rudaz, S. (2012) Capillary electrophoresis–electrospray ionization-mass spectrometry interfaces: fundamental concepts and technical developments. *Journal of Chromatography A*, **1267**, 17.

[142] Krenkova, J. and Foret, F. (2012) On-line CE/ESI/MS interfacing: recent developments and applications in proteomics. *Proteomics*, **12**, 2978.

[143] Thurman, E.M., Ferrer, I., and Barceló, D. (2001) Choosing between atmospheric pressure chemical ionization and electrospray ionization interfaces for the HPLC/MS analysis of pesticides. *Analytical Chemistry*, **73**, 5441.

[144] Titato, G.M., Bicudo, R.C., and Lanças, F.M. (2007) Optimization of the ESI and APCI experimental variables for the LC/MS determination of s-triazines, methylcarbamates, organophosphorous, benzimidazoles, carboxamide and phenylurea compounds in orange samples. *Journal of Mass Spectrometry*, **42**, 1348.

[145] Santos, T.C.R., Rocha, J.C., and Barceló, D. (2000) Determination of rice herbicides, their transformation products and clofibric acid using on-line solid-phase extraction

followed by liquid chromatography with diode array and atmospheric pressure chemical ionization mass spectrometric detection. *Journal of Chromatography A*, **879**, 3.

[146] De Almeida Azevedo, D., Lacorte, S., Vinhas, T. *et al.* (2000) Monitoring of priority pesticides and other organic pollutants in river water from Portugal by gas chromatography–mass spectrometry and liquid chromatography–atmospheric pressure chemical ionization mass spectrometry. *Journal of Chromatography A*, **879**, 13.

[147] Gardinali, P.R. and Zhao, X. (2002) Trace determination of caffeine in surface water samples by liquid chromatography–atmospheric pressure chemical ionization-mass spectrometry (LC–APCI-MS). *Environment International*, **28**, 521.

[148] Singh, S.P. and Gardinali, P.R. (2006) Trace determination of 1-aminopropanone, a potential marker for wastewater contamination by liquid chromatography and atmospheric pressure chemical ionization-mass spectrometry. *Water Research*, **40**, 588.

[149] Esparza, X., Moyano, E., de Boer, J. *et al.* (2011) Analysis of perfluorinated phosphonic acids and perfluorooctane sulfonic acid in water, sludge and sediment by LC–MS/MS. *Talanta*, **86**, 329.

[150] Castillo, M., Alonso, M.C., Riu, J. *et al.* (2001) Identification of cytotoxic compounds in European wastewaters during a field experiment. *Analytica Chimica Acta*, **426**, 265.

[151] Jin, M. and Yang, Y. (2006) Simultaneous determination of nine trace mono- and di-chlorophenols in water by ion chromatography atmospheric pressure chemical ionization mass spectrometry. *Analytica Chimica Acta*, **566**, 193.

[152] Jin, M.C. and Zhu, Y. (2006) Ion chromatography–atmospheric pressure chemical ionization mass spectrometry for the determination of trace chlorophenols in clam tissues. *Journal of Chromatography A*, **1118**, 111.

[153] Cherta, L., Portolés, T., Beltran, J. *et al.* (2013) Application of gas chromatography–(triple quadrupole) mass spectrometry with atmospheric pressure chemical ionization for the determination of multiclass pesticides in fruits and vegetables. *Journal of Chromatography A*, **1314**, 224.

[154] Portolés, T., Mol, J.G.J., Sancho, J.V., and Hernández, F. (2012) Advantages of atmospheric pressure chemical ionization in gas chromatography tandem mass spectrometry: pyrethroid insecticides as a case study. *Analytical Chemistry*, **84**, 9802.

[155] Badjagbo, K., Picard, P., Moore, S., and Sauvé, S. (2009) Direct atmospheric pressure chemical ionization-tandem mass spectrometry for the continuous real-time trace analysis of benzene, toluene, ethylbenzene, and xylenes in ambient air. *Journal of the American Society for Mass Spectrometry*, **20**, 829.

[156] Wu, L., Zheng, C., Ma, Q. *et al.* (2007) Chemical vapor generation for determination of mercury by inductively coupled mass spectrometry. *Applied Spectroscopy Reviews*, **42**, 79.

[157] Szlachciński, M. (2014) Recent achievements in sample introduction systems for use in chemical vapor generation plasma optical emission and mass spectrometry: from macro- to microanalytics. *Applied Spectroscopy Reviews*, **49**, 271.

[158] Aramendía, M., Resano, M., and Vanhaecke, F. (2009) Electrothermal vaporization-inductively coupled plasma-mass spectrometry: a versatile tool for tackling challenging samples – a critical review. *Analytica Chimica Acta*, **648**, 23.

[159] Vanhoe, H., Saverwijns, S., Parent, M. *et al.* (1995) Analytical characteristics of an inductively coupled plasma mass spectrometry coupled with a thermospray nebulization system. *Journal of Analytical Atomic Spectrometry*, **10**, 575.

[160] Durrant, S.F. and Ward, N.I. (2005) Recent biological and environmental applications of laser ablation inductively coupled plasma mass spectrometry (LA-ICP-MS). *Journal of Analytical Atomic Spectrometry*, **20**, 821.

[161] Mokgalaka, N.S. and Gardea-Torresdey, J.L. (2006) Laser ablation inductively coupled plasma mass spectrometry: principles and applications. *Applied Spectroscopy Reviews*, **41**, 131.

[162] Koch, J. and Günther, D. (2011) Review of the state-of-the-art of laser ablation inductively coupled plasma mass spectrometry. *Applied Spectroscopy*, **65**, 155A.

[163] Becker, J.S., Matusch, A., and Wu, B. (2014) Bioimaging mass spectrometry of trace elements – recent advance and applications of LA-ICP-MS: a review. *Analytica Chimica Acta*, **835**, 1.

[164] Becker, J.S. and Dietze, H.J. (1998) Inorganic trace analysis by mass spectrometry. *Spectrochimica Acta, Part B*, **53**, 1475.

[165] Brouwers, E.E.M., Tibben, M., Rosing, H. *et al.* (2008) The application of inductively coupled plasma mass spectrometry in clinical pharmacological oncology research. *Mass Spectrometry Reviews*, **27**, 67.

[166] Engelhard, C. (2011) Inductively coupled plasma mass spectrometry: recent trends and developments. *Analytical and Bioanalytical Chemistry*, **399**, 213.

[167] Balcerzak, M. (2011) Methods for the determination of platinum group elements in environmental and biological materials: a review. *Critical Reviews in Analytical Chemistry*, **41**, 214.

[168] Ammann, A.A. (2002) Speciation of heavy metals in environmental water by ion chromatography coupled to ICP-MS. *Analytical and Bioanalytical Chemistry*, **372**, 448.

[169] Casiot, C., Donard, O.F.X., and Potin-Gautier, M. (2002) Optimization of the hyphenation between capillary zone electrophoresis and inductively coupled plasma mass spectrometry for the measurement of As-, Sb-, Se- and Te-species, applicable to soil extracts. *Spectrochimica Acta, Part B*, **57**, 173.

[170] Ponce de Léon, C.A., Montes-Bayón, M., and Caruso, J.A. (2002) Elemental speciation by chromatographic separation with inductively coupled plasma mass spectrometry. *Journal of Chromatography A*, **974**, 1.

[171] Maher, W., Krikowa, F., Ellwood, M. *et al.* (2012) Overview of hyphenated techniques using an ICP-MS detector with an emphasis on extraction techniques for measurement of metalloids by HPLC–ICPMS. *Microchemical Journal*, **105**, 15.

[172] Michalski, R., Jabłońska-Czapla, M., Łyko, A., and Szopa, S. (2013) Hyphenated methods for speciation analysis, in *Encyclopedia of Analytical Chemistry*, (ed R.A. Meyers) John Wiley & Sons, Ltd. doi: 10.1002/9780470027318.a9291

[173] Jabłońska-Czapla, M., Szopa, S., Grygoyć, K. *et al.* (2014) Development and validation of HPLC–ICP-MS method for the determination inorganic Cr, As and Sb speciation forms and its application for Pławniowice reservoir (Poland) water and bottom sediments variability study. *Talanta*, **120**, 475.

[174] Parent, M., Vanhoe, H., Moens, L., and Dams, R. (1997) Investigation of HfO$^+$ interference in the determination of platinum in a catalytic converter (cordierite) by inductively coupled plasma mass spectrometry. *Talanta*, **44**, 221.

[175] Santos, M.C. and Nóbrega, J.A. (2006) Slurry nebulization in plasmas for analysis of inorganic materials. *Applied Spectroscopy Reviews*, **41**, 427.

[176] Baude, S., Broekaert, J.A.C., Delfosse, D. *et al.* (2000) Glow discharge atomic spectrometry for the analysis of environmental samples – a review. *Journal of Analytical Atomic Spectrometry*, **15**, 1516.

[177] Tanner, S.D., Baranov, V.I., and Bandura, D.R. (2002) Reaction cells and collision cells for ICP-MS: a tutorial review. *Spectrochimica Acta, Part B*, **57**, 1361.

[178] Vonderheide, A.P., Meija, J., Montes-Bayón, M., and Caruso, J.A. (2003) Use of optional gas and collision cell for enhanced sensitivity of the organophosphorus pesticides by GC–ICP-MS. *Journal of Analytical Atomic Spectrometry*, **18**, 1097.

[179] Koppenaal, D., Eiden, G.C., and Barinaga, C.J. (2004) Collision and reaction cells in atomic mass spectrometry: development, status, and applications. *Journal of Analytical Atomic Spectrometry*, **19**, 561.

[180] D'Ilio, S., Violante, N., Majorani, C., and Petrucci, F. (2011) Dynamic reaction cell ICP-MS for determination of total As, Cr, Se and V in complex matrices: still a challenge? A review. *Analytica Chimica Acta*, **698**, 6.

[181] Nonose, N. and Kubota, M. (2001) Non-spectral and spectral interferences in inductively coupled plasma high-resolution mass spectrometry. Part I. Optical characteristics of micro-plasmas observed just behind the sampler and the skimmer in inductively coupled plasma high resolution mass spectrometry. *Journal of Analytical Atomic Spectrometry*, **16**, 551.

[182] Nonose, N. and Kubota, M. (2001) Non-spectral and spectral interferences in inductively coupled plasma high-resolution mass spectrometry. Part 2. Comparison of interferences in quadrupole and high resolution inductively coupled plasma mass spectrometries. *Journal of Analytical Atomic Spectrometry*, **16**, 560.

[183] Balcerzak, M. (2009) Methods of elimination of hafnium interference in the determination of platinum in environmental samples by ICP-MS technique. *Chemical Analysis (Warsaw)*, **54**, 135.

[184] Zenobi, R. and Knochenmuss, R. (1998) Ion formation in MALDI mass spectrometry. *Mass Spectrometry Reviews*, **17**, 337.

[185] Stump, M.J., Fleming, R.C., Gong, W.H. *et al.* (2002) Matrix-assisted laser desorption mass spectrometry. *Applied Spectroscopy Reviews*, **37**, 275.

[186] El-Aneed, A., Cohen, A., and Banoub, J. (2009) Mass spectrometry, review of the basics: electrospray, MALDI, and commonly used mass analyzers. *Applied Spectroscopy Reviews*, **44**, 210.

[187] Batoy, S.M.A.B., Akhmetova, E., Miladinovic, S. *et al.* (2008) Developments in MALDI mass spectrometry: the quest for the perfect matrix. *Applied Spectroscopy Reviews*, **43**, 485.

[188] Hu, G.L., Xu, S.Y., Pan, C.S. *et al.* (2005) Matrix-assisted laser desorption/ionization time-of-flight mass spectrometry with a matrix of carbon nanotubes for the analysis of low-mass compounds in environmental samples. *Environmental Science and Technology*, **39**, 8442.

[189] Wang, H.Y., Chu, X., Zhao, Z.X. *et al.* (2011) Analysis of low molecular weight compounds by MALDI-FTICR-MS. *Journal of Chromatography B*, **879**, 1166.

[190] Lewis, J.K., Wei, J., and Siuzdak, G. (2000) Matrix-assisted laser desorption/ionization mass spectrometry in peptide and protein analysis, in *Encyclopedia of Analytical Chemistry* (ed R.A. Meyers), John Wiley & Sons Ltd, Chichester, p. 5880.

[191] Hunsucker, S.W., Watson, R.C., and Tissue, B.M. (2001) Characterization of inorganic coordination complexes by matrix-assisted laser desorption/ionization mass spectrometry. *Rapid Communications in Mass Spectrometry*, **15**, 1334.

[192] Cao, D., Wang, Z., Han, C. *et al.* (2011) Quantitative detection of trace perfluorinated compounds in environmental water samples by matrix-assisted laser desorption/ionization-time of flight mass spectrometry with 1,8-bis(tetramethylguanidino)-naphthalene as matrix. *Talanta*, **85**, 345.

[193] Zhang, J., Dong, X., Cheng, J. *et al.* (2011) Efficient analysis of non-polar environmental contaminants by MALDI-TOF MS with graphene as matrix. *Journal of the American Society for Mass Spectrometry*, **22**, 1294.

[194] Anderson, D.M.G., Carolan, V.A., Crosland, S. *et al.* (2010) Examination of the translocation of sulfonylurea herbicides in sunflower plants by matrix-assisted laser desorption/ionization mass spectrometry imaging. *Rapid Communications in Mass Spectrometry*, **24**, 3309.

[195] Mullen, A.K., Clench, M.R., Crosland, S., and Sharples, K.R. (2005) Determination of agrochemical compounds in soya plants by imaging matrix-assisted laser desorption/ionization mass spectrometry. *Rapid Communications in Mass Spectrometry*, **19**, 2507.

[196] Ruelle, V., El Moualij, B., Zorzi, W. *et al.* (2004) Rapid identification of environmental bacteria strains by matrix-assisted laser desorption/ionization time-of-flight mass spectrometry. *Rapid Communications in Mass Spectrometry*, **18**, 2013.

[197] Mugo, S.M. and Bottaro, C.S. (2004) Characterization of humic substances by matrix-assisted laser desorption/ionization time-of-flight mass spectrometry. *Rapid Communications in Mass Spectrometry*, **18**, 2375.

[198] Ivanova, B. and Spiteller, M. (2013) A novel UV-MALDI-MS analytical approach for determination of halogenated phenyl-containing pesticides. *Ecotoxicology and Environmental Safety*, **91**, 86.

[199] Niessen, W.M.A. (2003) Progress in liquid chromatography–mass spectrometry instrumentation and its impact on high-throughput screening. *Journal of Chromatography A*, **1000**, 413.

[200] Zwiener, C. and Frimmel, F.H. (2004) LC–MS analysis in the aquatic environment and in water treatment – a critical review. Part I: Instrumentation and general aspects of analysis and detection. *Analytical and Bioanalytical Chemistry*, **378**, 851.

[201] Hird, S.J., Lau, B.P.Y., Schuhmacher, R., and Krska, R. (2014) Liquid chromatography–mass spectrometry for the determination of chemical contaminants in food. *Trends in Analytical Chemistry*, **59**, 59.

[202] Petrović, M. and Barceló, D. (2006) Application of liquid chromatography/quadrupole time-of-flight mass spectrometry (LC–QqTOF-MS) in the environmental analysis. *Journal of Mass Spectrometry*, **41**, 1259.

[203] Llorca, M., Farré, M., Picó, Y., and Barceló, D. (2010) Study of the performance of three LC–MS/MS platforms for analysis of perfluorinated compounds. *Analytical and Bioanalytical Chemistry*, **398**, 1145.

[204] Cahill, M.G., Dineen, B.A., Stack, M.A., and James, K.J. (2012) A critical evaluation of liquid chromatography with hybrid linear ion trap-Orbitrap mass spectrometry for the determination of acidic contaminants in wastewater effluents. *Journal of Chromatography A*, **1270**, 88.

[205] Gros, M., Petrović, M., and Barceló, D. (2006) Multi-residue analytical methods using LC–tandem MS for the determination of pharmaceuticals in environmental and wastewater samples: a review. *Analytical and Bioanalytical Chemistry*, **386**, 941.

[206] Hao, C., Clement, R., and Yang, P. (2007) Liquid chromatography–tandem mass spectrometry of bioactive pharmaceutical compounds in the aquatic environment – a decade's activity. *Analytical and Bioanalytical Chemistry*, **387**, 1247.

[207] Gracia-Lor, E., Sancho, J.V., and Hernández, F. (2011) Multi-class determination of around 50 pharmaceuticals, including 26 antibiotics, in environmental and wastewater samples by ultra-high performance liquid chromatography–tandem mass spectrometry. *Journal of Chromatography A*, **1218**, 2264.

[208] Lopez-Serna, R., Petrović, M., and Barceló, D. (2011) Development of a fast instrumental method for the analysis of pharmaceuticals in environmental and wastewaters based on ultra high performance liquid chromatography (UHPLC)–tandem mass spectrometry (MS/MS). *Chemosphere*, **85**, 1390.

[209] Muz, M., Sönmez, M.S., Komesli, O.T. *et al.* (2012) Determination of selected natural hormones and endocrine disrupting compounds in domestic wastewater treatment plants by liquid chromatography electrospray ionization tandem mass spectrometry after solid phase extraction. *Analyst*, **137**, 884.

[210] Dorival-García, N., Zafra-Gómez, A., Camino-Sánchez, F.J. *et al.* (2013) Analysis of quinolone antibiotic derivatives in sewage sludge samples by liquid chromatography–tandem mass spectrometry: comparison of the efficiency of three extraction techniques. *Talanta*, **106**, 104.

[211] Gros, M., Rodríguez-Mozaz, S., and Barceló, D. (2013) Rapid analysis of multiclass antibiotic residues and some of their metabolites in hospital, urban wastewater and river water by ultra-high-performance liquid chromatography coupled to quadrupole-linear ion trap tandem mass spectrometry. *Journal of Chromatography A*, **1292**, 173.

[212] Tran, N.H., Hu, J.Y., and Ong, S.L. (2013) Simultaneous determination of PPCPs, endocrine-disrupting pesticides, and artificial sweeteners in environmental water samples using a single-step SPE coupled with HPLC–MS/MS and isotope dilution. *Talanta*, **113**, 82.

[213] Picó, Y., Blasco, C., and Font, G. (2004) Environmental and food applications of LC–tandem mass spectrometry in pesticide-residue analysis: an overview. *Mass Spectrometry Reviews*, **23**, 45.

[214] Alder, L., Greulich, K., Kempe, G., and Vieth, B. (2006) Residue analysis of 500 high priority pesticides: better by GC–MS or LC–MS/MS? *Mass Spectrometry Reviews*, **25**, 838.

[215] Kuster, M., López de Alda, M., and Barceló, D. (2006) Analysis of pesticides in water by liquid chromatography–tandem mass spectrometric techniques. *Mass Spectrometry Reviews*, **25**, 900.

[216] Hernández, F., Cervera, M.I., Portolés, T. *et al.* (2013) The role of GC–MS/MS with triple quadrupole in pesticide residue analysis in food and the environment. *Analytical Methods*, **5**, 5875.

[217] Kowal, S., Balsaa, P., Werres, F., and Schmidt, T.C. (2013) Fully automated standard addition method for the quantification of 29 polar pesticide metabolites in different water bodies using LC–MS/MS. *Analytical and Bioanalytical Chemistry*, **405**, 6337.

[218] Cervera, M.I., Portolés, T., López, F.J. *et al.* (2014) Screening and quantification of pesticide residues in fruits and vegetables making use of gas chromatography–quadrupole

time-of-flight mass spectrometry with atmospheric pressure chemical ionization. *Analytical and Bioanalytical Chemistry*, **406**, 6843.

[219] Barón, E., Eljarrat, E., and Barceló, D. (2012) Analytical method for the determination of halogenated norbornene flame retardants in environmental and biota matrices by gas chromatography coupled to tandem mass spectrometry. *Journal of Chromatography A*, **1248**, 154.

[220] Al-Odaini, N.A., Yim, U.H., Kim, N.S. *et al.* (2013) Isotopic dilution determination of emerging flame retardants in marine sediments by HPLC–APCI-MS/MS. *Analytical Methods*, **5**, 1771.

[221] Ens, W., Senner, F., Gygax, B., and Schlotterbeck, G. (2014) Development, validation, and application of a novel LC–MS/MS trace analysis method for the simultaneous quantification of seven iodinated X-ray contrast media and three artificial sweeteners in surface, ground, and drinking water. *Analytical and Bioanalytical Chemistry*, **406**, 2789.

[222] Shang, D.Y., Kim, M., and Haberl, M. (2014) Rapid and sensitive method for the determination of polycyclic aromatic hydrocarbons in soils using pseudo multiple reaction monitoring gas chromatography/tandem mass spectrometry. *Journal of Chromatography A*, **1334**, 118.

[223] Onghena, M., Moliner-Martinez, Y., Picó, Y. *et al.* (2012) Analysis of 18 perfluorinated compounds in river waters: comparison of high performance liquid chromatography–tandem mass spectrometry, ultra-high-performance liquid chromatography–tandem mass spectrometry and capillary liquid chromatography–mass spectrometry. *Journal of Chromatography A*, **1244**, 88.

[224] Nichols, D.S., Jordan, T.B., and Kerr, N. (2014) Determination of tributyltin in marine sediment and waters by pressurised solvent extraction and liquid chromatography–tandem mass spectrometry. *Analytical and Bioanalytical Chemistry*, **406**, 2993.

[225] Sekiguchi, Y., Mitsuhashi, N., Kokaji, T. *et al.* (2005) Development of a comprehensive analytical method for phosphate metabolites in plants by ion chromatography coupled with tandem mass spectrometry. *Journal of Chromatography A*, **1085**, 131.

[226] Gago-Ferrero, P., Díaz-Cruz, M.S., and Barceló, D. (2013) Liquid chromatography–tandem mass spectrometry for the multi-residue analysis of organic UV filters and their transformation products in the aquatic environment. *Analytical Methods*, **5**, 355.

[227] Lopez de Alda, M.J., Díaz-Cruz, S., Petrovic, M., and Barceló, D. (2003) Liquid chromatography–(tandem) mass spectrometry of selected emerging pollutants (steroid sex hormones, drugs and alkylphenolic surfactants) in the aquatic environment. *Journal of Chromatography A*, **1000**, 503.

[228] Prasain, J.K. (ed) (2012) *Tandem Mass Spectrometry – Applications and Principles*, InTech.

[229] Creaser, C.S. and Stygall, J.W. (1998) Recent developments in analytical ion trap mass spectrometry. *Trends in Analytical Chemistry*, **17**, 583.

[230] Douglas, D.J., Frank, A.J., and Mao, D. (2005) Linear ion traps in mass spectrometry. *Mass Spectrometry Reviews*, **24**, 1.

[231] Fernández, L.E.M. (2007) Introduction to ion trap mass spectrometry: application to the structural characterization of plant oligosaccharides. *Carbohydrate Polymers*, **68**, 797.

[232] March, R.E. (2009) Quadrupole ion traps. *Mass Spectrometry Reviews*, **28**, 961.

[233] Bueno, M.J.M., Aguera, A., Gómez, M.J. *et al.* (2007) Application of liquid chromatography/quadrupole-linear ion trap mass spectrometry and time-of-flight mass spectrometry to the determination of pharmaceuticals and related contaminants in wastewater. *Analytical Chemistry*, **79**, 9372.

[234] Andreu, V. and Picó, Y. (2005) Liquid chromatography–ion trap-mass spectrometry and its application to determine organic contaminants in the environment and food. *Current Analytical Chemistry*, **1**, 241.

[235] Gros, M., Rodríguez-Mozaz, S., and Barceló, D. (2012) Fast and comprehensive multi-residue analysis of a broad range of human and veterinary pharmaceuticals and some of their metabolites in surface and treated waters by ultra-high-performance liquid chromatography coupled to quadrupole-linear ion trap tandem mass spectrometry. *Journal of Chromatography A*, **1248**, 104.

[236] Price, D. and Williams, J.E. (eds) (1969) *Time-of-Flight Mass Spectrometry*, Pergamon Press, Oxford.

[237] Cotter, R.J. (1997) *Time-of-Flight Mass Spectrometry: Instrumentation and Applications in Biological Research*, American Chemical Society, Washington, DC 20036.

[238] Guilhaus, M., Selby, D., and Mlynski, V. (2000) Orthogonal acceleration time-of-flight mass spectrometry. *Mass Spectrometry Reviews*, **19**, 65.

[239] Balcerzak, M. (2003) An overview of analytical applications of time of flight-mass spectrometric (TOF-MS) analyzers and an inductively coupled plasma-TOF-MS technique. *Analytical Sciences*, **19**, 979.

[240] Mirsaleh-Kohan, N., Robertson, W.D., and Compton, R.N. (2008) Electron ionization time-of-flight mass spectrometry: historical review and current applications. *Mass Spectrometry Reviews*, **27**, 237.

[241] Yu, Q., Chen, L., Huang, R. *et al.* (2009) Laser ionization time-of-flight mass spectrometry for direct elemental analysis. *Trends in Analytical Chemistry*, **28**, 1174.

[242] Ferrer, I. and Thurman, E.M. (eds) (2009) *Liquid chromatography–Time of Flight Mass Spectrometry: Principles, Tools and Applications for Accurate Mass Analysis*, Wiley, New York, NJ.

[243] Lacorte, S. and Fernandez-Alba, A.R. (2006) Time of flight mass spectrometry applied to the liquid chromatographic analysis of pesticides in water and food. *Mass Spectrometry Reviews*, **25**, 866.

[244] Hernández, F., Sancho, J.V., Ibáñez, M., and Grimalt, S. (2008) Investigation of pesticide metabolites in food and water by LC–TOF-MS. *Trends in Analytical Chemistry*, **27**, 862.

[245] Lacina, O., Urbanova, J., Poustka, J., and Hajslova, J. (2010) Identification/quantification of multiple pesticide residues in food plants by ultra-high-performance liquid chromatography–time-of-flight mass spectrometry. *Journal of Chromatography A*, **1217**, 648.

[246] Hernández, F., Ibáñez, M., Bade, R. *et al.* (2014) Investigation of pharmaceutical and illicit drugs in waters by liquid chromatography–high-resolution mass spectrometry. *Trends in Analytical Chemistry*, **63**, 140.

[247] Vergeynst, L., van Langenhove, H., Joos, P., and Demeestere, K. (2014) Suspect screening and target quantification of multi-class pharmaceuticals in surface water based on large-volume injection liquid chromatography and time-of-flight mass spectrometry. *Analytical and Bioanalytical Chemistry*, **406**, 2533.

[248] Hernández, F., Portolés, T., Pitarch, E., and López, F.J. (2011) Gas chromatography coupled to high-resolution time-of-flight mass spectrometry to analyze trace-level organic compounds in the environment, food safety and toxicology. *Trends in Analytical Chemistry*, **30**, 388.

[249] Ferrer, I. and Thurman, E.M. (eds) (2013) Advanced techniques in gas chromatography–mass spectrometry (GC–MS-MS and GC–TOF-MS) for environmental chemistry. *Comprehensive Analytical Chemistry*, **61**, 2.

[250] Scigelova, M., Hornshaw, M., Giannakopulos, A., and Makarov, A. (2011) Fourier transform mass spectrometry. *Molecular and Cellular Proteomics*, **10** (7) M111.009431.

[251] Hu, Q., Noll, R.J., Li, H. *et al.* (2005) The Orbitrap: a new mass spectrometer. *Journal of Mass Spectrometry*, **40**, 430.

[252] Marshall, A.G., Hendrickson, C.L., and Jackson, G.S. (1998) Fourier transform ion cyclotron resonance mass spectrometry: a primer. *Mass Spectrometry Reviews*, **17**, 1.

[253] Heeren, R.M.A., Kleinnijenhuis, A.J., McDonnell, L.A., and Mize, T.H. (2004) A mini-review of mass spectrometry using high-performance FTICR-MS methods. *Analytical and Bioanalytical Chemistry*, **378**, 1048.

[254] Barrow, M.P., Burkitt, W.I., and Derrick, P.J. (2005) Principles of Fourier transform ion cyclotron resonance mass spectrometry and its application in structural biology. *Analyst*, **130**, 18.

[255] Perry, R.H., Cooks, R.G., and Noll, R.J. (2008) Orbitrap mass spectrometry: instrumentation, ion motion and applications. *Mass Spectrometry Reviews*, **27**, 661.

[256] Zubarev, R.A. and Makarov, A. (2013) Orbitrap mass spectrometry. *Analytical Chemistry*, **85**, 5288.

[257] van Agthoven, M.A., Delsuc, M.A., Bodenhausen, G., and Rolando, C. (2013) Towards analytically useful two-dimensional Fourier transform ion cyclotron resonance mass spectrometry. *Analytical and Bioanalytical Chemistry*, **405**, 51.

[258] Wang, J. and Gardinali, P.R. (2014) Identification of phase II pharmaceutical metabolites in reclaimed water using high resolution benchtop Orbitrap mass spectrometry. *Chemosphere*, **107**, 65.

[259] Cahill, M.G., Logrippo, S., Dineen, B.A. *et al.* (2015) Development and validation of a high-resolution LTQ Orbitrap MS method for the quantification of isoflavones in wastewater effluent. *Journal of Mass Spectrometry*, **50**, 112.

[260] Alder, L., Steinborn, A., and Bergelt, S. (2011) Suitability of an orbitrap mass spectrometer for the screening of pesticide residues in extracts of fruits and vegetables. *Journal of AOAC International*, **94**, 1661.

[261] Farré, M., Picó, Y., and Barceló, D. (2014) Application of ultra-high pressure liquid chromatography linear ion-trap orbitrap to qualitative and quantitative assessment of pesticide residues. *Journal of Chromatography A*, **1328**, 66.

[262] de Dominicis, E., Commissati, I., and Suman, M. (2012) Targeted screening of pesticides, veterinary drugs and mycotoxins in bakery ingredients and food commodities by liquid chromatography-high-resolution single-stage Orbitrap mass spectrometry. *Journal of Mass Spectrometry*, **47**, 1232.

3

HIGH-PERFORMANCE LIQUID CHROMATOGRAPHY COUPLED TO INDUCTIVELY COUPLED PLASMA MS/ELECTROSPRAY IONIZATION MS

JÜRGEN MATTUSCH

Department of Analytical Chemistry, Helmholtz Centre for Environmental Research-UFZ, Permoserstr. 15, Leipzig 04318, Germany

3.1 SEPARATION PRINCIPLES

Liquid chromatography includes a lot of specific modes to separate ionic and neutral compounds. Ionic compounds can be separated by ion-exchange interactions based on ion-exchange functionalities on stationary phases. An additional principle for separating ionic compounds is application of reversed-phase (RP) columns together with an ion-pairing reagent as mobile-phase modifier to create a more hydrophobic ion pair with the analyte ion. Common RP-HPLC using an aqueous buffer and an organic solvent can be applied for the separation of more hydrophobic molecule ions as well as hydrophilic interaction chromatography (HILIC) with organic buffer solution and water as eluent for the same group of molecule ions. Finally, for separation of mixtures of high-molecular-mass ions, size exclusion chromatography (SEC) is the method of choice because it provides a wide range for molar mass distribution analysis. A summary of the different options for chromatographic separation of ionic compounds is shown in Figure 3.1.

Application of IC-MS and IC-ICP-MS in Environmental Research, First Edition.
Edited by Rajmund Michalski.
© 2016 John Wiley & Sons, Inc. Published 2016 by John Wiley & Sons, Inc.

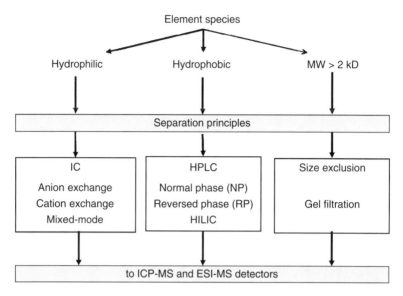

Figure 3.1 Overview about combinations of chromatographic principles with mass spectrometric detection.

3.1.1 Ion Chromatography (Anion/Cation Exchange, Mixed Mode)

In common ion chromatography, anions can interact with cationic functionalities immobilized on the stationary phase and vice versa [1–3]. With respect to ion-exchange interactions for retaining ionic analytes, two main directions of stationary phases and their corresponding eluents were developed and applied successfully in the past. Briefly, on the one hand, stationary phases with very low exchange capacities combined with low concentrations of weakly dissociated organic acid salts (phthalate, benzoate) could be applied for common anion separation and detection in the nonsuppressed mode of IC. The low equivalent conductivity of the organic acids could be suppressed electronically. A second possibility to achieve a good separation of ions could be realized by columns with a high exchange capacity combined with strong eluting ions such as carbonate or hydroxide. In these cases, the conductivity of the eluent ions exceeds in a high extent the conductivity of the analyte ions. To decrease drastically the conductivity of the mobile phase, a chemically or an electrochemically working suppressor has to be used.

To separate both anionic and cationic analytes simultaneously, multidimensional chromatography was applied in the off-line and online mode. Especially in the speciation of arsenic, combinations of anion- and cation-exchange columns are necessary to determine a multitude of inorganic, methylated, and organic arsenic compounds. A variety of arsenic speciation methods [4–8] based on IC-ICPMS analyses was summarized in specialized reviews and is also presented in the chapter "Applications." An example for the application of an anion-exchange and a cation-exchange column for the separation of anionic and cationic arsenic species is shown in Figure 3.2.

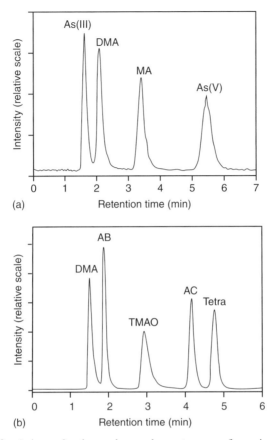

Figure 3.2   Anion and cation exchange chromatograms of arsenic species [9].

Additionally, the analytes can also bind to the polymeric resins of the stationary phase to achieve a more unspecific retention. If both are the case, then a mixed-mode mechanism conducts to the separation of different charged analytes. An example for a mixed-mode separation of different charged arsenic species is shown in Figure 3.3. Under the conditions of an acidic eluent (e.g., nitric acid gradient), the arsenic species existing as anionic, neutral, and cationic compounds can be separated on an anion-exchange stationary phase with a polymeric core responsible for additional reversed-phase interactions.

3.1.2   High-Performance Liquid Chromatography (Reversed-Phase Mode, HILIC)

HPLC with a reversed-phase separation principle or HILIC [10, 11] is suitable for separation of polar and less polar ionic compounds. The polarity of the analytes can vary over a wide range that the mobile phase has to modify with organic

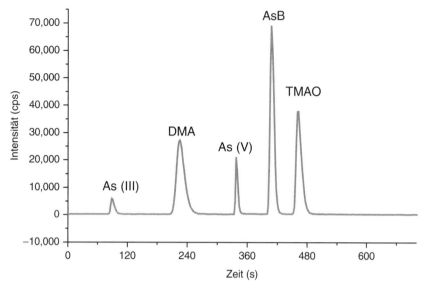

Figure 3.3 Example for mixed-mode ion chromatography for the speciation of arsenic with IC-ICP-MS. All arsenic species are commercially available for their identification. Analytical column: IonPacAS7+AG 7, Eluent: pH-Wert <5 (HNO_3 – gradient) [12].

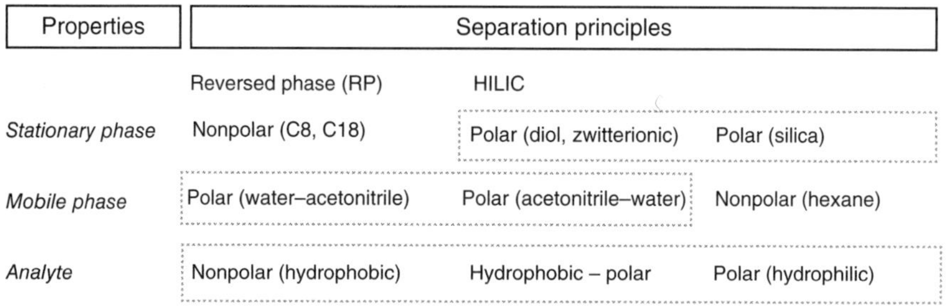

Properties	Separation principles		
	Reversed phase (RP)	HILIC	
Stationary phase	Nonpolar (C8, C18)	Polar (diol, zwitterionic)	Polar (silica)
Mobile phase	Polar (water–acetonitrile)	Polar (acetonitrile–water)	Nonpolar (hexane)
Analyte	Nonpolar (hydrophobic)	Hydrophobic – polar	Polar (hydrophilic)

Figure 3.4 Separation principles and properties of stationary, mobile phase and analyte.

solvents as eluting agent. Besides isocratic elution, gradient elution with increasing organic solvent content (RP-LC) or increasing water content (HILIC) enables the reduction of retention time and solvent consumption. A comparison of normal-phase, reversed-phase, and HILIC is shown in Figure 3.4.

3.1.3 Size Exclusion Chromatography (SEC) (Gel Filtration Chromatography, GFC)

Larger charged molecules, biomolecules (proteins, enzymes), and synthetic polymers can be separated by size exclusion chromatography (SEC) also known as gel filtration chromatography (GFC). The principles, columns, detectors, calibration of these separation techniques can be reviewed in several articles [13–16]. If we focus our attention on the separation of water-soluble high-molecular-weight compounds such as biomolecules and metal biomolecules, then, in most cases, the mobile-phase compositions (most aqueous) and buffer concentrations (millimolar range) get along well with the compatibility of the ICP-MS. It should be considered that cations of alkaline and alkaline earth elements in buffer components or modifiers should be substituted by volatile ammonium cations to inhibit metal oxide deposits on the cones. The eluent flow rates are in the range of $0.1–1\,ml\,min^{-1}$ and therefore in the same order of magnitude that the outlet tubing of the columns can be connected online with conventional nebulizers.

Generally, the separation conditions have to be individually adapted to the performance of the element specific detectors with respect to organic solvent tolerance and buffer composition. Flow rates $<100\,\mu L\,min^{-1}$ require special nebulizers with optimal performance at reduced flow. Also, the nebulizer gas flow should be tuned carefully for highest sensitivity.

3.2 DETECTION PRINCIPLES

3.2.1 Common Detection in IC: Conductivity, UV–Vis, Electrochemical Detection

Beyond a reasonable doubt, the conductivity detection dominates the ion chromatography applied for common anions and aliphatic carboxylic acids. However, this detection is unspecific and needs standards to identify the analytes by their retention times. More selective is the UV–Vis detection because only a few analytes show specific absorption signals for identification, such as nitrate, nitrite, bromide. Metal ions can also be detected as complexes after postcolumn derivatization. The electrochemical detection is limited to ions that undergo direct redox reactions such as carbohydrates, amino acids, or influences indicator reactions by potential shift resulting from complex formation such as sulfide, cyanide on an Ag/Ag^+ electrode.

3.2.2 Element Specific Detection

3.2.2.1 Inductively Coupled Plasma Atomic Emission Spectrometry (ICP-AES)
This spectrometric technique based on the same inductively coupled argon plasma as ionization source such as ICP-MS but for the quantification and identification of the analytes serves the characteristic emission radiation of the excited element ions. The detection of the characteristic wavelength emitted can be performed by a monochromator for a single-wavelength detection at a time or by using a polychromator to capture emissions of numerous wavelengths simultaneously combined with photodetectors such as charge-coupled devices. Nowadays, the instruments and software allow detecting fast transient signals for coupling with separation techniques. The limit of detection (LOD) for a multielement analysis ranges from milligram per

liter (ppm) to microgram per liter (ppt) depending on the ionization degree of the elements and their optimal wavelength for interference-free detection. Often, wavelengths with lower intensity have to be chosen for quantification because of the overlapping with lines of diverse intensities of other elements. The detectable wavelength range lies between 130 nm (Vacuum UV (VUV)) and 700 nm (Vis). Especially, the VUV is important for the detection of halides, which are more difficult to analyze with ICP-MS.

3.2.2.2 *Inductively Coupled Plasma Mass Spectrometry (ICP-MS)*   The instruments of choice for trace element determination are ICP-MSs as powerful analytical instruments for the ultrasensitive and selective detection of the most elements and isotopes of the Periodic Table of Elements [17].

The element-specific detection results from the ionization through a hot argon plasma radio frequency generated. During the ionization process in the plasma, the molecules are completely destroyed and the resulting elements are ionized by argon ions to element cations M^+ (M^{2+}, MO^+). Solutions to be analyzed are nebulized before by pneumatic microconcentric nebulizers combined with spray chambers of different types or by direct injection nebulizer (DIN). Subsequently, the formed aerosol is transferred to the plasma torch by an argon gas flow. The optimal nebulization performance depends mainly on the flow rate of the eluent and the corresponding argon gas nebulizer flow rates. After ionization in the plasma, the cations reach the high-vacuum part of the mass spectrometer via sample and skimmer cones. Before they are separated with a nominal mass resolution in a quadrupole mass analyzer and detected by an electron multiplier, they will pass through a collision/reaction cell (CRC) installed in many modern ICP-MS between the lenses and the quadrupole. Numerous element cations can be detected directly as M^+ without CRC, and for others, it is necessary to use the CRC to (I) remove interfering ions by an inert collision gas (CH_4, He) or (II) it has to react with a reactive gas (O_2, H_2, NH_3) to form, for example, MO^+ molecule detectable interference-free by the mass shift on the mass-to-charge ratio $(M+16)^+$ [18]. Using both of these variants, the most isobaric interferences can be drastically reduced. A full description of collision/reaction cell technology in ICP-MS was given by Tanner *et al.* (2002) and of the dynamic reaction cell (ICP-DRC-MS) by Tanner and Baranov (1999) and by Bandura *et al.* (2002) [19–21].

Recently, further progress in ICP-MS technology could be introduced in analytical laboratories. Instead of CRC and one quadrupole or equivalent configurations, the new generation of ICP-MS contains two-quadrupole mass analyzer (QMA) and a collision/reaction cell between them. What are the advantages of this concept (Figure 3.5)?

The main benefit with the first QMA is that the analyte ions of interest indicated by the dotted line can be selected by their nominal mass-to-charge ratio and spectral interferences and high matrix load can be excluded effectively as illustrated on the parabolic dashed lines. Then, only the selected ions can reach the CRC to be manipulated similarly to the principles described earlier. The second QMA operates in this way that the mass-to-charge ratio of the original analyte cation can pass (dotted line)

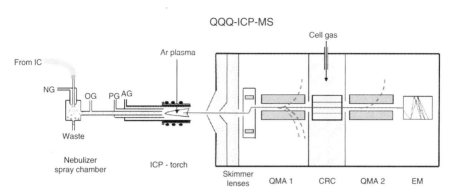

Figure 3.5 ICP-MS with a triple quadrupole arrangement. NG: nebulizer gas (Ar), OG: optional gas (Mix Ar/O$_2$), PG: plasma gas (Ar), AG: auxiliary gas (Ar), QMA 1: first quadrupole mass analyser, CRC: collision/reaction cell (octopole, hexapole), QMA 2: second quadrupole mass analyser, EM: electron multiplier.

if the interference could be removed (parabolic dashed line) or the analyte cation reacts with the cell gas to get a mass-to-charge ratio of a newly formed molecule ion (mass shift) whose mass-to-charge ratio can select on the second QMA for passing. The separated ions can be detected with an electron multiplier in an analogue (high ion density) or digital (low ion density) mode that an extremely wide concentration range of 10 orders of magnitude can be recorded. The optional gas (OG) is added to the aerosol stream and serves to support the complete oxidation of organic solvents of the mobile phase in the plasma. According to previous experience, the aerosol stream should be doped with 10–15 vol% OG that consists of approximately 80 mass% Ar and 20 mass% O$_2$ (Figure 3.6).

Alternative to the employment of a quadrupole-ICP-MS (Q-ICP-MS), also high-resolution (HR) ICP-MS instruments can be used in the determination of elements and isotopes [22].

The high-resolution MS allows the separation of spectral interferences from the analyte cation by higher mass resolution and measuring its accurate mass (mass resolution <10,000). However, associated with a higher mass resolution is the loss of sensitivity that a compromise between interference removal and sensitivity must be found. In general, they are also approximately five times slower than a Q-ICP-MS, so a transient signal analysis necessary for coupling with chromatography (gas chromatography, liquid chromatography (IC)) and capillary electrophoresis is unsatisfactory and only suitable for required speedy chromatographic signal generations. HR-ICP-MS was successfully online coupled to field-flow fractionation (FFF) device to identify the transport of toxic metals by nanominerals and mineral nanoparticles. FFF separations are comparatively slow that HR-ICP-MS can be applied effectively [23].

HR-ICP-MS instruments can be found as sector field (SF) ICPMS in different modifications, for instance, as double-focusing magnetic sector field or magnetic/electric sector field. A second type of HR-ICP-MS named multicollector

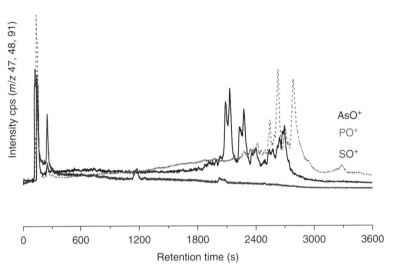

Figure 3.6   RP-HPLC chromatograms with ICP-MS detection of a fish extract with simultaneous detection of As, P, and S after reaction with O_2 in the CRC.

(MC) ICP-MS uses multiple detectors in combination with a double-focusing mass spectrometer and an electrostatic zoom lens approach. It is mainly applied for the accurate isotope and isotope ratio analysis, and concerning the fast mass scanning, it can also be combined with a laser ablation (LA) sampling. The disadvantages are that these HR-ICP-MSs are more expensive (three to four times) than Q-ICP-MS instruments and the operation is more complicated and extensive [24, 25].

3.2.2.3 Molecule-Specific Detection – Electrospray Ionization Mass Spectrometry (ESI-MS) Since the development and application of the electrospray ionization as a "soft" ion source in mass spectrometry of large biomolecules by J. B. Fenn *et al.*, this technique is now widespread in organic analytical mass spectrometry for polar and fragile polar compounds. In most applications, ESI-MS is combined with liquid chromatographic techniques outlined in Figure 3.1. In electrospray ionization processes, the intact molecule can be ionized in the positive or negative mode and can enter the transfer capillary as protonated $(M+H)^+$ or deprotonated $(M-H)^-$ ions in the simplest case to reach the high-vacuum part of the MS. In addition, depending on the properties of the compounds of interest, other atmospheric pressure ionization techniques such as atmospheric pressure chemical ionization (APCI) or atmospheric pressure photoionization (APPI) can also be used as interface for LC-MS. For the mass separation, also a broad spectrum of techniques are available, such as linear quadrupole (Q)-MS, Tandem-MS, Time-of-flight (TOF)-MS, Q-TOF-MS, Q-Ion Trap MS, OrbiTrap-MS, and FT-ICR-MS. The MS techniques, their basic concepts, and applications are described very informatively and detailed in a textbook [26, 27].

3.3   HYPHENATED TECHNIQUES

3.3.1   HPLC(IC)–ICP-MS

3.3.1.1   Analytes of Interest   The coupling of ICP-MS as detector for liquid chromatography is especially attractive for (i) speciation of metals, metalloids, and nonmetals and (ii) identification and quantification of heterorganic compounds with respect to the detectable heteroatoms (sulfur, phosphorous, silicon, bromine, iodine) in environmental and biological samples. An example for the separation and detection of gadolinium complexes by IC-ICP-MS is shown in Figure 3.7.

The advantages of this detector can be characterized by its

- ability for a multielemental analysis, for example, heavy metal complexes;
- high sensitivity (nanogram per liter);
- high selectivity (nominal- to high-resolution masses, CRC modes);
- detection stability with complex matrices (8000 K plasma temperature).

The sensitivity of elements is governed by their first ionization potentials with respect to the argon value (15.8 eV). Most of the metals and metalloids can be ionized with efficiencies of >80%. The advantage of ICP-MS is also that nonmetals such

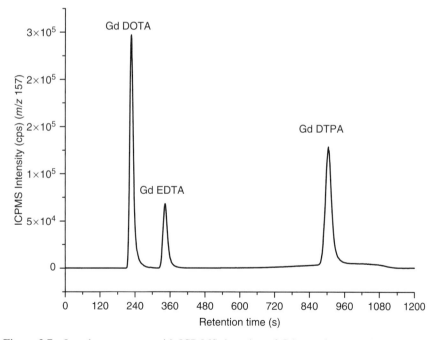

Figure 3.7   Ion chromatogram with ICP-MS detection of Gd complexes used as MRI contrast agents. Detection: m/z 157 (Gd); separation: anion-exchange column, ammonium acetate gradient, flow rate 1 ml min^{-1} [28].

as the elements mentioned earlier are sufficiently ionized and therefore sensitively detectable. Only fluorine cannot be ionized because the first ionization potential is higher (17.4) than that of argon. The ICP-MS detector is predestined for quantification of the separated element species. Theoretically, all the compounds are showing the same detector response (slop of the calibration curve) if they contain the same element and the same number of this element. Taking into account this fact then, only one calibration is necessary to quantify all compounds detected under constant aqueous elution conditions.

This behavior was evident for arsenic compounds and can be applied for compounds with a low molecular weight (<1 kD). In the case of high-molecular-weight compounds, the sensitivity can be influenced by the accompanying carbon atoms, restricting the ionization efficiency. The consideration of retention-time-dependent sensitivities is described in Section 3.3.1.2.

Unfortunately, with the combination of liquid chromatography and ICP-MS detection, only two parameters are provided for identification of the detected compounds:

(I) The m/z identifies the compound by its mass.
(II) The retention time of the signal stated something about ion charge (IC) or polarity (HPLC) of the compound.

With these two parameters, the detected compound cannot be identified unless one is in the possession of commercially available or synthesized standard compounds. An example for a multimode IC coupled with ICP-MS of commercial available arsenic species is shown in Figure 3.8 [29–31].

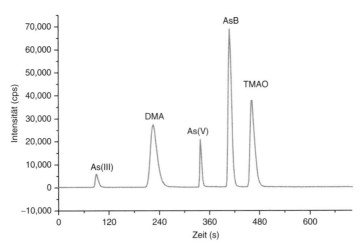

Figure 3.8 Example for mixed-mode ion chromatography for the speciation of arsenic with IC–ICP-MS. All arsenic species are commercially available for their identification. Analytical Column: IonPacAS7+AG 7, Eluent: pH-Wert < 5 (HNO$_3$ – gradient) [12].

3.3.1.2 Mobile-Phase Effects The sensitivity depends strongly on the organic composition of the mobile phase; therefore, retention-time-dependent sensitivity factors are required. Especially in the case of arsenic speciation by HPLC with a high content of organic solvents (methanol, acetonitrile) in the eluent or by gradient elution, the sensitivity is strongly affected [12, 32] and has to be corrected externally or internally. External correction can be performed by flow injection analysis (FIA) by injecting serially the arsenic species under the conditions of increasing organic solvent concentrations in the mobile phase. The increasing peak heights with increasing organic solvent percentage at the same analyte concentration indicate the sensitivity factor for each retention time. These factors can then be transmitted in the real chromatogram to calculate the response factor for each compound. This procedure is very helpful to calculate the concentration of unknown arsenic species such as arsenolipids [33].

High organic content in the eluent (>20%) can cause an instable plasma or can also put out the plasma. To reduce the influence of organics on plasma performance, three steps for prevention can be proposed:

- Addition of a gas mixture of argon/oxygen (80/20) to the nebulizer aerosol
- Application of a plasma torch with reduced aerosol channel (normal: 1.5–2 mm, for organic: ≤1 mm)
- Reduction of the flow rate of the eluent (<100 µl min^{-1}) or postcolumn splitting the eluent.

To protect the sample and skimmer cones of the ICP-MS from plaque, the mobile phase should contain components that are volatile such as ammonium salts of acetate, carbonate, nitrate or aliphatic carbonic acids such as formic acid, acetic acid. Sodium, potassium salts, and their hydroxides have to be avoided because they form the appropriate metal oxide deposits on the orifice of the cones and thus constipate them.

3.3.1.3 Critical Estimation of LC-ICP-MS

Advantages	Disadvantages
Very selective and high sensitive	Organic solvents can interfere
Suitable for quantification	Nonvolatile buffer unsuitable
The most elements detectable	Standard compounds necessary
Easy to couple with LC via nebulizer	No identification
Collision/reaction cell to remove *m/z* interferences	
HR-ICP-MS for reduction of *m/z* interferences	

3.4   HPLC(IC)–ICP-MS/ESI-MS

3.4.1   Fundamentals

Not only the numerous samples, for example, of biological origin contain the expected analyte, but the ICP-MS chromatogram also contains signals from unknown components. In those cases, the in-parallel simultaneous detection of element-specific signals by ICP-MS and molecular selective information by ESI-MS can support the elucidation of those unknown components. It is important to underline the simultaneity of both detections because the identification is much more difficult with two separately working instrument combinations such as LC-ICP-MS and LC-ESI-MS. In these cases, the signals of both detectors cannot correlate so well that the information recorded by ICP-MS can assist in finding the accurate molecular weight in ESI-MS especially if monoisotopic analytes have to be determined.

3.4.2   Methodology of Data Evaluation

The comparison of both comprehensive chromatograms shows that, in principle, well-defined peaks are obtained by the element-specific detection. Through the selective detection, the ICP-MS chromatogram is almost matrix independent and the selected masses can be only influenced by polyatomic or double-charged interferences. Against it, the ESI-MS detects all components of the probe unselectively, resulting in an extreme peak-rich chromatogram. The components to be identified are overlapped by the major components in the sample and cannot be seen as a well-defined peak. Only after searching of suitable accurate masses and their retention times by the extracted ion chromatogram (EIC), a definite peak becomes visible which should have the same retention time like in the ICP-MS chromatogram. With the confirmation of correct compound formula, the search for a suitable compound structure can begin. To support this process, the retention time (capacity factor) in reversed-phase LC as indicator for the polarity of the compound can be correlated with the assessment of polarity after structure modeling and optimization (MM2). In contrast to organic substances, the structure modeling/optimization of metal(loide) organic compounds is more problematic because of the lack of thermodynamic data for metals and metalloids in those calculation programs.

The final available data set includes a minimum of five important parameters to facilitate quantification and identification of unknown analytes stemming from chromatographic results, as well as the appropriate ICP-MS and ESI-MS data:

1. LC: retention time, capacity factor
2. ICP-MS: element information about target analytes and accompanying elements
3. ESI-MS: polarity switching (positive, negative) for optimization of detection and as a rough option for identification

4. HR-MS: accurate masses depending on the resolution provided by the instrument, isotopic pattern

5. HR-MS/MS: fragmentation of precursor ions to product ions with accurate mass resolution.

3.4.3   Technical Requirements

For the technical implementation of this instrument arrangement, only an additional T-piece is to be mounted in the capillary connecting LC and ICP-MS. An additional capillary connects the ESI-MS with the T-piece. With the length and inner diameter of the capillaries after splitting the eluent flow, the split ratio can be regulated roughly. The split ratio also depends on the flow and pressure conditions dominating the nebulizers in the spray chamber of ICP-MS and the ESI source. Another splitting of the eluent with different split ratios can be performed by using a fixed-flow splitter with a constant split ratio or an adjustable-flow splitter commercially available with different minimum and maximum split ratios between 1:1 and 20,000:1, respectively. Generally, the split ratio should be chosen with respect to the sensitivity of both detectors. Normally, the ICP-MS is more sensitive than the ESI-MS that a higher portion of the eluent should flow to the latter to enhance the ability for identification and fragmentation. The change of the split ratio involves an alteration of the time shift between the ICP-MS and ESI-MS chromatograms and leads to a readjustment of the retention time by analyzing a standard compound and comparing the retention times obtained by both detectors.

The element-specific detector can also be substituted by an HR-ICP-MS to improve the interference-free detection of the elements (attention of limitation: loss of sensitivity) or by an MC-ICP-MS with accurate isotopic analysis [34, 35].

On the site of the molecule-specific detectors, also Tandem MS, Ion Trap, and OrbiTrap MS instruments summarized in Table 3.1 are suitable for coupling with (U)PLC and ion chromatography.

High-resolution mass spectrometer on the element-specific (double-focusing magnetic sector field ICP-MS, Element-2^{TM}, Thermo Scientific) and molecule-specific (OrbiTrap-MS, Discovery, Thermo Scientific) site of detection allows a universal performance in sensitivity (quantification) and identification feasibilities (accurate mass and MS/MS of choice). After RP-HPLC separation, the speciation of numerous lipophilic arsenic compounds divisible into three classes (As-phospholipids, As-fatty acids, and As-hydrocarbons) in brown alga (*Saccharina latissima*) is shown in Figure 3.9 [36, 37]. As-lipids in fish samples shown as example in Figure 3.10 [38, 39] and free and As-phytochelatines in rice plants [40] could be achieved in an excellent manner.

3.5   APPLICATIONS AND CONCLUSION

The varied applications of HPLC(IC)–ICP-MS/ESI-MS have been continuously increasing in the past two decades and now include mainly the analysis of metal(loid)

TABLE 3.1   Advantages and Disadvantages of Mass Spectrometer and Their Mass
Resolutions and Mass Ranges

Principle	Mass resolution	Mass range	Advantages	Disadvantages
Quadrupole	0.7–1	<2,000		Nominal mass
Tandem MS	1	<2,000	High-sensitivity MS/MS	Nominal mass
Time-of-flight	20,000–40,000	<3,200	Accurate mass	No MS/MS
Q-TOF	20,000–40,000	<3,200	MS/MS	
Ion Trap	1	<6,000	MS(n)	Nominal mass
OrbiTrap	<100,000 at 1 Hz[a]	<4,000	MS/MS	
FT-ICR-MS	<600,000 at 1 Hz[a]	Unlimited	MS/MS	

[a] Values for coupling with LC at 1 Hz duty cycle.

organic compounds in complex biological samples. It has entered into the analysis
of organic compounds with heteroatoms and their metabolites as well as in nonmetal
speciation. For a more clear overview, the applications are therefore classified into
four groups: (i) speciation of metals including rare earth elements, (ii) speciation of
metalloids, (iii) speciation of nonmetals, and (iv) organic compounds with ICP-MS
detectable heteroatoms. There are also excellent reviews, which should be read
for more details on these emerging subjects of analytical chemistry, biochemical
analysis up to proteomics and metabolomics. Finally, many interesting fact sheets
and expertises of research groups around the world were collected in the last
15 years by Michael Sperling *et al.* and presented in the very expedient "European
Virtual Institute for Speciation Analysis – EVISA" [41]. Some selected milestone
papers also reflecting the fast methodological development are summarized in
Table 3.2.

The list of references that would include all papers in this research field shows
this oversized interest and the evolution in terms of analytes and instrumentation.
Two decades ago, the first steps were undertaken to separate, for example, inorganic and methylated arsenic species by FIA-ICP-MS and shortly afterward with
IC-ICP-MS. Today, the newest inorganic and organic mass spectrometer with respect
to fascinating ability for high mass resolution and ultratrace sensitivity is coupled
with high-efficient separation equipments such as (U)HPLC, gas chromatography,
or capillary electrophoresis. FT-ICR-MS with an enormous resolution power and
an unlimited mass range might supersede chromatographic separations and dual MS
detections in future. Only the price and the high effort for maintenance of the equipments available on the market are the limiting factors for the feasibility of this technique in element speciation and organic analysis.

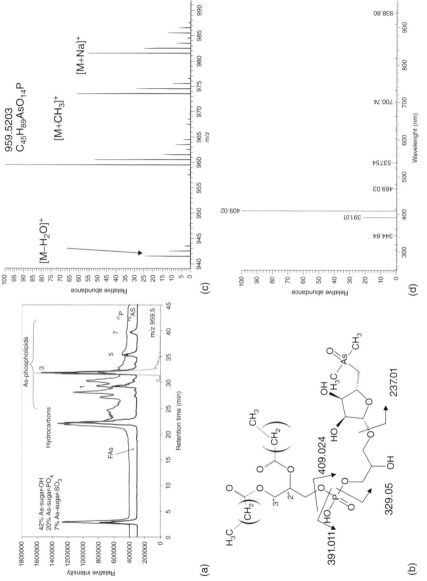

Figure 3.9 HPLC–HR-ICP-MS/HR-ESI-MS analysis of a brown alga (*Saccharina latissima*) [36].

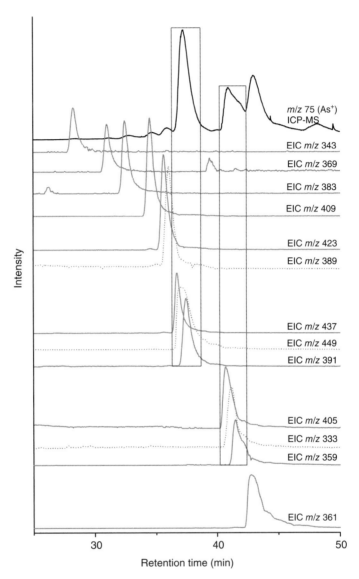

Figure 3.10 HPLC–ICPMS/ESIQTOFMS of arsenolipids detected in canned cod liver extract and indicated by the extracted ion chromatograms (EIC) in the ESI-MS Chromatogram. As[+] signal (m/z 75) was compared simultaneously with the extracted ion chromatogram (EIC) recorded by ESI-MS [39].

TABLE 3.2 Reviews and Applications

Groups	Analytes	Method	Applications	Matrix	References
Review	Trace elements	LC (CE)-ICPMS	Speciation in large biomolecules	Proteomics, biological samples	[40]
	Selenium	LC with ICPMS or ESIMS detection	Speciation	Yeast, plants, human serum, nutritional supplements	[42]
	Halides, metals, metalloids	IC with ICPMS and MS detection	Numerous elements	Numerous environmental samples	[43]
	As, Se, Sn, S, Pt, heavy metals	Different LC methods, LC-ICPMS coupling devices	Element speciation, environmental application, biological materials and clinical samples, pharmaceuticals	Water, soil extract, plant extracts, body fluids, food extracts, animal tissue extract	[31]
	Arsenic speciation	Coupled techniques	Food and drinking water	Rice, fruits and vegetables, algae and freshwater plants, marine organisms	[44]
	Arsenic speciation	Hyphenated techniques	Inorganic arsenic and methylated species	Water samples	[6]
	Arsenic speciation	IC-ICPMS	As(III), As(V), thioarsenic species	Sulfidic waters, groundwaters	[45]
	Speciation	Small-bore HPLC-ICPMS	Metals and metalloids	Applications in all fields	[46]
	Explosives	IC-MS	Low-order explosives-related ionic species (chlorate, perchlorate, nitrate)	Forensic and environmental analysis	[47]
	Numerous elements	Ion chromatography	IC with different detectors	Natural waters with high ionic strength	[48]

(continued)

95

TABLE 3.2 (*Continued*)

Groups	Analytes	Method	Applications	Matrix	References
Metals					
	Aluminum complexes	IC-ICPAES	Al(III)-citrate species	Decomposition studies	[49]
	Aluminum	CE-ICPMS, CE-ESIMS	Free Al, Al-F-complexes	Natural water samples	[50]
	Chromium	CE-ICPMS	Chromium speciation (Cr(III), Cr(VI) and Cr-complexes)	Nutritional supplement	[51]
	Chromium, cobalt, molybdenum	Asymmetric FFF (AF(4))-UV-multiangle light scattering-ICPMS	Micro- to nanosized metal wear particles	Interaction with transport proteins	[52]
	Cobalt	SEC-ICPMS	Large molecular cobalt (e.g., albumin-Co), small molecular Co (e.g., glutathione-Co)	Studies for kinetic and toxicology of Co compounds in human serum	[53]
	Gold, silver	Micellar electrokinetic chromatography-ICPMS	Au and AgNP	Au speciation in dietary supplements	[54]
	Heavy metals	IC-ICPMS	Speciation of Fe, Cu, Mo, Co, Z, Mn with aminopolycarboxylates (e.g., EDTA)	Ecotoxicological studies with algae, mobility of MeEDTA in river water into groundwater, Cu species in river water	[30]
	Heavy metals	IC-ICPMS	Heavy metals (Fe, Co, Ni, Cu, Zn, Cd, Pb) and their complexes (e.g., NTA, EDTA, DTPA, citrate, etc.)	Groundwater infiltrated by a polluted water	[55]

Element	Technique	Species	Application	Reference
Iron	IC-ICPMS	Redox species of Fe	Synthetic aqueous standard solution and beverage samples	[56]
Mercury species	HPLC-ICPMS	$Hg(II)$, $MeHg^+$	Fish samples	[57]
Mercury species	IC-ICPMS	$Hg(II)$, $MeHg^+$, $EtHg^+$, $PhHg^+$	CRMs of seawater and marine fish, seawater and marine fish samples	[58]
Mercury species	HPLC-ICPMS	$Hg(II)$, $MeHg^+$, $EtHg^+$	Seawater	[59]
Metal complexes	FIA-ESIMS, IC-ICPMS	EDTA complexes of Mn, Fe, Co, Ni, Zn, Cu, and Pb	Wastewaters	[29]
Molybdenum	IPC-ICP-MS	Thiomolybdate speciation	Natural waters	[60]
Molybdenum, rhenium	AEC-ICPMS	Mo and Re speciation in sulfidic water, thiometalates	Geothermal tracers in oxic and anoxic waters	[61]
Paladium	HPLC-ICPMS/ESIMS	Pd – complexes and reaction products	Interaction of Pd – complexes with guanylic acid	[62]
Platinum	ZIC-HILIC-ID-ICPMS	Speciation of cisplatin	Hospital wastewater samples	[63]
Silver	Flow field–flow fractionation	Ag nanoparticles (AgNP) – plasma protein associates, AgNP incubated with proteins, determination of association rates and binding stoichiometry	Interaction between bovine serum albumin, globulin, and fibrinogen and AgNP	[64]

(continued)

TABLE 3.2 (*Continued*)

Groups	Analytes	Method	Applications	Matrix	References
	Silver	HPLC-UV-ICPMS	Ag^+ and AgNP	Method development for analysis of AgNP – containing products	[65]
	Silver	HPLC-ICPMS	Ag^+ and AgNP	Clothing (e.g., sport socks)	[66]
	Tin	IC-ICPMS	Inorganic Sn as DTPA complexes	On-column complexation	[67]
	Transition metals	LC-ICPMS	Mn, Fe, Cu, Zn complexes with masses of 600–5000 Da, low-molecular-mass transition metal complexes (LMMTMC)	In mitochondria isolated from fermenting yeast cells, human Jurkat cells, and mouse brain and liver	[68]
	Vanadium	IC-ICPMS	V speciation (V(IV), V(V))	Groundwater, drinking water, exposure to children	[69]
	Vanadium species	IC-ICPMS	V(IV), V(VI)	Groundwater	[70]
Metalloids					
	Antimony	HPLC-ICPMS	Sb(III), Sb(V), TMSb	Speciation in soil samples	[71]
	Arsenic species, lipophilic	RP-HPLC-ICPMS/ESIMS	As-sugar-phosphate, As-fatty acids, As-hydrocarbons	Brown alga (*Saccharina latissima*)	[36]
	Arsenic species	IC-ICPMS, microwave extraction	Inorganic arsenic and methylated species	Soil extracts (Bangladesh), NIST SRM 2711 (Montana soil)	[72]

Arsenic species	HPLC-ICPMS	Inorganic arsenic and methylated species	Toxicological biotests with frog embryos	[73]
Arsenosugar	HPLC-ICPMS (ion pairing)	Pentavalent and trivalent arsenosugars	Influence of arsenosugar on DNA nicking assay and cytotoxicity array, human risk assessment	[74]
Arsenic, antimony, thallium	HPLC-ICPMS	inorganic ionic forms of As(III), As(V), Sb(III), Sb(V), Tl(I), and Tl(III)	Analysis in river water	[75]
Arsenic, sulfur	HPLC-ICPMS/ESIMS	Metabolites of dimethylarsinothioyl glutathione	Cellular extracts of treated human multiple myeloma cell lines	[76]
Selenium species	HPLC-ICPMS	Se-containing antioxidative peptides	Enzymatic hydrolysates of Se-enriched brown rice protein	[77]
Selenium species	LC-ICPMS, LC-ESI-Q-TOFMS	Inorganic and organic Se species	Biological samples	[78]
Selenium species	High temperature LC-ICPMS	Selenosugars, trimethylselenonium ion	Analysis of urine samples, Brazil nuts, selenium supplement	[79]
Selenium species	HPLC-ICPMS	Inorganic Se, organic Se compounds	Biotransformation of Se in onion (*Allium cepa* L.)	[80]

(continued)

Groups	Analytes	Method	Applications	Matrix	References
	Selenium	IP-HPLC-ICPMS	Inorganic Se, organic Se compounds	Analysis of selenium enriched pakchoi (*Brassica chinensis Jusl var parachinensis* (Bailey) Tsen & Lee)	[81]
	Silicon	IEC-MC-ICPMS	Silicon isotope ratios	Commercial Si standards and natural waters	[82]
	Silicon	IEC-ID-ICPMS	Dissolved silica	Seawater	[83]
Anions/non-metals					
	Amino-polycarboxylic acids (APCA)	IEC-ICPMS	Determination of APCA after complexation with Cd^{2+}	Method development, comparison with existing DIN methods	[84]
	Bromine	IC-ICPMS, IC-APIMS	Bromate	Surface water samples, round robin test	[85]
	Bromine	Low pressure LC-ICPMS	Bromate	Drinking waters	[86]
	Iodine-129	IC-SF-ICPMS	Ultra trace concentrations of I-129	Greenland snow	[87]
	Iodine	IP-IC-ICPMS	IO_3^-, I^-, and five iodo amino acids	Speciation analysis in real urine samples	[88]

Element	Technique	Analyte/Speciation	Sample	Ref.
Iodine	HPLC-ICPMS	IO^{3-}, I$^-$, organic iodine in low content	Iodine-rich Groundwater in North China	[89]
Iodine (Uranium)	SEC-ICPMS	Speciation of I and U	Groundwater, radioactive waste disposal	[90]
Phosphorus, sulfur	HPLC-ICPMS	Phosphorothioate oligonucleotide	Method development using hexafluoroiso-propanol/trimethylamine based eluent	[91]
Phosphorus, sulfur	UPLC-ICPMS	PO$^+$, SO$^+$ (DRC)	Complex biological samples (blood plasma/serum, tissues)	[92]
Phosphorus	IC-ICPMS	Glyphosate, glufosinate, reaction products	Biological samples: human serum and urine	[93]
Phosphorus	HPIC-ICPMS	Inorganic and organic P compounds	Extracts of Soil and plant samples (*Brassica napus* L.)	[94]
Phosphorus	IC-ICPMS	Phosphite, phosphate	Plant extracts	[95]
Sulfur, bromine, chlorine	IC-MC-ICPMS	Isotope analysis in anionic species	Environmental samples, seawater	[96]
Phosphorus, silicon	IEC-SF-ICPMS	Dissolved phosphate, silicate	Seawater	[97]
Phosphorus	IC-ICPMS	Phosphonates	River water affected by discharge of municipal or industrial wastewater treatment plants	[98]

3.6   REFERENCES

[1] Haddad, P.R., Nesterenko, P.N., and Buchberger, W. (2008) Recent developments and emerging directions in ion chromatography. *Journal of Chromatography A*, **1184**, 456.

[2] Weiss, J. (2008) *Ion Chromatography*, 2nd edn, VCH, Weinheim.

[3] Fritz, J.S. and Gjerde, D.T. (2009) *Ion Chromatography*, 4th edn, Wiley-VCH Verlag GmbH &Co. KGaA, Weinheim.

[4] Gong, Z., Lu, X., Ma, M. *et al.* (2002) Arsenic speciation analysis. *Talanta*, **58**, 77.

[5] B'Hymer, C. and Caruso, J.A. (2004) Arsenic and its speciation analysis using high-performance liquid chromatography and inductively coupled plasma mass spectrometry. *Journal of Chromatography A*, **1045**, 1.

[6] Terlecka, E. (2005) Arsenic speciation analysis in water samples: a review of the hyphenated techniques. *Environmental Monitoring and Assessment*, **107**, 259.

[7] Komorowicz, I. and Barałkiewicz, D. (2011) Arsenic and its speciation in water samples by high performance liquid chromatography inductively coupled plasma mass spectrometry-last decade review. *Talanta*, **84**, 247.

[8] Ammann, A.A. (2011) Arsenic speciation analysis by ion chromatography – a critical review of principles and applications. *American Journal of Analytical Chemistry*, **2**, 27.

[9] Francesconi, K., Visoottiviseth, P., Sridokchan, W., and Goessler, W. (2002) Arsenic species in an arsenic hyperaccumulating fern, *Pityrogramma calomelanos*: a potential phytoremediator of arsenic-contaminated soils. *Science of the Total Environment*, **284**, 27.

[10] Tang, D.-Q., Zou, L., Yin, X.-X., and Ong, C.N. (2014) HILIC–MS for metabolomics: an attractive and complementary approach to RPLC–MS. *Mass Spectrometry Reviews*. doi: 10.1002/mas.21445, 1–27.

[11] Jandera, P. (2011) Stationary and mobile phases in hydrophilic interaction chromatography: a review. *Analytica Chimica Acta*, **692**, 1.

[12] Mattusch, J. and Wennrich, R. (1998) Determination of anionic, neutral, and cationic species of arsenic by ion chromatography with ICPMS detection in environmental samples. *Analytical Chemistry*, **70**, 3649.

[13] Hong, P., Koza, S., and Bouvier, E.S.P. (2012) A review: size exclusion chromatography for the analysis of protein biotherapeutics and their aggregates. *Journal of Liquid Chromatography and Related Technologies*, **35**, 2923.

[14] Barth, H.G. and Saunders, G.D. (2012, Mar 31) Fundamentals and properties of size exclusion chromatography: packings and columns. *LC-GC The Magazine of Separation Sciences*, **30(S4)**, 46.

[15] Konstanski, L.K., Keller, D.M., and Hamielec, A.E. (2004) Size-exclusion chromatography: a review of calibration methodologies. *Journal of Biochemistry and Biophysical Methods*, **58**, 159.

[16] Berek, D. (2010) Size exclusion chromatography – a blessing and a curse of science and technology of synthetic polymers. *Journal of Separation Science*, **33**, 315.

[17] Gross, J.H. (2013) *Anorganische Massenspektrometrie*, Springer, Berlin, Heidelberg, p. 731.

[18] Palcic, J.D., Jones, J.S., Flagg, E.L., and Donovan, S.F. (2015) Lewisite exposure biomarkers in urine by liquid chromatography–inductively coupled plasma tandem mass

spectrometry: with an accelerated matrix-matched stability study. *Journal of Analytical Atomic Spectrometry*, **30**, 1799.

[19] Tanner, S.D., Baranov, V.I., and Bandura, D.R. (2002) Reaction cells and collision cells for ICP-MS: a tutorial review. *Spectrochimica Acta, Part B*, **57**, 1361.

[20] Tanner, S.D. and Baranov, V.I. (1999) A dynamic reaction cell for inductively coupled plasma mass spectrometry (ICP-DRC-MS). II. Reduction of interferences produced within the cell. *Journal of the American Society for Mass Spectrometry*, **10**, 1083.

[21] Bandura, D.R., Baranov, V.I., Litherland, A.E., and Tanner, S.D. (2006) Gas-phase ion-molecule reactions for resolution of atomic isobars: AMS and ICP-MS perspectives. *Journal of Mass Spectrometry*, **255**, 312.

[22] Douthitt, C.B. (2008) Commercial development of HR-ICPMS, MC-ICPMS and HR-GDMS. *Journal of Analytical Atomic Spectrometry*, **23**, 685.

[23] Plathe, K.L., von der Kammer, F., Hassellöv, M. *et al.* (2013) The role of nanominerals and mineral nanoparticles in the transport of toxic trace metal: filed-flow fractionation and analytical TEM analyzes after nanoparticle isolation and density separation. *Geochimica et Cosmochimica Acta*, **102**, 213.

[24] Yang, L. (2009) Accurate and precise determination of isotopic ratios by MC–ICP-MS: a review. *Mass Spectrometry Reviews*, **28**, 990.

[25] Baskaran, M. (2012) Environmental isotope geochemistry: past, present and future, in *Handbook of Environmental Isotope Geochemistry*, Springer, Berlin, Heidelberg, p. 3.

[26] Gross, J.H. (2011) *Mass Spectrometry: A Textbook*, 2nd edn, Springer-Verlag, Berlin, Heidelberg, New York, ISBN 978-3-642-10711-5.

[27] Fenn, J.B., Mann, M., Meng, C.K. *et al.* (1989) Electrospray ionization for mass spectrometry of large biomolecules. *Science*, **246**, 64.

[28] Yun, W., (2014) Master's thesis, University of Applied Sciences, Merseburg, Germany.

[29] Chen, Z., Sun, Q., Xi, Y., and Owens, G. (2008) Speciation of metal–EDTA complexes by flow injection analysis with electrospray ionization mass spectrometry and ion chromatography with inductively coupled plasma mass spectrometry. *Journal of Separation Science*, **31**, 3796.

[30] Ammann, A.A. (2002) Speciation of heavy metals in environmental water by ion chromatography coupled to ICP-MS. *Analytical and Bioanalytical Chemistry*, **372**, 448.

[31] Montes-Bayón, M., DeNicola, K., and Caruso, J.A. (2003) Liquid chromatography–inductively coupled plasma mass spectrometry. *Journal of Chromatography A*, **1000**, 457.

[32] Larsen, E.A. and Stürup, S. (1994) Carbon-enhanced inductively coupled plasma mass spectrometric detection of arsenic and selenium and its arsenic speciation. *Journal of Analytical Atomic Spectrometry*, **9**, 1099.

[33] Amayo, K.O., Petursdottir, A., Newcombe, C. *et al.* (2011) Identification and quantification of arsenolipids using reversed-phase HPLC coupled simultaneously to high-resolution ICPMS and high-resolution electrospray MS without species-specific standards. *Analytical Chemistry*, **83**, 3589.

[34] Günther-Leopold, I., Waldis, J.K., Wernli, B., and Kopajtic, Z. (2005) Measurement of plutonium isotope ratios in nuclear fuel samples by HPLC–MC–ICP-MS. *Journal of Mass Spectrometry*, **242**, 197.

[35] Santamaria-Fernandez, R., Hearn, R., and Wolff, J.C. (2008) Detection of counterfeit tablets of an antiviral drug using δ 34 S measurements by MC–ICP-MS and confirmation

by LA-MC–ICP-MS and HPLC-MC–ICP-MS. *Journal of Analytical Atomic Spectrometry*, **23**, 1294.

[36] Raab, A., Newcombe, C., Pitton, D. *et al.* (2013) Comprehensive analysis of lipophilic arsenic species in a brown alga (*Saccharina latissima*). *Analytical Chemistry*, **85**, 2817.

[37] García-Salgado, S., Raber, G., Raml, R. *et al.* (2012) Arsenosugar phospholipids and arsenic hydrocarbons in two species of brown macroalgae. *Environmental Chemistry*, **9**, 63.

[38] Amayo, K.O., Petursdottir, A., Newcombe, C. *et al.* (2011) Identification and quantification of arsenolipids using reversed-phase HPLC coupled simultaneously to high-resolution ICPMS and high-resolution electrospray MS without species-specific standards. *Analytical Chemistry*, **83**, 3589.

[39] Arroyo-Abad, U., Lischka, S., Piechotta, C. *et al.* (2013) Determination and identification of hydrophilic and hydrophobic arsenic species in methanol extract of fresh cod liver by RP-HPLC with simultaneous ICP-MS and ESI-Q-TOF-MS detection. *Food Chemistry*, **141**, 3093.

[40] Sanz-Medel, A., Montes-Bayonand, M., and Fernandes Sanchez, M.L. (2003) Trace element speciation by ICP-MS in large biomolecules and ist potential for proteomics. *Analytical and Bioanalytical Chemistry*, **377**, 236.

[41] http://www.speciation.net/index.html

[42] Pedrero, Z. and Madrid, Y. (2009) Novel approaches for selenium speciation in foodstuffs and biological specimens: a review. *Analytica Chimica Acta*, **634**, 135.

[43] Michalski, R., Jablonska, M., Szopa, S., and Lyko, A. (2011) Application of ion chromatography with ICP-MS or MS detection to the determination of selected halides and metal/metalloids species. *Critical Reviews in Analytical Chemistry*, **41**, 133.

[44] Sadee, B., Foulkes, M.E., and Hill, S.J. (2015) Coupled techniques for arsenic speciation in food and drinking water: a review. *Journal of Analytical Atomic Spectrometry*, **30**, 102.

[45] Stucker, V.K., Williams, K.H., Robbins, M.J., and Ranville, J.F. (2013) Arsenic geochemistry in a biostimulated aquifer: an aqueous speciation study. *Environmental Toxicology and Chemistry*, **32**, 1216.

[46] Grotti, M., Terol, A., and Todoli, J.L. (2014) Speciation analysis by small-bore HPLC coupled to ICP-MS. *Trends in Analytical Chemistry*, **61**, 92.

[47] Barron, L. and Gilchrist, E. (2014) Ion chromatography-mass spectrometry: a review of recent technologies and applications in forensic and environmental explosives analysis. *Analytica Chimica Acta*, **806**, 27.

[48] Gros, N. (2013) Ion chromatographic analyses of sea waters, brines and related samples. *Water*, **5**, 659.

[49] Peukert, A. and Seubert, A. (2009) Characterization of an aluminium(III)-citrat species by means of ion chromatography with inductively coupled plasma-atomic emission spectrometry detection. *Journal of Chromatography, A*, **1216**, 7946.

[50] Nakamoto, D. and Tanaka, M. (2014) Speciation of aluminum by CE-ESI-MS and CE-ICP-MS. *Bunseki Kagaku*, **63**, 383.

[51] Cheng, Y., Cheng, J., Xi, Z. *et al.* (2015) Simultaneous analysis of Cr(III), Cr(VI), and chromium picolinate in foods using capillary electrophoresis coupled to inductively coupled plasma mass spectrometry. *Electrophoresis*, **36**, 1208.

[52] Loeschner, K., Harrington, C.F., Kearney, J.L. *et al.* (2015) Feasibility of asymmetric flow field-flow fractionation coupled to ICP-MS for the characterization of wear metal particles and metalloproteins in biofluids from hip replacement patients. *Analytical and Bioanalytical Chemistry*, **407**, 4541.

[53] Kerger, B.D., Gerads, R., Thuett, H.K.A. *et al.* (2013) Cobalt speciation assay for human serum, Part I. Method for measuring large and small molecular cobalt and protein-binding capacity using size exclusion chromatography with inductively coupled plasma-mass spectroscopy detection. *Toxicological and Environmental Chemistry*, **95**, 687.

[54] Franze, B. and Engelhard, C. (2014) Fast separation, characterization, and speciation of gold and silver nanoparticles and their ionic counterparts with micellar electrokinetic chromatography coupled to ICP-MS. *Analytical Chemistry*, **86**, 5713.

[55] Ammann, A.A. (2002) Determination of strong binding chelators and their metal complexes by anion-exchange chromatography and inductively coupled plasma mass spectrometry. *Journal of Chromatography, A*, **947**, 205.

[56] Wolle, M.M., Fahrenholz, T., Rahman, G.M.M. *et al.* (2014) Method development for the redox speciation analysis of iron by ion chromatography-inductively coupled plasma mass spectrometry and carryover assessment using isotopically labeled analyte analogues. *Journal of Chromatography, A*, **1347**, 96.

[57] Döker, S. and Bosgelmez, I.I. (2015) Rapid extraction and reverse phase-liquid chromatographic separation of mercury(II) and methylmercury in fish samples with inductively coupled plasma mass spectrometric detection applying oxygen addition into plasma. *Food Chemistry*, **184**, 147.

[58] Chen, X., Han, C., Cheng, H. *et al.* (2013) Rapid speciation analysis of mercury in seawater and marine fish by cation exchange chromatography hyphenated with inductively coupled plasma mass spectrometry. *Journal of Chromatography, A*, **1314**, 86.

[59] Cheng, H., Wu, C., Liu, J., and Xu, Z. (2015) Thiol-functionalized silica microspheres for online preconcentration and determination of mercury species in seawater by high performance liquid chromatography and inductively coupled plasma mass spectrometry. *RSC Advances*, **5**, 19082.

[60] Lohmayer, R., Reithmaier, G.M.S., Bura-Nakic, E., and Planer-Friedrich, B. (2015) Ion-pair chromatography coupled to inductively coupled plasma-mass spectrometry (IPC–ICP-MS) as a method for thiomolybdate speciation in natural waters. *Analytical Chemistry*, **87**, 3388.

[61] Vorlicek, T.P., Chappaz, A., Groskreutz, L.M. *et al.* (2015) A new analytical approach to determining Mo and Re speciation in sulfidic waters. *Chemical Geology*, **403**, 52.

[62] Liu, D.Y., Zhu, F., Ma, Y.-J. *et al.* (2015) Analysis of reaction products of (ethylenediamine) palladium(II) Chloride and 5 '-deoxyguanylic acid in aqueous solution by liquid chromatography. Inductively coupled plasma mass spectrometry and electrospray ionization mass spectrometry. *Chinese Journal of Analytical Chemistry*, **43**, 193.

[63] Vidmar, J., Martinic, A., Milacic, R., and Scancar, J. (2015) Speciation of cisplatin in environmental water samples by hydrophilic interaction liquid chromatography coupled to inductively coupled plasma mass spectrometry. *Talanta*, **138**, 1.

[64] Wimuktiwan, P., Shiowatana, J., and Siripinyanond, A. (2015) Investigation of silver nanoparticles and plasma protein association using flow field-flow fractionation coupled with inductively coupled plasma mass spectrometry (FlFFF-ICP-MS). *Journal of Analytical Atomic Spectrometry*, **30**, 245.

[65] Hanley, T.A., Saadawi, R., Zhang, P. *et al.* (2014) Separation of silver ions and starch modified silver nanoparticles using high performance liquid chromatography with ultraviolet and inductively coupled mass spectrometric detection. *Spectrochimica Acta, Part B*, **100**, 173.

[66] Soto-Alvaredo, J., Montes-Bayon, M., and Bettmer, J. (2013) Speciation of silver nanoparticles and silver(I) by reversed-phase liquid chromatography coupled to ICPMS. *Analytical Chemistry*, **85**, 1316.

[67] Huang, L., Yang, D., Guo, X., and Chen, Z. (2014) Speciation analysis of inorganic tin by on-column complexation ion chromatography with inductively coupled plasma mass spectrometry and electrospray mass spectroscopy. *Journal of Chromatography, A*, **1368**, 217.

[68] McCormick, S.P., Moore, M.J., and Lindahl, P.A. (2015) Detection of labile low-molecular-mass transition metal complexes in mitochondria. *Biochemistry*, **54**, 3442.

[69] Arena, G., Copat, C., Dimartino, A. *et al.* (2015) Determination of total vanadium and vanadium(V) in groundwater from Mt. Etna and estimate of daily intake of vanadium(V) through drinking water. *Journal of Water and Health*, **13**, 522.

[70] Gamage, S.V., Hodge, V.F., Cizdziel, J.V., and Lindley, K. (2010) Determination of vanadium (IV) and (V) in Southern Nevada groundwater by ion chromatography–inductively coupled plasma mass spectrometry. *Chemical and Biochemical Methods Journal*, **3**, 10.

[71] Ge, F. and Wei, C. (2013) Simultaneous analysis of Sb-III, Sb-V and TMSb by high performance liquid chromatography–inductively coupled plasma-mass spectrometry detection: application to antimony speciation in soil samples. *Journal of Chromatographic Science*, **51**, 391.

[72] Rahman, M.M., Cheng, Z., and Naidu, R. (2009) Extraction of arsenic species in soils using microwave-assisted extraction detected by ion chromatography coupled to inductively coupled plasma mass spectrometry. *Environmental Geochemistry and Health*, **31**, 93.

[73] Koch, I., Zhang, J., Button, L.A. *et al.* (2015) Arsenic(+3) and DNA methyltransferases, and arsenic speciation in tadpole and frog life stages of western clawed frogs (Silurana tropicalis) exposed to arsenate. *Metallomics*, **7**, 1274.

[74] Andrews, P., Demarini, D.M., Funasaka, K. *et al.* (2004) Do arsenosugars pose a risk to human health? The comparative toxicities of a trivalent and pentavalent arsenosugar. *Environmental Science and Technology*, **38**, 4140.

[75] Szopa, S. and Michalski, R. (2015) Simultaneous determination of inorganic forms of arsenic, antimony, and thallium by HPLC–ICP-MS. *Spectroscopy*, **30**, 54.

[76] Yehiayan, L., Stice, S., Liu, G. *et al.* (2014) Dimethylarsinothioyl glutathione as a metabolite in human multiple myeloma cell lines upon exposure to Darinaparsin. *Chemical Research in Toxicology*, **27**, 754.

[77] Liu, K., Zhao, Y., Cheng, F., and Fang, Y. (2015) Purification and identification of Se-containing antioxidative peptides from enzymatic hydrolysates of Se-enriched brown rice protein. *Food Chemistry*, **187**, 424.

[78] Anan, Y., Nakajima, G., and Ogra, Y. (2015) Complementary use of LC–ICP-MS and LC–ESI-Q-TOF-MS for selenium speciation. *Analytical Sciences*, **31**, 561.

[79] Terol, A., Ardini, F., Basso, A., and Grotti, M. (2015) Determination of selenium urinary metabolites by high temperature liquid chromatography–inductively coupled plasma mass spectrometry. *Journal of Chromatography, A*, **1380**, 112.

[80] Michalska-Kacymirow, M., Kurek, E., Smolis, A. *et al.* (2014) Biological and chemical investigation of Allium cepa L. response to selenium inorganic compounds. *Analytical and Bioanalytical Chemistry*, **406**, 3717.

[81] Thosaikham, W., Jitmanee, K., Sittiput, R. *et al.* (2014) Evaluation of selenium species in selenium-enriched pakchoi (Brassica chinensis Jusl var parachinensis (Bailey) Tsen & Lee) using mixed ion-pair reversed phase HPLC–ICP-MS. *Food Chemistry*, **145**, 736.

[82] Yang, L., Zhou, L., Hu, Z., and Gao, S. (2014) Direct determination of Si isotope ratios in natural waters and commercial Si standards by ion exclusion chromatography multi-collector inductively coupled plasma spectrometry. *Analytical Chemistry*, **86**, 9301.

[83] Nonose, N., Cheong, C., Ishizawa, Y. *et al.* (2014) Precise determination of dissolved silica in seawater by ion-exclusion chromatography isotope dilution inductively coupled plasma mass spectrometry. *Analytica Chimica Acta*, **840**, 10.

[84] Nette, D. and Seubert, A. (2015) Determination of aminopolycarboxylic acids at ultra-trace levels by means of online coupling ion exchange chromatography and inductively coupled plasma-mass spectrometry with indirect detection via their Pd^{2+}-complexes. *Analytica Chimica Acta*, **884**, 124.

[85] Seubert, A., Schmincke, G., Nowak, M. *et al.* (2000) Comparison of on-line coupling of ion chromatography with atmospheric pressure ionization mass spectrometry and with inductively coupled plasma mass spectrometry as tools for the ultra-trace analysis of bromate in surface water samples. *Journal of Chromatography, A*, **884**, 191.

[86] Tirez, K., Brusten, W., Beutels, F. *et al.* (2013) Determination of bromate in drinking waters using low pressure liquid chromatography/ICP-MS. *Journal of Analytical Atomic Spectrometry*, **28**, 1894.

[87] Ezerinskis, Z., Spolaor, A., Kirchgeorg, T. *et al.* (2014) Determination of I-129 in Arctic snow by a novel analytical approach using IC–ICP-SFMS. *Journal of Analytical Atomic Spectrometry*, **29**, 1827.

[88] Han, C., Sun, J., Cheng, H. *et al.* (2014) Speciation analysis of urine iodine by ion-pair reversed-phase liquid chromatography and inductively coupled plasma mass spectrometry. *Analytical Methods*, **6**, 5369.

[89] Tang, Q., Xu, Q., Zhang, F. *et al.* (2013) Geochemistry of iodine-rich groundwater in the Taiyuan Basin of central Shanxi Province, North China. *Journal of Geochemical Exploration*, **135**, 117.

[90] Kozai, N., Ohnukiand, T., and Iwatsuki, T. (2013) Characterization of saline groundwater at Horonobe, Hokkaido, Japan by SEC–UV-ICP-MS: speciation of uranium and iodine. *Water Research*, **47**, 1570.

[91] Studzinska, S., Mounicou, S., Szpunar, J. *et al.* (2015) New approach to the determination phosphorothioate oligonucleotides by ultra high performance liquid chromatography coupled with inductively coupled plasma mass spectrometry. *Analytica Chimica Acta*, **855**, 13.

[92] Thompson, D.F., Micholoulos, F., Smith, C.J. *et al.* (2013) Phosphorus and sulfur metabonomic profiling of tissue and plasma obtained from tumour-bearing mice using ultra-performance liquid chromatography/inductively coupled plasma mass spectrometry. *Rapid Communications in Mass Spectrometry*, **27**, 2539.

[93] Kazui, Y., Seto, Y., and Inoue, H. (2014) Phosphorus-specific determination of glyphosate, glufosinate, and their hydrolysis products in biological samples by liquid

chromatography inductively coupled plasma mass spectrometry. *Forensic Toxicology*, **32**, 317.

[94] Rugova, A., Puschenreiter, M., Santner, J. *et al.* (2014) Speciation analysis of orthophosphate and myo-inositol hexakisphosphate in soil- and plant-related samples by high-performance ion chromatography combined with inductively coupled plasma mass spectrometry. *Journal of Separation Science*, **37**, 1711.

[95] Elguera, J.C.T., Barrientos, E.Y., Wrobel, K., and Wrobel, K. (2013) Monitoring of phosphorus oxide ion for analytical speciation of phosphite and phosphate in transgenic plants by high-performance liquid chromatography–inductively coupled plasma mass spectrometry. *Journal of Agriculture and Food Chemistry*, **61**, 6622.

[96] Zakon, Y., Halicz, L., and Gelman, F. (2014) Isotope analysis of sulfur, bromine, and chlorine in individual anionic species by ion chromatography/multicollector-ICPMS. *Analytical Chemistry*, **86**, 6495.

[97] Yang, L., Paglianoand, E., and Mester, Z. (2014) Direct determination of dissolved phosphate and silicate in seawater by ion exclusion chromatography sector field inductively coupled plasma mass spectrometry. *Analytical Chemistry*, **86**, 3222.

[98] Schmidt, C.K., Raue, B., Brauch, H.J., and Sacher, F. (2014) Trace-level analysis of phosphonates in environmental waters by ion chromatography and inductively coupled plasma mass spectrometry. *International Journal of Environmental and Analytical Chemistry*, **94**, 385.

4

APPLICATION OF IC-MS IN ORGANIC ENVIRONMENTAL GEOCHEMISTRY

KLAUS FISCHER

Faculty VI – Regional and Environmental Sciences, Department of Analytical and Ecological Chemistry, University of Trier, Behringstr. 21, Trier 54296, Germany

4.1 INTRODUCTION

Organic environmental geochemistry is a subdiscipline of organic geochemistry, dealing with the dynamics of organic compounds in the environment. Here, the term "dynamics" represents a network of physical, biological, and chemical processes responsible for the occurrence, distribution, lifetime, reaction, and transformation of organic substances in the ecosphere and in its various compartments. Different from its parent discipline, properties and fate of fossil organic matter as well as processes in the lower lithosphere and in deeper earth zones are usually not subject to organic environmental geochemistry. The research focus stretches over many scales in space and time, for example, from microscale rhizosphere processes at the tip of a tiny root hair to global substance trajectories and from fast photochemical reactions to long-term entrapment of low-molecular-weight organic substances (LMWOS) in organoclay complexes of soils. On the environmental compartment level, research subjects and methods are often identical to those of the chemical branch of the compartment-specific scientific discipline, that is, soil science, hydrology, sedimentology, and atmospheric research. A summary of relevant environmental compartments and related matrices is given in Table 4.1.

Application of IC-MS and IC-ICP-MS in Environmental Research, First Edition.
Edited by Rajmund Michalski.
© 2016 John Wiley & Sons, Inc. Published 2016 by John Wiley & Sons, Inc.

TABLE 4.1   Main Environmental Compartments and Related Matrices (Except Organisms) Relevant as Sources and Reaction Media for Natural Organic Compounds

Compartment	Phase/Matrix
Atmosphere	Gas phase
	(Bio)aerosols, fine particulate matter
	Smoke, soot
	Ice/snow particles
	Rain/fog droplets
Hydrosphere	Free water phase
	Suspended matter/colloids
	Organic detritus, sediment, sediment pore water
Pedosphere	Solum including soil organic matter
	Organic litter horizon, rhizosphere
	Soil water (percolate)
	Plant root, and microbial exudates

Hence, in practice, a differentiation between these disciplines, for example, soil organic chemistry and organic environmental geochemistry, is often not possible or not meaningful. Nevertheless, a characteristic feature of environmental geochemistry is its integrative approach, linking together processes inside and across boundaries of various (sub)compartments into the framework of global models for element and substance cycles. These cycles are plotted not only for the present but also for the past, reading historical records and nature archives of, for example, ice or sediment cores. A specific task is the identification of the influence of humans on natural substance cycles and ecosystem functions. With regard to the occurrence and effects of environment-contaminating chemicals, especially organic xenobiotics, there is no clearly defined delimitation from the field of environmental chemistry. However, this chapter concentrates on the analysis of natural organic substances.

According to the principles of ion chromatography, used as a generic term for ion exchange chromatography (I-EC), ion exclusion chromatography (IEC), ion-pair chromatography (IPC), and zwitterionic chromatography (ZIC), an organic molecule subjected to this separation technique must possess one or more polar functional groups being able to dissociate or to take up protons at a certain pH value or pH range, defined by the eluent composition and separation temperature. Further criteria are at least moderate hydrophilicity and a low-to-moderate molecular mass, but their relevance for the separation process depends on the applied technique and separation conditions. Different from the general organic geochemistry, where diagenetically formed lipophilic substances dominate, various groups of polar organic compounds, for example, carbohydrates, carboxylic acids, amino acids, peptides,

TABLE 4.2 Main Topics of Environmental Geochemical Research on Carboxylic Acids (CA)

Soil and Agriculture	Hydrosphere	Atmosphere
• Influence of soil properties, climatic conditions, vegetation, and agricultural practice on organic acid production and release	• Seasonal dynamics of CA • Influence of trophic levels • Formation during degradation of detritus and suspended particulate matter	• Emission into the gas phase from biological sources and biomass burning (source identification)
• Rhizosphere processes (interaction between plants, soil microbes, and fungi)	• Interrelations with refractory organic substances	• Distribution between gas and particle phase
• Mineral nutrition of plants, plant response to nutrient deficiency	• Formation during photochemical degradation of humic and fulvic acids	• Formation and transformation during photooxidation reactions
• Carbon turnover and sequestration	• Metal mobilization and speciation	• Interaction with atmospheric radicals
• Pedodynamics of nutritional elements including speciation and bioavailability	• Concentration profiles in sediments and sediment pore waters	• Organic signature of atmospheric particles
• Formation and degradation of humic substances	• Contribution to the regulation of sediment pH and redox potential	• Contribution to the acidity of precipitation
• Podzolization and acidification	• Participation in geochemical sediment conversion processes	
• Mineral weathering		
• Allelopathy and phytotoxicity		
• Molecular signaling		

amines, have attracted considerable attention in environmental geochemical research. For example, current topics of environmental research on carboxylic acids are summarized in Table 4.2.

Similar research fields being slightly modified exist for carbohydrates. Noteworthy specific tasks in environmental carbohydrate research are the formation of mono- and oligosaccharides during biodegradation or hydrolysis of polysaccharides or lignocellulosic materials; the applicability of monosaccharide anhydrosugars, that is, levoglucosan, mannosan, and glucosan, as biomarkers for biomass burning; and the determination of the mass ratios of selected pentoses to hexoses as indication for abundance ratios of different microbial subgroups in soils and sediments. Clearly, all of the mentioned polar compounds are part of the analytical portfolio of modern, that is, "high-performance" ion chromatography.

The development of efficient methods for the separation of polar organic compounds was and still is a continuous challenge for instrumental IC since its introduction in the late 1960s. It has initiated some great inventions and effected several breakthroughs. First presentations of prototypes of instrumental "high-resolution" or high-pressure anion-exchange chromatography (HPAEC) devices demonstrated their capabilities to separate more than 100 physiological compounds including many urinary acids within one run, which could take 20 h [1, 2]. These prototypes were also able to analyze carbohydrates in physiological fluids [3, 4]. Katz *et al.* made first attempts to use this IC instrumentation for the determination of organic compounds in municipal and industrial sewage effluents [5]. Their high-resolution IC analyzer featured some advanced technical attributes, for example, an eluent gradient generator and a two-wavelength, dual-beam UV photometer. In the 1970s, carboxylic acid analyses put demand on the development of improved ion exclusion chromatographic methods, since anion-exchange resins provided insufficient selectivity and retention for organic acids at that time [6]. To separate complex mixtures of inorganic and organic acids, a further enhancement of chromatographic resolution was necessary. To achieve this aim, two-dimensional chromatographic techniques were already introduced in the 1970s. The first approach in this direction was a combination of an ICE column (first dimension) with an IEC column (second dimension) [7, 8]. The organic acids were separated in the first dimension, whereas the fraction of the nearly unresolved, highly dissociated inorganic acids were passed to an anion concentrator column by means of a column-switching valve. Afterward, the concentrator was flushed by the IEC eluent, transferring the inorganic analytes into the IEC column, where they were separated. To avoid the effort of column switching and the usage of two separation channels, techniques employing two columns in series were also invented. These techniques require the compatibility of the eluent with both columns. In the 1980s, mainly two alternative analytical strategies were developed. The first one combined two different anion-exchange columns, for example, Dionex (now Thermo Scientific) IonPac AS4 and AS4-SC, applying an $NaHCO_3/Na_2CO_3$ eluent [9]. The second one hyphenated an ICE column (Aminex HPX-87H) with an RP-C_{18} column, percolating diluted sulfuric acid through both columns [10]. In addition, the demands of organic acid analyses gave rise to the invention of new types of chemical eluent suppressors. Despite some attempts to use hollow fiber suppressors to minimize background conductivity after the HPAEC separation of carboxylic acids [11, 12], it was found that this suppressor type offers no wide applicability, since its suppression capacity was not sufficient for the eluent concentrations often required [13]. The problem could be solved with the introduction of micromembrane suppressors.

Carbohydrate analysis has stimulated column and detector development simultaneously. In the 1970s, HPAEC played a minor role compared with the separation on cation-exchange resins [14, 15] and with amino phases bonded to silica gels [16, 17]. This situation changed with the introduction of HPAEC columns providing an improved selectivity for mono- and oligosaccharides, for example, Dionex IonPac

AS6 (CarboPac PA1) [18]. At the same time, the invention of pulsed amperometric detection (PAD) made a new dimension in carbohydrate detection sensitivity available, replacing the formerly often used refractive index detector (RID), especially in cases where high sensitivity is required or gradient elution is performed [19, 20].

As for carbohydrates, for every group of organic analytes, routine detection systems for liquid chromatographic separation exist, some of them accompanying HPLC and IC through their whole developmental progress. Conventional IC detection modes for organic acids are conductivity detection and, of minor importance, UV–Vis absorption. Amino acids are mainly analyzed by means of UV/Vis absorption or fluorescence measurement in combination with pre- or postcolumn derivatization. Some newer applications make use of the PAD as well. Despite the great progress in terms of detection sensitivity and structural analyte characterization achieved by the hyphenation of IC with MS, it should not be ignored that the great majority of IC analysis of natural polar organic compounds is still performed with conventional detection systems, at least in the field of environmental geochemistry.

IC-MS hyphenation for the determination of small polar organic molecules started to flourish with the launch of the atmospheric pressure ionization technique, in particular with technically matured electrospray interface (ESI) devices in the mid-1990s. Surely, the history of organic trace analysis by IC-MS began before the introduction of ESI. Pacholec *et al.* [21] were among the first connecting an ion exclusion column (Dionex IonPac ICE-AS1) with an MS system by a thermospray interface. They used the discharge ionization mode of the interface and positive ion scanning of the $[M+H]^+$ ions of the analytes and of their water adducts to identify glycolic, acetic, propionic, and butyric acids. They did not record mass fragmentation spectra. Before its replacement by ion spray (pneumatically assisted ESI), the thermospray interface was widely used in the HPAEC-MS analysis of oligosaccharides during the 1990s [22–26].

One of the few attempts to perform organic acid analysis by ion exclusion chromatography–particle beam mass spectrometry was made by Alexander and Quinn [27]. They combined a Waters Fast Fruit Juice ICE column (mobile phase 1.0 mM HCl, flow rate 0.5 ml min^{-1}) with an HP 5988 quadrupole MS by an HP particle beam interface Model 59980A. Chemical ionization with methane to determine molecular mass and electron impact ionization to attain structural information of various aliphatic di- and tricarboxylic acids were applied. Replacing the ICE column by an anion-exchange column (Ringwood SGE Model 250 GL), the same HP particle beam MS system was applied for the determination of acidic organic pollutants in fruit juice, surface waters, and hazardous waste [28, 29]. In another investigation, a mixed-mode anion-exchange/reversed-phase column (Dionex OmniPac PAX-500) was combined with the HP MS system to analyze aromatic and polycyclic sulfonic acids in hazardous waste samples [30].

The work of Conboy and Henion [30] was a milestone in the hyphenation of HPAEC with MS by an ESI interface. Their aim was to analyze N-linked glycans from glycoproteins. The paper of Xiang *et al.* [31] pioneers in the

HPAEC-ESI-MS/MS analysis of low-molecular-weight organic acids (LMWOA) together with inorganic anions. This investigation already combined many aspects of modern IC-MS techniques. A Dionex IonPac AS5A-5μ column was used together with a sodium hydroxide gradient at a flow rate of 1 ml min^{-1}. The column effluent was passed through a Dionex Anion Self-Regenerating Suppressor-1 (ASRS), operated in the autosuppression recycle mode. Before entering the MS interface, the eluent flow was reduced to approximately its half by means of a tee piece. The MS unit was a Finnigan TSQ-7000 triple-stage quadrupole mass spectrometer, applying nitrogen as sheath and auxiliary gas and xenon as collision gas for collision-induced dissociation (CID) fragmentation.

An early example for the combination of HPICE with ESI-MS is provided by Johnson *et al.* [32]. They employed a Dionex IonPac ICE-AS6 column with 0.4 mM trifluoroacetic acid as mobile phase. The mass spectrometer was an API/1 single quadrupole apparatus (PerkinElmer SCIEX). Several fundamental aspects and problems of ICE-MS hyphenation were addressed by their method development, for example, the need for postcolumn pH adjustment and organic solvent addition to facilitate eluent evaporation, analyte ionization, and droplet formation and to achieve a minimization of unwanted electrical discharge (ion suppression) of the analytes in the ion spray, caused by matrix components and remaining eluent anions. Even effects of sheath gas type and flow on spray formation and of the eluent split ratio on detection sensitivity were tested.

This treatise does not follow the intention to provide a deeper insight into specific chromatographic phenomena, which are not of general importance for the considered analytical procedures. Relevant chromatographic separation principles are briefly mentioned in the following text. The description of specific technical features of different MS instruments is also beyond the scope of this review. Instead, it aims at familiarizing the reader with the various IC-(ESI)-MS methods developed within the last 15 years to analyze small ionizable organic compounds of natural origin in environmental media and to provide a broad overview of their applications. Besides the general analytical performance criteria, that is, selectivity, sensitivity, and achievable structural information, this method presentation points to practical aspects, for example, method versatility, matrix compatibility, robustness, and ease of use, as far as they are noticed.

4.2 CARBOXYLIC ACIDS

4.2.1 Molecular Structure, Molecular Interaction Potential, and Chromatographic Retention

Correlations between the molecular structure of carboxylic acids, kinds of intermolecular forces, and retention in anion-exchange and anion exclusion chromatography are briefly outlined. Details can be found in Section 1-1, in several monographs about ion chromatography, and elsewhere [33–38]. Clearly, the chromatographic distribution of analytes is influenced by all parts of the chromatographic process, that is, by the properties and composition of the chromatographic phases and by the process conditions, for example, temperature. Hence, a few general rules are outlined in the following text, but this does not mean that they will apply in every chromatographic situation.

Electrostatic forces, that is, electrostatic attraction in exchange chromatography and electrostatic repulsion in exclusion chromatography, are the main interaction mechanisms for simple mono- and oligoprotic aliphatic carboxylic acids, followed by RP (van der Waals) interactions of the hydrocarbon units. Since HPAEC separations of organic acids are usually made with strongly alkaline eluents, the carboxylic groups are completely dissociated independently from their pK_a values. Retention increases with increasing analyte charge. Within subgroups of analytes with identical charge, retention increases with carbon chain length and unbranched molecules are more strongly retained than their branched isomers. In ICE, retention decreases with increasing charge. Here, the adjustment of the (acidic) eluent pH is the master variable to regulate the dissociation and charge formation of the analytes, depending on their pK_a values. If the proportion of the undissociated structure on total species distribution is more or less the same for a group of acids, retention is mainly ruled by RP interactions. Additional functional groups such as hydroxyl and keto groups enhance molecule polarity and give rise to polar interactions and H-bonding but at the expense of diminished RP interactions. Thus, retention is often weaker compared to the nonfunctionalized analogues even with columns offering specific polar interaction sites, for example, the Dionex anion exclusion column IonPac ICE-AS6 [35]. In ICE, polar functional analyte groups might influence the chromatographic process by a modification of the pK_a values as well. Olefinic groups and, particularly, aromatic and polycyclic moieties offer π-electron bonding potential and, in the latter case, charge-transfer interactions together with increased van der Waals bonding forces. Hence, retention by columns synthesized through polymerization of aromatic monomers, for example, PSDVB, is strongly enhanced. Size exclusion effects may contribute to chromatographic selectivity in HPIEC and HPICE resulting in a reduced retention with increasing molecular volume.

Due to their large number of hydroxyl groups, sugar mono- and dicarboxylic acids, for example, aldonic, aldaric, and uronic acids, are strongly polar and highly hydrated compounds with an insignificant RP interaction potential. Principally, separation by HPICE and HPAEC is possible. In the latter case, separation on columns designed for carbohydrate analysis provides an alternative to general-purpose anion-exchange columns. Carbohydrate separation columns provide higher retention than common anion-exchange columns do for sugar acids at recommended operation conditions. For instance, C_6 sugar monocarboxylic acids (hexonic acids) are poorly retained at the Dionex IonPac AS11-HC column even at low hydroxide ion concentration (capacity factors < 1) due to missing RP interactions and additional size exclusion effects. Diastereoisomers are poorly resolved. Uronic acids are more strongly retained on both types of HPAEC columns, and separation of some diastereoisomers is possible.

4.2.2   Environmental Analysis of Carboxylic Acids by Ion Exclusion Chromatography–Mass Spectrometry (HPICE-MS)

The analysis of soil solutions, root exudates, and plant tissues is a main field in the environmental analysis of carboxylic acids by ICE-MS. Here and in other applications of ICE-MS and IEC-MS, hyphenation of the chromatographic unit with MS is realized by the ESI interface. Organic acid analysis is made with negative ionization throughout. Bylund *et al.* developed one of the first methods for this purpose [39]. They combined a Supelcogel C610H column with an MDS SCIEX API 3000 triple quadrupole MS detector. The essentials of this procedure and of the following methods are listed in Table 4.3.

The isocratic eluent contained 10% (v/v) methanol and 90% (v/v) of a diluted aqueous formic acid solution with a volume concentration of formic acid of 0.01%. The replacement of the recommended eluent for this column (0.2% H_3PO_4) by the newly introduced one was necessary to meet the requirements of the MS system related to the volatility of the mobile phase. The addition of methanol served to improve the stability of the electrospray. The eluent replacement led to a somewhat lower chromatographic resolution. From the 17 selected reference compounds, that is, multifunctional mono-, di-, and tricarboxylic acids, two fractions, containing four acids each, were formed, eluting in retention time intervals with spans of less than 1 min. Reasons for the reduced chromatographic resolution might be the higher pH value of the formic acid eluent, allowing for a higher dissociation degree of the analytes. Consequently, analyte exclusion is enhanced and retention reduced. Furthermore, the addition of methanol, lowering hydrophobic interactions between the analytes and the column resin, might have contributed to a suboptimal chromatographic selectivity. Nevertheless, due to the multiple reaction monitoring mode (MRM) of the MS apparatus, all acids were unambiguously identified. MRM helped to distinguish between various isomers, for example, citrate and isocitrate due to different mass spectra. Relatively large differences in analyte detection sensitivity were stated, ranging from a LOD ($3 \times S/N$) of 1 nM (*cis*-aconitic acid and malic acids) to 300 nM (pyruvic and lactic acids). They are attributed both to differences in precursor ion formation and transmission and to differences in fragmentation behavior. Moreover, differences in background noise levels were noticed. The practical application of the method revealed the occurrence of 12 organic acids in a soil solution. The sampling and the sample preparation details are summarized in Table 4.4.

Oxalic acid had highest concentrations, followed by lactic acid. Amounts of both compounds were nearly identical in stream water. Many of the other acids were found in concentrations $\leq 0.05\,\mu M\,l^{-1}$. Ali *et al.* utilized this method to determine carboxylic acids in extracts of podzolic boreal forest soils [41]. Among other aspects, they focused on the influence of the sample preparation method on extraction efficiency. For that purpose, they compared soil solution obtained by drainage centrifugations with soil extraction by different buffer solutions, conducted in an ultrasonic bath or in a rotary shaker. As expected, the acid concentration in the organic (O) horizon was essentially higher than in the eluvial (E) horizon featuring a factor of 30 related to the respective centrifugate. Extraction with 10 mM K_2HPO_4

TABLE 4.3 Environmental Analysis of Carboxylic Acids by Ion Exclusion Chromatography–ESI-Mass Spectrometry

Matrix	Column	Eluent[a]	Eluent Flow (ml min^{-1})	Eluent Split Ratio	MS[b]	References	Publishing Year
Aerosol particles	LC × LC: 1st dim.: Hamilton SCX 10 μm, 2nd dim.: Waters C$_{18}$ XBridge	ACN/HAc- gradient	1st dim.: 0.04, 2nd dim.: 2.00	1:30	Bruker Daltonics Micro-TOF	[40]	2006
Soil extracts, soil solution, surface water	Sigma Aldrich Supelcogel C610H	Methanol/0.01% formic acid	0.40	≈1:7 and 1:3	MDS SCIEX API 3000 MS/MS, MRM[c] mode	[39, 41]	2007, 2011
Plant nutrition solution with root exudates	Bio-Rad HP-87H	0.1% HAc	0.40	—	Agilent Single Quadrupole MS, Full Scan	[42]	2008
Plant tissues, root exudates	Phenomenex Rezex RMH- Monosaccharide H^{+}	0.1% HAc	0.50	—	MDS SCIEX QTRAP 3200 MS/MS, MRM	[43]	2009

(*continued*)

TABLE 4.3 *(Continued)*

Matrix	Column	Eluent[a]	Eluent Flow (ml min^{-1})	Eluent Split Ratio	MS[b]	References	Publishing Year
Root tips [44], sugar beet leaves, roots and xylem sap [45], sugar beet leaves, tomato leaves, tomato xylem sap [46], roots and leaves of *Lotus corniculatus* [47]	Sigma–Aldrich Supelcogel H	0.1% formic acid or 5% methanol with 0.1% formic acid or 5% 2-propanol with 0.1% formic acid	0.20	—	Bruker Daltonics Micro-TOF II	[44–47]	2011, 2011, 2011, 2012

[a]HAc: acetic acid, ACN: acetonitrile, % relate to volume concentrations of aqueous solutions.
[b]Negative ionization mode.
[c]Multiple reaction monitoring analytes listed in Table 4.4.

118

TABLE 4.4 Sample Matrix, Sampling, and Sample Preparation Techniques for the Analysis of Carboxylic Acids

Sampling and Sample Preparation	Matrix	Detected Acids	References
Soil solution by suction lysimeter, MF[a] (0.45 μm)	Soil solution, surface water	17 MFA[b]	[39]
Soil solution: water percolation through laboratory soil columns; filtration through glass fiber and membrane (0.4 μm) filters	Soil leachates, drinking water	18 (poly)hydroxy acids	[36]
Extraction with 10 mM K_2HPO_4 (pH 7.2) or with 1:1 (v/v) K_2HPO_4/methanol mixture; agitation in ultrasonic bath or rotary shaker, MF (0.45 μm)	Soil extracts, soil centrifugates	17 MFA	[41]
Collection of percolated plant nutrition solution, MF (0.45 μm)	Plant nutrition solution with root exudates	Aconitic, citric, lactic, maleic, malonic, oxalic, pyruvic, succinic, tartaric	[42]
Tissues: homogenization in a mill, extraction with water; exudates: immersion of roots into aerated trap solutions of 0.5 mM $CaSO_4$ $2H_2O$ for 2 h, MF (0.45 μm)	Root exudates, roots, and shoots	12 MFA	[43]
Freeze crushing, extraction with aqueous methanol at 80 °C, centrifugation, evaporation of methanol, MF (0.22 μm)	Root tips	cis-Aconitic, citric, lactic, malic, quinic, succinic	[45]

(continued)

119

TABLE 4.4 *(Continued)*

Sampling and Sample Preparation	Matrix	Detected Acids	References
Leaves and roots: extraction with 4% (w/v) metaphosphoric acid, centrifugation, MF (0.22 μm)	Leaves, roots and xylem sap of sugar beet	Citric, lactic, malic, succinic	[46]
Leaves as in [46], MF (0.22 μm)	Sugar beet leaves, tomato xylem sap	*cis*-Aconitic, ascorbic, citric, fumaric, malic, oxalic, 2-oxoglutaric, quinic, shikimic, succinic	[44]
Wet effluent diffusion denuder/aerosol collector (WEDD/AC) and high-volume sampler with quartz fiber filters; WEDD/AC effluents passed through anion concentrator column (Dionex TAC-LP1); ultrasonic extraction of fiber filters with water	Gas phase, aerosol particles	Up to 20 C_1–C_6 mono-, di-, and tricarboxylic acids	[48–51]
High-volume sampler and virtual impactor (fraction < 1.3 μm), static sonication assisted solvent extraction (3 × 15 ml methanol)	Aerosol particles	≈35 mid- and long-chain aliphatic mono- and dicarboxylic acids, terpene-derived acids	[40]

Sample preparation	Matrix	Analytes	Reference
Quartz filter, water extraction in ultrasonic bath (15 min), MF (0.45 μm)	Organic aerosol particles from biomass burning	19 aliphatic (hydroxy) acids	(Fischer, K., Höffler, S. and Meyer, A. Unpublished Results)
Sampling by bag houses, sieving (<63 μm), extraction with water, MF (0.20 μm)	Urban dust: NIST SRM 1649b standard material	34 mono- and multifunctional acids	[52]
Sampling with HDPE bottles, MF (0.22 μm)	Surface waters	Acetic, citric, formic, glycolic, lactic, malic, oxalic, succinic, tartaric	[53, 54]
Preparation from synthetic sea salt	Synthetic sea water	Acetic, butyric, formic, propionic, pyruvic, valeric	[55]
Collection from fermentation vessels, centrifugation, water addition, homogenization	Microalgae (biofuel feedstock)	17 MFA	[56]
Liquors from laboratory hydrolysis experiments: dilution with water, MF (0.20 or 0.45 μm)	Alkaline wood hydrolysates (kraft black liquor)	12 acids including sugar acids	[57]

[a]Membrane filtration.
[b]Multifunctional aliphatic acids.

buffer (pH 7.2) led to higher release efficiencies for the E horizon, but not for the O horizon. The addition of methanol to the buffer solution provoked an extreme increase in acid amounts liberated from the O horizon and had a moderate stimulating effect on acid mobilization from the E horizon. All of the targeted 17 organic acids were detected in quantifiable concentrations in at least one sample. Citric acid prevailed in all centrifugates and in most of the extracts. Highest extraction efficiencies, compared to the soil centrifugates, were achieved for isocitric and fumaric acids.

Chen et al. hyphenated a Bio-Rad HP-87H column with an Agilent 1100 single quadrupole MS to determine multifunctional aliphatic acids (MFAs) in plant nutrition solutions containing root exudates [42]. 0.1% (v/v) acetic acid with a flow rate of 0.40 ml min^{-1} served as eluent. The separation of the nine reference compounds needed 8 min. The detection limits, based on a signal-to-noise ratio of 3, ranged from 10 to 30 µg l^{-1}. Compared with 0.1% (v/v) formic acid, acetic acid yielded better MS sensitivity. This effect was already observed by Rosenberg [58]. Relative standard deviations of peak areas (100 µg l^{-1}, $n = 3$) were less than 3.5%, and spike recoveries were between 88.7% and 95.4%. The concentrations of pyruvic, aconitic, and citric acids in nutrition solutions containing root exudates from *Brassica juncea* (Indian mustard) were 32.4, 24.4, and 44.3 µg l^{-1}, respectively. These data matched results from measurements with a suppressed ion-exchange chromatography–conductivity detection system very well.

Twelve MFAs were analyzed in trapped root exudates and in water extracts of plant tissues by a combination of the Phenomenex Rezex RHM-Monosaccharide-H-ICE column with the MDS SCIEX Q TRAP 3200 triple quadrupole MS system, operated in the MRM mode [43]. A 0.1% (v/v) acetic acid solution with a flow rate of 0.5 ml min^{-1} was used as eluent. To achieve an optimal separation, other liquid chromatographic techniques (RP-HPLC with C_{18} and C_8 columns, HILIC) were tested as well, but best results were achieved with the ICE column. Separation was accomplished within 7 min, but citric and isocitric peaks were not totally resolved. Limits of detection ($3 \times S/N$) for standards and exudates ranged from 0.5 with *cis*-aconitic and maleic acids to 30.0 µg l^{-1} with pyruvic acid. LODs for plant tissues were between 0.4 and 12.0 mg kg^{-1}. Recoveries in roots and shots spanned between 72% and 115%. Matrix effects were neglectable. Total acid concentrations were highest in the shoots. Exudated amounts were one to a few percent of the organic acid pool in roots.

Another examination of organic acids in plant tissue extracts utilized a Supelco H column together with a Bruker Daltonics Micro-TOF II MS [44]. Eluents were 0.1% (v/v) formic acid or a mixture of formic acid with 5% (v/v) methanol. The flow rate was 0.2 ml min^{-1}. LODs ($3 \times S/N$) were below 0.1 µM l^{-1} for the majority of the 10 standard compounds. The inferior LOD of oxalic acid amounting to 12.8 µM l^{-1} was traced back to a poor ionization efficiency of this analyte. The elevated LOD of ascorbic acid amounting to 1.25 µM l^{-1} is explained by its fast oxidation, which prevents its ESI-MS detection. Repeatabilities of retention time and peak area were <5% with the exception of oxalic acid. The recoveries of the analytes from the plant tissues were between 93% and 110% mainly. Considerably lower values were measured for oxalic acid (46.5% in tomato xylem sap), 2-oxoglutaric acid (mean of 67.5% for sugar beet leaves and tomato xylem sap), and ascorbic acid (21.5% in sugar beet

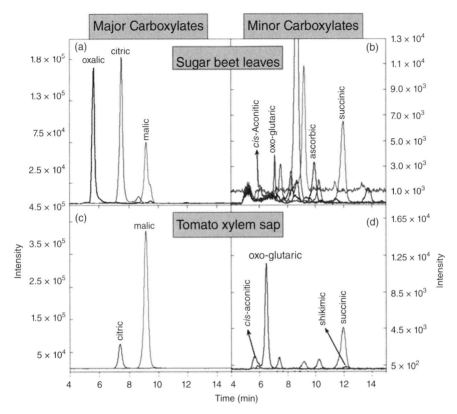

Figure 4.1 ICE-ESI-TOF-MS chromatograms of major (a and c) and minor (b and d) carboxylates found in sugar beet leaves and tomato xylem sap. Chromatogram traces correspond to the $[M-H]^- \pm 0.03$ *m/z* of each organic acid. For details of the analytical method, refer to Table 4.3. Source: From [44] with permission. Copyright (2011), American Chemical Society.

leaves). With respect to these low recoveries, the authors stated that "some organic acid/plant matrix combinations may require specific extraction procedures and/or the use of ^{13}C labeled internal standards" [44]. In tomato xylem sap, major carboxylates were citric and malic acids, whereas *cis*-aconitic, 2-oxoglutaric, shikimic, and succinic acids were minor components. The dominating carboxylates in sugar beet leaf extracts were oxalic, citric, and malic acids. The aforementioned minor acids were also detectable. Figure 4.1 depicts the ICE-MS chromatograms of major and minor carboxylates in both plant samples.

The technique and instrumentation described earlier were applied with very small modifications (altered eluent composition, Table 4.3) for the investigation of metabolic responses of plants to reduced or elevated supplies with essential (Fe) and potentially toxic (Zn, Al) elements. To study the effect of iron deficiency (iron chlorosis) on organic acid concentrations in root tips, various genotypes of *Prunus* rootstocks were hydroponically grown in iron-sufficient and -deficient conditions during 2 weeks [45]. It was found that malic, citric, and succinic acid concentrations increased in roots of plants grown under iron-deficient conditions, independently from their genotype. The results of the ICE-MS analyses were confirmed by quantitative ^{1}H-NMR measurements, which revealed increases of sucrose and amino acid concentrations as well.

Concentrations of oxalic, citric, malic, and succinic acids were monitored in roots, leaves, and the xylem sap of sugar beet plants subjected to different Zn concentrations in nutrition solutions [46]. The concentration of citric acid, but not of the other acids, increased in roots with increasing Zn concentration. The total acid amounts were highest in xylem sap and leaves at moderately elevated Zn levels. Similar aspects were addressed in another study concerned with effects of different Al concentrations on growth and metabolic processes of the forage legume *Lotus corniculatus* (bird's-foot trefoil) [47]. The increased root contents of malic and 2-isopropylmalic acids indicated the participation of these Al chelators in Al detoxification. Acidic components in atmospheric aerosols were analyzed by means of comprehensive two-dimensional liquid chromatography–time-of-flight mass spectrometry [40]. The reason for this enhanced analytical effort was the purpose to determine mid- and long-chain mono- and dicarboxylic acids together with terpene-derived acids simultaneously. Separation in the first dimension was accomplished by the ion exclusion column Hamilton SCX 10 μm in capillary format. The mobile phase was an ACN/acetic acid gradient with a flow rate of 40 μl min^{-1}. The effluent of the first column was sampled in a loop of a modulating valve and completely transferred to the second column within short and constant periods. The second column was a Waters C_{18} XBridge (50×3 mm I.D.) with 2.5 μm packing material. To realize a high sampling rate in the second dimension, the short RP column was operated with a high flow rate (2.0 ml min^{-1}, the eluent composition was the same as in the first dimension), performing a complete chromatographic run within 120 s. The effluent of the second column was split with a ratio of 1:30 before transfer into the ESI–micro-TOF (Bruker Daltonics). Injecting a standard mixture, it was ascertained that tartaric acid was least retained in both dimensions, followed by other low-molecular-mass dicarboxylic acids and by dicarboxylic acids with a keto group. Monocarboxylic acids were more strongly retained in the first dimension. The gradient elution caused some irregularities in peak shapes and even peak splitting in some cases. Limits of detection ($3 \times S/N$) were in the range of 2–200 μg l^{-1}. Average reproducibility of peak areas was 8% (10 mg l^{-1}, $n = 3$). Up to 35 organic acids were detected in methanol extracts of urban and rural aerosol samples, but only a few were quantified. The rural aerosols differed from the urban ones in comparably high amounts of oxidized terpene derivatives, especially pinonic acid, indicating high contributions of volatile organic carbon emissions from coniferous forests to aerosol formation.

4.2.3   Environmental Analysis of Carboxylic Acids by Ion-Exchange Chromatography–Mass Spectrometry (HPI-EC-MS)

Reviewing published methods on environmental geochemical analysis of carboxylic acids by HPIEC-MS, a few general features become obvious. Chromatography was exclusively performed choosing Dionex (Thermo Scientific) anion-exchange columns. Most of the methods made use of the Dionex IonPac AS11-HC column. This column belongs to the group of latex-agglomerated anion exchangers with ethylvinylbenzene/divinylbenzene (EVB/DVB) core material and alkanol quaternary ammonium functions fixed at the latex particle surface, which serve as anion-exchange sites. Particle diameter is 9 µm, and the degree of cross-linking is 6% [59]. This column is designated for operation with a hydroxide gradient. An anion self-regenerating suppressor, usually the Dionex ASRS ultra, is inserted in the effluent stream of the column to remove the eluent ions before entering the ESI interface. As far as no further information is provided, all of the applications presented in this chapter were based on this chromatographic design (Table 4.5).

Various aspects of atmospheric gas-phase and particle reactions, for example, the transformation of volatile organic compounds, the generation of organic acids, and their contents in aerosols from different origins, were addressed by several studies of the Laboratory of Atmospheric Chemistry, Paul Scherrer Institute, Switzerland, applying HPIEC-ESI-MS, equipped with a Dionex MSQ ELMO single quadrupole instrument [48–51]. The method was presented for the first time in the context of the identification of organic acids in secondary organic aerosol and in the corresponding gas phase [48]. The aerosol and the gas phase stemmed from smog chamber experiments with 1,3,5-trimethylbenzene, photooxidized in the presence of propene and NO_x. Sampling was achieved by means of a wet effluent diffusion denuder/aerosol collector (WEDD/AC) coupled to an IC with conductivity detection (analytical column: Dionex IonPac AS17). The detector cell outlet was connected with a fraction collector, and IC-MS was consulted to identify unresolved or unidentified compounds in the fractions. Twenty molar masses, presumingly belonging to organic acids, were registered by IC-MS. A tentative identification was possible for half of the compounds. High concentrations of formic, lactic, pyruvic, acetic, and methyl maleic acids were characteristic of the gas phase, whereas lactic acid was dominating in the particle phase, followed by formic and acetic acids.

Furthermore, this analytical technique was applied to determine water-soluble carboxylic acids in aerosols from Zurich, Switzerland [49], and to identify anthropogenic sources for organic acids in an Alpine valley [50]. About 8 mono-, 10 di-, and 2 tricarboxylic acids were quantified in aerosol particles from Zurich, with oxalic acid as main component, followed by malonic, malic, and formic acids. Most of the quantified acids contained additional functional groups. Hydroxy carboxylic acids were the most abundant subgroup of dicarboxylic acids. Pyruvic acid was the only keto acid identified. Most of the analytes occurring in higher concentrations had carbon chains composed of 2–5 carbon atoms. In addition, more than 20 unknown organic acids with m/z values up to 249 were also detected. Concentrations of nine of these acids were estimated on the basis of response factors of closely eluting calibrated compounds. The estimated average concentrations of two of the unknown, presumingly dicarboxylic acids (m/z 148 and 164, respectively) surpassed oxalic acid. The sum of the carboxylic acids contributed on average 2% to the water-soluble organic carbon (WSOC). The fraction of dicarboxylic acids to the WSOC was higher in summer compared to winter, suggesting that dicarboxylic acids are mainly a result

TABLE 4.5 Environmental Analysis of Carboxylic Acids by Ion-Exchange Chromatography–ESI-Mass Spectrometry

Matrix	Column[a]	Eluent	Eluent Flow (ml min⁻¹)	Suppressor[b]	Postcolumn Modifications	MS Apparatus[c]	References	Publishing Year
Gas phase and aerosol	IonPac AS11-HC	NaOH gradient	n. d.[d]	ASRS Ultra	—	Dionex MSQ ELMO SQ[e]	[48–51]	2004, 2006, 2008, 2012
Drinking water, soil leachates, organic aerosol particles	IonPac AS11-HC	KOH gradient	0.30	ASRS Ultra	—	Dionex MSQ ELMO SQ	[36] (Fischer, K., Höffler, S. and Meyer, A. Unpublished Results)	2007
Alkaline wood hydrolysates	IonPac AS11-HC	NaOH gradient	1.00	ASRS Ultra	Eluent split ratio 3:10	HP Series 1100 SQ, SIM and SCAN modus	[57]	2008
Fermenting microalgae	IonPac AS11-HC with Anion Trap Column (CR-ATC)	KOH gradient	0.38	ASRS 300	Continuous ACN addition to eluent (0.20 ml min⁻¹)	Dionex MSQ Plus SQ	[56]	2009
Surface water	IonPac AS11	NaOH gradient	n.d.	ASRS Ultra	—	AB/MDS SCIEX Q-TRAP 2000 MS/MS	[53, 54]	2010, 2010

Synthetic sea water	LC×LC: 1st dim.: IonPac AS24, 2nd dim.: IonPac AS11-HC + IonPac Ultra Trace Anion Concentrator (UTAP-ULP-1)	KOH gradient	0.40 (both dimensions)	2 × ASRS 300	Continuous ACN addition to eluent (0.40 ml min^{-1})	Dionex MSQ Plus SQ	[55]	2012
Urban dust	IonPac AS17-C	KOH gradient	0.25	ASRS	—	Agilent 6410 MS/MS, MRM and SRM mode	[52]	2014

[a] IonPac is a trademark of Dionex (Thermo Scientific).
[b] Suppressors from Dionex (Thermo Scientific).
[c] Negative ionization. single quadrupole detection in SIM modus.
[d] No data.
[e] Single quadrupole.

of photochemical reactions in summer, whereas in winter, they principally result from primary sources. Later on, a 3-week measuring campaign was performed in early winter in an Alpine valley [50]. Here, the emission situation was coined by residential wood burning. Other emission sources that are of great importance for Zurich, that is, traffic, played a minor role. Employing the WEDD/AC sampling system, the alternating analysis of the denuder and the particle collection system resulted in an overall temporal resolution of 2 h. Eleven carboxylic acids were detected by IC-MS and six were identified, namely, formic, acetic, glutaric, succinic, oxalic, and methyl-maleic acids. Roughly 40% of the total amount of identified acids was acetic acid. As far as comparable, the acid concentrations were essentially lower, as in Zurich.

Liu *et al.* made smog chamber experiments to investigate the aqueous-phase processing of secondary organic aerosol formed during isoprene photooxidation [51]. For that purpose, the organic particles were collected on filters, extracted by water, and subsequently oxidized in the aqueous phase either by H_2O_2 under dark conditions or by OH radicals in the presence of light, using a photochemical reactor. Several advanced analytical techniques, including IC-MS, were arranged in an extensive analytical program to detect online and off-line different polarity classes of organic compounds in all atmospheric phases generated during the experiments (Figure 4.2). Besides IC-MS, other MS techniques such as HPLC-APCI-MS/MS and a high-resolution time-of-flight aerosol mass spectrometer (HR-AMS) were part of the analytical platform. The ion chromatographic analysis revealed a significant increase in concentrations of formic, glyoxylic, glycolic, butyric, and oxalic acids

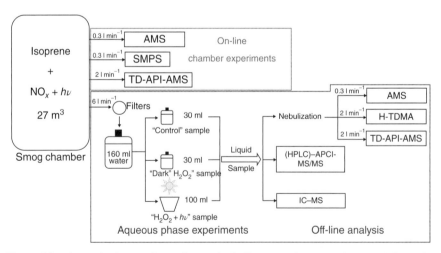

Figure 4.2 General scheme of experiments including secondary organic aerosol formation in the smog chamber, particles sampling on filters and subsequent extraction in pure water, aqueous-phase processing, and nebulization, together with the applied analytical instruments. AMS: aerosol mass spectrometer, H-TDMA: humidified tandem differential mobility analyzer, SMPS: scanning mobility particle sizer, TD: thermodesorption. For details of the IC-MS method, refer to Table 4.5. Source: From [51] with permission.

by the aqueous processing of the organic aerosol particles, both by interaction with H_2O_2 under light exclusion and with irradiation. Highest formation rates were found for formic acid. The IC-MS data were in a good agreement with the increase of the mass fragment m/z 44, measured with the HR-AMS after nebulization. This fragment is indicative of organic acids.

Organic aerosol particles formed by biomass burning, for example, rice straw, are often enriched with saccharides and their dehydrated derivatives, for example, levoglucosan [60, 61]. The concomitant carbohydrate oxidation generates organic acids. Monosaccharides are often oxidized at their terminal C-atoms, mainly yielding sugarmono- and diacids, for example, gluconic and glucaric (saccharic) acids. Further oxidation proceeds via cleavage of the carbon chain forming small hydroxyl acids, for example, glyceric and glycolic acids. These compounds are separable on the IonPac AS-11HC as well. Figure 4.3 depicts SIM-MS traces (Dionex MSQ ELMO single quadrupole detector) of organic acids including sugar acids, extracted from organic aerosol particles by water in an ultrasonic bath (Fischer, K., Höffler, S. and Meyer, A. Unpublished Results). The particles were sampled on quartz filter stripes during a biomass burning event at Mt. Bei Tungyen, Taiwan (courtesy of G. Engling, Academia Sinica, Taiwan). Figure 4.3 illustrates the grouping of the analytes according to the number of their carboxyl groups into three coarse fractions under chosen chromatographic conditions (Table 4.5). Monocarboxylic acids eluted within 10 min, dicarboxylic acids eluted between 30 and 35 min, and the tricarboxylic citric acid

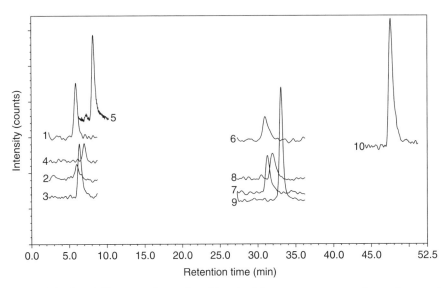

Figure 4.3 SIM-MS traces of organic acids solved in aqueous extracts of organic aerosol particles, sampled during a biomass burning event at Mt Bei Tungyen, Taiwan (Fischer, K., Höffler, S. and Meyer, A. Unpublished Results). SIM signals (acids): 1: gluconic, 2: threonic, 3: lactic, 4: shikimic, 5: formic, 6: saccharic (glucaric), 7: malic, 8: tartaric, 9: tartronic, 10: citric. Analytical details are provided in Table 4.5.

was retained for 47.5 min. The SIM mode allowed for the determination of several analytes despite their incomplete chromatographic separation. Eleven out of thirteen detected compounds were quantifiable. Formic acid possessed highest concentration, followed by citric acid and gluconic acids. Besides the last one, further detected sugar acids were saccharic, xylonic, tartaric, and threonic acids.

To characterize the content of water-soluble organic acids in an air particulate standard reference material (SRM 1649b, urban dust) from the US National Institute of Standards and Technology (NIST), Brent *et al.* applied an IC system (column: Dionex IonPac AS-17C), interfaced with an Agilent 6410 mass spectrometer. The triple quadrupole MS instrument was equipped with an ESI source [52]. Compared to an IonPac AS-11HC, the IonPac AS-17C, a hydroxide-selective low-capacity anion-exchange column, provided the advantage of a baseline resolution between nitrate and organic diacids. Compound identification was made through MRM in the negative ionization mode. The MS/MS conditions for each reference compound were individually determined by selection of an appropriate precursor ion, followed by collision-induced dissociation to determine the most abundant product ions. The two most intense precursor-to-product ion transitions were selected for peak assignment. Quantitative results were obtained using selective reaction monitoring (SRM)-based calibrants, provided that baseline resolution of the respective compound was achieved and CID transition intensity ratio matched with reference standards. In total, 36 organic acids were spotted including aliphatic mono-, di-, and tricarboxylic acids and some aromatic (hydroxy) acids. The following compounds were quantified in the extracts and their average concentrations in the reference material were calculated: glycolic ($20.3 \, \text{mg kg}^{-1}$), citric and isocitric ($7.2 \, \text{mg kg}^{-1}$), quinic ($4.2 \, \text{mg kg}^{-1}$), galacturonic ($3.8 \, \text{mg kg}^{-1}$), malic ($3.6 \, \text{mg kg}^{-1}$), and tartaric ($0.14 \, \text{mg kg}^{-1}$) acids.

Meyer *et al.* developed an HPAEC-ESI single quadrupole MS method for the environmental analysis of aliphatic polyhydroxy carboxylic acids [36]. They considered 18 reference compounds, mainly sugar acids. One aim was to modulate potential polar interactions between the IonPac AS11-HC column and the hydroxy acids to optimize analyte separation. Various eluent modifiers such as borate, ion-pairing reagents, carbohydrates, alkylglycosides, and zwitterionic reagents were tested for this purpose. Some of the modifiers influenced the retention behavior of specific analyte subgroups. The addition of borate generally enhanced retention by complex formation, which selectively boosted retention of unbranched sugar monocarboxylic acids and of shikimic acid. The zwitterionic reagents enhanced the retention of monocarboxylic acids and reduced the retention of the polycarboxylates. None of these modifiers was able to generate a baseline resolution of all compounds. Thus, a general use of a specific modifier could not be recommended despite the achievable improved resolution of some peak groups. The findings indicate that the IonPac AS11-HC does not offer great potential to adapt polar interactions for specific separation needs. Furthermore, the postcolumn addition of methanol prior to the ESI interface considerably increased sensitivity for some of the more hydrophobic analytes, but strongly depressed sensitivity for C_6 sugar acids. Figure 4.4 combines SIM-MS traces of (poly)hydroxy carboxylic acids detected in a soil leachate (B4, Table 4.6).

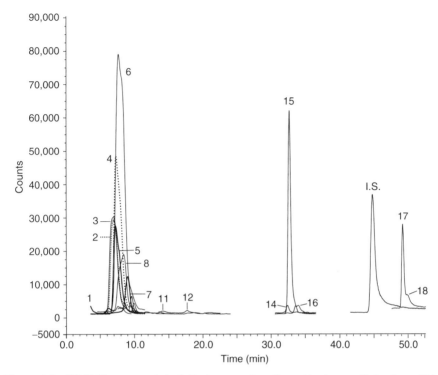

Figure 4.4 SIM-MS traces of (poly)hydroxy carboxylic acids in a soil leachate (B4, Table 4.6). Analytical conditions as in Figure 4.3. SIM signals (acids): 1: isosaccharinic, 2: D-xylonic, 3: D-gluconic, 4: 4-hydroxybutyric, 5: D-glyceric, 6: lactic, 7: shikimic, 8: glycolic, 11: D-galacturonic, 12: D-glucuronic, 14: D-glucaric, 15: malic, 16: tartaric, I.S. (internal standard): terephthalic acid-d_4, 17: citric, 18: isocitric. Source: From [36] with permission.

The clustering of the acids according to their charge number is obvious. The essentially higher retention of the internal standard terephthalic acid-d_4 as the aliphatic dicarboxylic acids points to the additional retardation by π-electron interactions.

Concentration and recovery rates for hydroxy acids in several soil leachates and in a drinking water sample are shown in Table 4.6. Composition and acid concentrations of the various soil leachates differed considerably. In total, 6 out of the 18 reference compounds, that is, 4-hydroxybutyric, D-glyceric, lactic, shikimic, glycolic, and malic acids, were detected in all soil leachates. C_5- and C_6-sugar acids were not detectable by IC-MS in two samples. On average, highest concentrations were found for 4-hydroxybutyric and lactic acids, based on IC-MS quantification. Due to incomplete separation and interferences with inorganic ions, especially chloride, conductivity detection could not be used for the quantification of monocarboxylates eluting before 10 min. For most of the analytes, the recovery rates (spike with 2.5 mg l^{-1}) were within the 90–110% interval in the SIM mode. Most of the recoveries for drinking water were slightly below 100%. Due to coelutions, recoveries determined with conductivity detection were often above 100%.

TABLE 4.6 (Poly)hydroxy Carboxylic Acids in Drinking Water and Soil Leachates (B1–B4): Concentrations and Recovery Rates with MS and Conductivity (CD) Detection

Analyte (acid)	Concentration (mg l⁻¹)										Recovery Rate (%)			
	Drinking Water		B1		B2		B3		B4		Drinking Water		B1	
	MS [a]	CD	MS	CD	MS	CD	MS	CD	MS	CD	MS	CD	MS	CD
Isosaccharinic	—	—	—	—	—	—	—	n.q.	0.15	—	100	99	101	100
D-Xylonic	—	n.q.	0.45	n.q.	d	n.q.	d	n.q.	7.51	n.q.	99	n.d.	103	n.d.
D-Gluconic	—	n.q.	0.41	n.q.	—	n.q.	—	n.q.	6.80	n.q.	101	n.d.	106	n.d.
4-Hydroxybutyric	—	n.q.	2.44	n.q.	0.48	n.q.	0.52	n.q.	44.5	n.q.	97	n.d.	104	n.d.
D-Glyceric	—	n.q.	0.65	n.q.	0.08	n.q.	0.09	n.q.	9.61	n.q.	99	n.d.	107	n.d.
Lactic	0.7	n.q.	3.08	n.q.	2.21	n.q.	1.89	n.q.	29.7	n.q.	75	n.d.	93	n.d.
Shikimic	—	n.q.	0.65	n.q.	1.42	n.q.	1.30	n.q.	3.43	n.q.	93	n.d.	83	n.d.
Glycolic	0.05	n.q.	1.31	n.q.	0.47	n.q.	0.38	n.q.	12.2	n.q.	95	n.d.	114	n.d.
2-Hydroxybutyric	1.42	—	—	—	—	—	—	—	—	—	100	105	115	55
2-Hydroxyisovaleric	—	—	—	—	—	0.07	—	—	—	—	98	115	103	113
D-Galacturonic	—	—	—	—	—	—	—	—	0.09	—	91	97	103	130
D-Glucuronic	—	—	0.06	—	—	—	—	—	0.15	—	81	—	96	—
2-Hydroxycaproic	—	—	—	—	—	—	—	—	—	—	68	124	91	126
D-Glucaric	—	—	—	—	d	0.27	—	0.29	0.29	11.2	81	79	87	101
D,L-Malic	—	—	0.58	—	0.29	0.12	0.29	0.15	6.01	9.99	99	94	101	85
L-Tartaric	—	—	—	—	—	—	—	—	0.17	—	99	—	95	—
Citric	—	—	1.42	—	d	0.09	—	0.08	2.22	2.11	91	96	97	98
D,L-Isocitric	—	—	—	—	—	0.14	0.12	0.24	0.45	—	99	100	77	114

CD: conductivity detection; n.q.: not quantifiable; n.d.: not determined; d: detectable; concentration < LOQ.

[a] Spike concentration for determination of recovery rates: 2.5 mg l⁻¹; $N = 3$; MS conditions: cone voltage: 50 V; temperature: 600 °C; needle voltage: −3 kV. B: soil leachates.

Source: From [36] with permission.

The alkaline delignification of wood is an important technical procedure for the pulp industry. During the treatment, roughly half of the wooden raw material is degraded and dissolved into the process liquor. Simultaneously, a large number of hydroxy mono- and dicarboxylic acids are formed together with formic and acetic acids. By means of an HPAEC-ESI-MS method, applying an HP 1100 Series single quadrupole detector in the SCAN mode, various sugar acids (α- and β-glucosisosaccharinic acid, xyloisosaccharinic acid, glucoisosaccharinaric acid) and other hydroxy acids, for example, lactic, glycolic, 2-hydroxybutyric, and malic acids were detected [57]. Quantifications were made with conductivity detection.

The fermentation of biomass, for example, of microalgae, produces a multitude of carboxylic acids. To control the fermentation process with regard to optimizing biomass transformation into biofuel, monitoring of organic acid concentration is essential. To provide the adequate analytical tool, Wang *et al.* refined the separation of 32 carboxylic acids on the IonPac AS11-HC [56]. With respect to MS detection, they tried to improve the resolution of acids with identical or close molecular masses such as butyrate and pyruvate ($m/z = 87.05$ and 87.02, respectively) and maleate and fumarate ($m/z = 115.01$) especially. To assist matrix desolvation during ion spray formation, ACN was mixed with the eluent before entering the ESI interface. Method detection limits, achieved with the Dionex MSQ Plus single quadrupole detector in the SIM mode, were in the range of 1.0–$5.0\,\mu g\,l^{-1}$ typically. The application of the method to identify carboxylic acids in microalgae subjected to fermentation revealed 16 compounds. A corresponding SIM-MS chromatogram is reproduced together with a suppressed conductivity chromatogram in Figure 4.5. Succinic acid was present in biomass as the most prominent organic acid with concentrations larger than $200\,mg\,l^{-1}$. Mucic, α-ketoglutaric, oxalic, and citric acids were also present in large amounts ($>5\,mg\,l^{-1}$).

In unproductive freshwater systems, bacterial metabolism might rely on the import of terrestrially derived dissolved organic matter to a great extent. To test this hypothesis, the amounts of several groups of LMWOS, transported during spring flood events from terrestrial areas into small forestral streams and lakes, were calculated and their degradability by aquatic microorganisms was measured [53, 54]. IC-MS with an AB/MDS SCIEX QTRAP 2000 linear ion trap MS/MS system established increased levels of acetic, oxalic, lactic, and formic acids in the receiving water bodies. Formic, acetic, lactic, and citric acids were quickly biodegraded, whereas glycolic acid and various dicarboxylic acids showed higher recalcitrance. Acetic acid was the most important compound for bacterial growth, representing 45% of the total bacterial consumption of all low-molecular-weight compounds.

The determination of traces of organic acids in sea water is challenging because of the high inorganic ion content. One approach to solve this analytical problem is to apply two-dimensional IC. Miller and Schnute presented a method that enables the separation of the early eluting organic acids from the stronger retained inorganic anions on an IonPac AS24 column as first chromatographic dimension [55]. This fraction is trapped by a Dionex IonPac UTAC-ULP1 column. Afterward, the analytes are

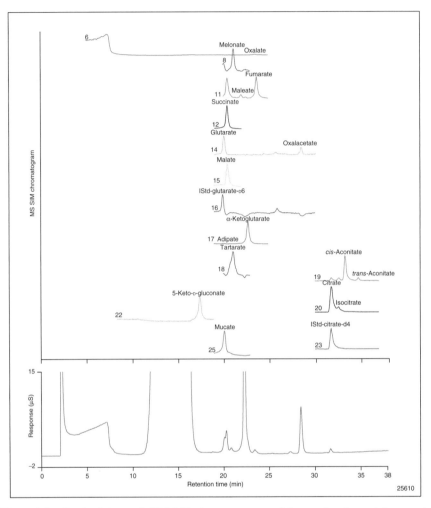

Figure 4.5 Conductivity and SIM-MS chromatograms of low-molecular-weight organic acids in microalgae biomass. The peak sizes of the MS chromatogram are not uniformly scaled. Analytical conditions listed in Table 4.5. Source: From [56] with permission.

mobilized from the trap column and transferred on an IonPac AS11-HC column, the second chromatographic dimension. Both columns are operated with KOH gradients at the same flow rate of $0.4\,\mathrm{ml\,min^{-1}}$. The conductivity is suppressed after separation in both dimensions by a Dionex ASRS in the external water mode. Detection was accomplished by the Dionex MSQ Plus single quadrupole detector achieving method detection limits between $3.8\,\mathrm{\mu g\,l^{-1}}$ (valeric acid) and $200\,\mathrm{\mu g\,l^{-1}}$ (formic acid).

4.3   CARBOHYDRATES

4.3.1   Structural Diversity and Ion Chromatographic Behavior

Carbohydrates constitute a large and diverse class of organic compounds, sub-divided into various groups by specific molecular features. On the level of the monomeric units, the number of carbon atoms, steric arrangement of hydroxyl groups (diastereoisomeric, anomeric, and enantiomeric forms), keto- or aldehyde functionality, types of cyclic forms, and potential presence of further functional groups, for example, carboxylic acid, amine, ether, or ester groups, are the classification criteria. Oligo- and polysaccharides are characterized, among other criteria, by type and number of the monomeric units, type and position of linkages between these units, three-dimensional molecule shape, and bindings with structurally different molecules such as phosphoric acid, proteins, or lipids. On the one hand, this structural diversity offers many starting points for the development of chromatographic separation strategies. On the other hand, it makes the separation of very complex carbohydrate mixtures by one chromatographic procedure impossible.

The primary factor of carbohydrate separation by HPAEC (HPICE is a possible but seldom applied alternative to HPAEC for this purpose) is their charge at given eluent pH, ruled by their related dissociation degree, depending on their pK_a values. Thus, all molecular features acting on the pK_a or, strictly speaking, on the dissociation constant of the most acidic hydroxyl group, influence the retention in HPAEC systems as well. Small pK_a differences of diastereoisomers, for example, glucose and galactose, are sufficient to allow for separation under suited HPAEC conditions. The following sequence of retention of C_6 carbohydrates on the Dionex CarboPac PA20 column, which matches the increase in acidity of compounds very well, might illustrate the importance of carbohydrate charge for anion-exchange chromatographic separation:

Sugar alcohol $\approx$ anhydro sugar $<$ desoxysugar $<$ aminosugar $<$ neutral sugar

$\ll$ amino sugar acid $<$ sugar monoacid $<$ uronic acid $<$ sugar diacid

Clearly, this row and the following selectivity sequences provide a coarse orientation only and might be altered for different columns, eluents, and separation temperatures. Restricted to neutral monosaccharides, increased retention with an increased number of carbon atoms and subsequently hydroxyl groups is often found, that is, tetroses $<$ pentoses $<$ hexoses. Given a distinct type of linkage between the monomeric units, retention of linear oligo- and polysaccharides increases with an increasing degree of polymerization.

The environmental geochemical analysis of carbohydrates with HPAEC-MS is focused on mono- and disaccharides including anhydro sugars, especially levoglucosan, sugar alcohols, and amino sugars. For this purpose, Dionex CarboPac columns (PA1, PA10, and PA20) are almost exclusively used. The corresponding separation resins have the same general architecture in common, that is, they are latex ("MicroBead™") agglomerated, hydrophobic, pellicular anion-exchange polymers with exchange capacities of $100\,\mu mol_c$ (PA1) and $90\,\mu mol_c$ (PA10) per 4×250 mm column or 65 μmol_c per 3×150 mm column, respectively. They differ in the composition of the polymer core material, degree of cross-linking, diameter of resin and of latex particle, and type of the functional groups attached to the latex particles [62, 63]. The unique selectivity of every column is shaped by these parameters, but the manufacturer, as many others, does not disclose the logic between targeted chromatographic selectivity and column design. According to the manufacturer, the CarboPac PA10 and PA20 columns are recommended for monosaccharide composition analysis, whereas the PA1 is a general-purpose column for the separation of mono-, di-, and some oligosaccharides. As mentioned in the previous chapter, acidic carbohydrates might also be separated together with aliphatic carboxylic acids, applying the IonPac AS-11HC or similar columns.

4.3.2 Environmental Analysis of Carbohydrates by Various IC-MS Methods

Aerosol particles are preferred objects for the environmental analysis of carbohydrates. Monosaccharide anhydrides (MA) play a central role, that is, levoglucosan (1,6-anhydro-β-glucopyranose), mannosan, and galactosan. It has been shown that these three MA isomers are exclusively generated as thermal degradation products of plant polysaccharides, that is, starch, cellulose, and hemicellulose [64]. Due to their low vapor pressures, MAs are enriched in aerosol particles [65]. In particular, levoglucosan is produced in large amounts in biomass burning processes and therefore used as a tracer for such events [60, 66–68]. Besides GC-MS, HPAEC-PAD is the method of choice for the determination of MAs, but some newer investigations preferred HPAEC-MS. Saarnio *et al.* coupled a Dionex CarboPac PA10 column with a Dionex MSQ single quadrupole MS [69, 70]. The KOH gradient was provided by a Dionex eluent generator (Table 4.8). KOH was removed from the column effluent before entering the ESI interface by a Dionex ASRS 300 suppressor. The authors underlined the necessity to separate the MAs before MS detection, since they are isobaric isomers, forming MS signals with identical *m/z* values. The applied method met this requirement but failed to separate the target compounds from some sugar alcohols. This limitation of the CarboPac PA10 separation efficiency was even earlier noted [77]. To compensate for the ion suppression caused by coelution of sugar alcohols, signal corrections for various MA–sugar alcohol combinations were made. Detection limits of 1 and $2 \mu g \, l^{-1}$, respectively, were attained and recoveries from quartz filter samples spanned from 94% to 103%. The highest MA concentrations in ambient fine particle samples from urban (Helsinki) and rural background were typically found in weekend samples [69]. Taking MAs as biomass burning tracers, the same analytical approach was used to estimate the impact of wood combustion on fine particles in the Helsinki Metropolitan Area [70]. Particles were sampled by means of a Berner low-pressure impactor, and a fraction <1 μm was collected on quartz filters (Table 4.7).

The estimated average contribution of wood combustion to fine particles ranged from 18% to 29% at the urban sites and from 31% to 66% at the suburban sites in the cold season. The fine particle measurements showed that the proportions of the

TABLE 4.7 Sample Matrix, Sampling, and Sample Preparation Techniques for the Environmental Analysis of Carbohydrates (CH)

Matrix	Detected Carbohydrates	Sampling and Sample Preparation	References
Aerosol particles	Levoglucosan, mannosan, galactosan	Berner low-pressure impactor, collection of <1 µm particles on quartz filters, filter extraction with water, MF[a] (0.45 µm)	[69, 70]
Aerosol particles	Levoglucosan	Collection on quartz filters, online analysis with particle-into-liquid sampler (PILS)	[71]
Aerosol particles, soil	Levoglucosan, mannosan, galactosan	Aerosol: high-volume sampler, two filter stages (PM_{10} and $PM_{2.5}$) soil: surface collection, air-drying, sieving (<2 mm); water extraction (both sample types) under Vortex agitation	[72]
Surface waters	Glucose, fructose, sucrose, mannose, mannitol	Sampling with HDPE bottles, MF (0.22 µm)	[53, 54]
Lake water	18 CH (mono- and disaccharides, acetylated amino sugars, amino sugar acids)	Sampling with glass bottles, glass fiber filtration (0.20 µm)	[73]
Lake sediments	Levoglucosan, mannosan, galactosan	Drilling of sediment cores, freeze-drying, milling, and homogenization. MeOH-extraction of freeze-dried sediment using pressurized solvent extraction (PSE). Evaporation to dryness, dissolution in water, centrifugation	[74]
Popular leaves	18 CH including oligosaccharides and sugar alcohols	Harvesting of leaves, freezing, and crushing in liquid nitrogen, extraction with ethanol/water (80/20 v/v), centrifugation; evaporation of the supernatant, dissolution of the residue in water, MF (0.45 µm)	[75]
Biomass hydrolysates	Mono- and oligosaccharides	Samples collected from chemically generated biomass hydrolysates and from laboratory fermentation processes, MF (0.22 µm)	[76]

[a]Membrane filtration.

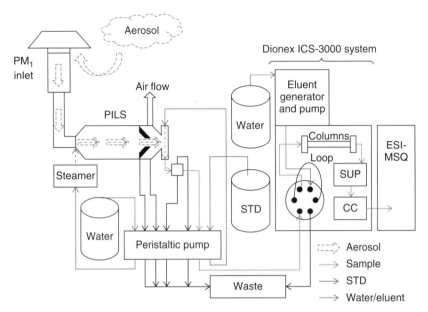

Figure 4.6 Schematic representation of the sampling line and the online coupled PILS-HPAEC-MS apparatus. Arrows represent the flow of the aerosol sample and the liquids within the system. STD: standard solution containing internal standard and standard addition of levoglucosan, SUP: suppressor, CC: conductivity cell. The figure is not to scale. For details of the HPAEC-MS method, refer to Table 4.8. Source: From [71] with permission.

three MAs can be exploited to distinguish between wildfire particles and particles from residential wood combustion.

Recently, an online technique for the determination of carbohydrates in aerosol particles was invented to achieve highly time-resolved data [71]. For this purpose, a particle-into-liquid sampler (PILS) was coupled with HPAEC-MS. A schematic representation of the sampling line and the PILS-HPAEC-MS apparatus is depicted in Figure 4.6. Briefly, fine particles were sampled from ambient air through a size-selective inlet and directed to the PILS, where the particles were grown with saturated water vapor to droplets and impacted on the impaction plate, which was flushed with a continuous stream of water, thereby extracting the water-soluble analytes from the particles. A partial stream of the water extract was directed to the HPAEC sample loop and analyzed. With respect to the required chromatographic run time being optimized for levoglucosan, this semicontinuous procedure attained a time resolution of roughly 8 min, whereas conventional discontinuous filter sampling techniques work with time resolutions from several hours to a few days. A good agreement between the semicontinuous and the offline method was established. The practical application of the method to monitor levoglucosan in aerosol particles from the region of Helsinki indicated the presence of imported particles, originating from biomass burning in Baltic countries.

Piot *et al.* determined MAs in water extracts of atmospheric and soil particles by means of the Dionex CarboPac PA1 column in combination with a tandem MS system [72]. The isocratic eluent was composed of 0.5 mM KOH with a flow rate of 1.20 ml min^{-1}, reduced by a split valve to 0.8 ml min^{-1} before entering the ESI interface (Table 4.8).

A suppressor was not used. MS parameters were optimized to achieve best CID fragmentation efficiency with respect to the formation of MA daughter ions m/z 113 and m/z 101, respectively. The HPAEC-MS/MS method was proven to be superior to a GC-MS procedure in terms of detection sensitivity and analysis time. The results of both hyphenation techniques for levoglucosan matched very well.

In the context of the studies of Berggren *et al.* mentioned earlier, addressing the importance of terrestrially derived dissolved organic matter for bacterial metabolism in unproductive freshwater systems, concentrations of various neutral mono- and disaccharides and of sugar alcohols were determined in small aquatic systems during spring flood events [53, 54]. For this purpose, the IC-MS/MS system was equipped with a Dionex CarboPac PA20 column, guarded by a Dionex AminoTrap column (Table 4.8). Glucose and fructose in nearly identical proportions were the main constituents of the dissolved carbohydrate fraction. Minor compounds were sucrose, mannose, and mannitol. The share of the carbohydrate fraction on the total amount of the LMWOS was essentially lower than the carboxylic acid fraction and comparable to the amino acids, ranging from 127 µg C l^{-1} to almost 0.

Hornak and Pernthaler made the uncommon attempt to exploit ion exclusion chromatography for the determination of mono- and disaccharides, amino sugars, and amino sugar acids in lake water [73]. The chromatographic unit was hyphenated to an AB/MDS SCIEX API 5000 triple quadrupole MS. The negatively ionized neutral carbohydrates and the positively ionized amino sugar (acids) were detected in the MRM mode (Table 4.8). Separation was performed on a Supelcogel C610H column with a 0.1% (v/v) formic acid/water gradient (flow rate 0.7 ml min^{-1}). Eighteen carbohydrates including various mono- and disaccharides and acetylated amino sugars were chosen to determine and to optimize selectivity and efficiency of the chromatographic system. With a few exceptions, the retention of the carbohydrates increased following the sequence disaccharides < hexoses < pentoses ≈ desoxyhexoses, quite different from HPAEC selectivity. The retention increase with decreasing molecule size and with decreasing number of hydroxyl groups (increasing lipophilicity) indicated the participation of size exclusion and reversed-phase interactions in the separation process. For instance, fucose eluted after galactose and the triple chlorinated, synthetic disaccharide sucralose eluted after sucrose. All of the 18 reference compounds eluted between 7 and 13 min, leading to several partial and complete coelutions. As far as the analytes were determined at different mass transitions in the MRM mode, MS detection compensated for insufficient chromatographic resolution. Obviously, this was not possible in the case of the coeluting isobaric carbohydrates galactose, mannose, and fructose. Nevertheless, the method was applied to record depth profiles of glucose and *N*-acetylglucosamine in the water body of Lake Zurich.

MAs in lake sediments were screened by HPAEC-MS, based on the optimization of the method of Saarnio *et al.* mentioned earlier [74]. Sampling and sample

TABLE 4.8 Environmental Analysis of Carbohydrates by IC-ESI-MS

Matrix	Column[a]	Eluent	Eluent Flow (ml min^{-1})	Suppressor[b]	Postcolumn Modifications	MS Apparatus[c]	References	Publishing Year
Aerosol particles	CarboPac PA10	KOH gradient	0.20	ASRS 300		Dionex MSQ SQ	[69–71]	2010, 2012, 2013
Aerosol particles, soil	CarboPac PA1	0.5 mM KOH	1.20	—	Eluent split (0.8 ml min^{-1})	Thermo Fisher Scientific LCQ Fleet Ion Trap	[72]	2012
Surface water	CarboPac PA20, with Amino Trap guard column	NaOH gradient	n.d.[d]	ASRS Ultra	—	AB/MDS SCIEX Q-TRAP 2000 linear ion trap MS/MS	[53, 54]	2010
Lake water	Supelcogel C610H	0.1% (v/v) formic acid/water gradient	0.70	—	—	AB/MDS SCIEX API 5000 triple quadrupole. MRM mode; neutral carbohydrates: negative ionization; amino sugar (acids): positive ionization	[73]	2014

	Column	Eluent	Flow rate	Suppressor	Modification	Detector	Ref.	Year
Lake sediments	CarboPac PA1 and PA10 in series with AminoTrap guard column	NaOH gradient	0.25	ASRS 300	Addition of methanol/NH_{3aq} (0.02 ml min^{-1})	Dionex MSQ SQ	[74]	2014
Poplar leaves	CarboPac PA20	KOH gradient	0.25	ASRS Ultra	Addition of 0.5 mM LiCl (0.05 ml min^{-1})	Finnigan MSQ SQ, positive ionization	[75]	2005
Biomass hydrolysates	CarboPac PA1	NaOH/Na acetate gradient	0.215	ASRS 300	—	Thermo LTQ LT – 1000 linear ion trap, positive ionization	[76]	2013

[a] CarboPac is a trademark of Dionex (Thermo Scientific).
[b] All suppressors from Dionex (Thermo Scientific).
[c] As far as not stated otherwise, ESI with negative ionization, SIM mode for single quadrupole detector (SQ).
[d] No data.

preparation are summarized in Table 4.7. Two Dionex CarboPac columns (PA1 and PA10) were placed in series to improve MA resolution and separation from minor compounds (Table 4.8). To enhance ionization of the aqueous eluent (NaOH gradient, suppressed), methanol/aqueous NH_3 solution was added postcolumn. By means of a single quadrupole MS detector, LODs $(3 \times S/N)$ of 0.9–1.8 µg l^{-1} and LOQs $(10 \times S/N)$ of 3.2–5.8 µg l^{-1} were achieved. The method accuracy based on the relative recovery of the MAs from spiked blank samples $(n = 5)$ was 99.8% (±1.3%) for levoglucosan, 97.7% (±6.4%) for mannosan, and 94.1% (±7.6%) for galactosan. The investigation of lake sediments revealed a decreasing MA concentration with increasing sediment depth and a significant correlation with macroscopic charcoal concentration. The calculation of concentration ratios of individual MAs might be helpful to determine not only when biomass burning occurred but also if changes occurred in the primary burned vegetation.

The content of various mono- and oligosaccharides including sugar alcohols in native biomass (poplar leaves) was characterized by employing a Dionex CarboPac PA20 column coupled to a single quadrupole detector [75]. The postcolumn addition of a diluted aqueous LiCl solution to enhance MS sensitivity via the formation of $[M+Li]^+$ and $[M+Li+H_2O]^+$ ions is one of the specific features of the method. Cone voltages were adapted to each compound individually. It was found that the optimal cone voltage increased with increasing carbohydrate mass. Compared with PAD detection, MS quantification yielded lower linearity and sensitivity. Consequently, HPAEC-MS was used for qualitative characterization of the carbohydrates only. It allowed the confirmation of peak attribution and the identification of salicin as a major compound in the ethanol/water extracts of poplar leaves.

Coulier *et al.* developed an HPAEC-MS method to identify (oligo)saccharides in hydrolysates of lignocellulosic biomass, generated by chemical, fermentative, or enzymatic treatments [76]. Structural information was gained by MS^n experiments, applying a Thermo LTQ LT-1000 linear ion trap mass detector (Table 4.8). For quantification, the positive ionization method was selected to target $[M+Na]^+$ ions. Separation was accomplished on a Dionex CarboPac PA1 column, operated with a sodium hydroxide/sodium acetate gradient. Response factors for MS carbohydrate detection sensitivity relative to glucose were determined ranging from 0.3 (glucuronic acid) to 2.1 (xylobiose). Ten carbohydrates were putatively identified in the hydrolysates, and it was possible to correlate product patterns with hydrolyzation conditions. The degree of polymerization and the carbon number of the monomeric units were determined for several otherwise unidentified compounds.

Although not in the focus of this treatise, it should be mentioned that hyphenation of IC with MS is not restricted to mass detectors supplying structural information about the organic analytes. In several cases, knowledge of the ratios of stable isotopes of various elements, especially carbon, is of high importance for environmental geochemical research. Isotope ratios might offer the key to the understanding of element cycles, substance transformations in ecosystems, metabolic activities, and interactions of different groups of (micro)organisms and of paleoclimatic processes. Here, the method of choice is isotope ratio MS (IRMS). For a long time, GC was the unique chromatographic technique combinable to IRMS. The first commercially available

LC-IRMS interface was introduced roughly 10 years ago [78]. Today, hyphenation remains a challenging task. Meanwhile, various applications of HPAEC-IRMS for the environmental analysis of carbohydrates were presented [79–82]. Separation was based on Dionex CarboPac columns. Carbonate contamination of the hydroxide eluents had to be strongly avoided due to possible interferences with the carbon isotope ratio determination of the analytes. Thus, the utilization of an anion trap column to remove carbonate from the eluent stream and/or the installation of a hydroxide eluent generator is recommended, especially for gradient applications [81]. Experiences with this new analytical technique were gathered for neutral pentoses, hexoses, and deoxysugars [79, 81], for acidic monosaccharides [79], and for amino sugars [80, 82]. The analyzed matrices comprise plant materials, soils, and sediments.

4.4   AMINES AND AMINO ACIDS

Short-chain aliphatic amines might be present in trace concentrations in ambient air. Often, emissions can be traced back to anthropogenic sources including waste incineration, sewage treatment, cattle breeding, and vehicle exhaust gases [83]. However, natural sources such as the biodegradation of nitrogen-containing biomolecules such as proteins and amino acids might also contribute to environmental concentrations of amines. To monitor natural and synthetic amines in air samples, Verriele *et al.* coupled a Dionex cation-exchange column (IonPac CS14) with a conductivity detector followed by an MSQ Plus single quadrupole MS with ESI interface, operated in the positive ionization mode [84]. Elution was performed by a gradient mixing variable proportions of water, methylsulfonic acid, and acetonitrile (flow rate: $0.25 \, \text{ml} \, \text{min}^{-1}$). Conductivity was suppressed by a Dionex CSRS 300 suppressor. The target analytes comprised primary, secondary, and tertiary mono- and diamines together with cyclic amines and amino alcohols. To scavenge amines from air, samples were pumped through midget impingers filled with ultrapure water. The chromatographic performance was limited by several coelutions both between amines and of amines with ammonia and metal ions. Coelution problems and interferences with large ammonia excess were also addressed in other IC-MS studies [85, 86]. With a few exceptions, MS detection sensitivity (LOQ, $3 \times S/N$) surpassed conductivity by 1–3 orders of magnitude, ranging from $0.02 \, \mu\text{g} \, \text{l}^{-1}$ (dipropylamine) to $26.0 \, \mu\text{g} \, \text{l}^{-1}$ (trimethylamine), allowing the quantification of amines at levels of microgram per cubic meter in air with a good accuracy for most compounds (RSD < 10%). The great superiority of MS over conductivity in amine quantification was also stated by Hermans *et al.*, who applied a Dionex IonPac CS18 column in combination with a single quadrupole MS [86]. They ascertained a MS sensitivity 100-fold higher on average.

   IC-MS applications for the environmental analysis of amino acids are very scarce. HPAEC-IRMS helped to determine the natural ^{13}C abundances of amino acids in proteins. The main components of the analytical device were a Dionex AminoPac PA10 column, a PAD, a Liquiface interface, and an IRMS_r instrument (both Isoprime Ltd) with an electron ionization ion source [87]. Basic and neutral amino acids were resolved with 35 mM NaOH in isocratic mode. To elute aromatic and acidic amino acids, increased elution strength is required, generated by a combination of a higher hydroxide concentration with a second displacing ion, usually acetate. Since carbon isotope ratio determination necessitates a carbon-free eluent, acetate was replaced by nitrate. The method was able to demonstrate that anion-exchange chromatography is a viable alternative to cation chromatography, especially where analysis of basic amino acids is important.

4.5 TRENDS AND PERSPECTIVES

In the past years, IC-MS has demonstrated its suitability as an analytical tool for organic environmental geochemical research. Main application fields are organic acid and carbohydrate analysis. Primarily, single quadrupole MS detectors were coupled with IC systems, but some examples exist for applications of ion trap, triple quadrupole, or TOF instruments. Compared with IC without mass-selective detection, benefits of the hyphenated systems are the ability to resolve and to quantify coeluting compounds, the potential to confirm the purity of peaks and the identity of compounds, recorded by common detectors, and a higher sensitivity, depending on the MS type. Despite these merits, IC-MS has not found widespread application in geochemical research in contrast to IC-conductivity detection or IC-PAD. The following reasons might be blamed for this situation:

- In several cases, elucidation of structural information in combination with IC is of minor importance due to the reproducible and already known sample composition and/or analyte identification by alternative methods, for example, GC-MS.
- Expenditure of operation and maintenance of MS detectors are higher, robustness and operational reliability are lower than common, that is, conductivity, detectors.
- Hyphenation of IC with MS is not trivial, and several interfering effects, for example, ion suppression, insufficient ion spray formation, low analyte transfer rates, salt deposition, have to be controlled and avoided.
- In some cases, single quadrupole MS detection offers lower sensitivity and a narrower linear detection range than conductivity or PAD [75].

On the other hand, the experiences made with IC and IC-MS provide some starting points for an enhancement of the performance of IC-based analytical systems and for an increased exploitation of IC-MS. Advanced environmental geochemical research needs further improvements of stationary phases for organic acid and carbohydrate analysis. The use of an MS detector to compensate for insufficient chromatographic resolution might be an interim state rather than a final goal of analytical performance development. Several investigations point to the relevance of coelutions of various multifunctional acids on Dionex IonPac-AS11HC [36, 49, 57]. Another chromatographic challenge is the separation of mixtures from neutral monosaccharides, sugar alcohols, and monosaccharide anhydrides. Capillary IC might offer some advantages in sensitivity, but not necessarily in selectivity and separation performance [88–90]. To improve separation efficiency and to widen the polarity range of detectable compounds, principally the same measures were taken as 40 years ago, that is, increase of length of the stationary phase by column coupling and implementation of two-dimensional chromatography realized by various column switching techniques. Today, several achievements, for example, the possibility to combine columns of different physical dimensions and the built-in capacity for column switching methods of modern, integrated IC systems provide improved prerequisites for the implementation of 2D methods. With respect to the increased diversity of HPLC tools and techniques (zwitterionic IC, HILIC, mixed-mode phases) and to the almost unmanageable variety of available stationary phases, 2D or even multidimensional chromatography seems to have a great potential even in organic environmental geochemical research. To obtain quick and reliable information about the attained fractionations in multidimensional LC, MS is an indispensable tool. To promote the elaboration of new 2D chromatographic methods, a practical guide for suited combinations of chromatographic dimensions would be very helpful.

A further stimulus to increase the utilization of IC-MS in geochemical analysis is expected to come from the MS side. Many researchers tackle the various challenges of IC-MS hyphenation mentioned earlier with a "trial-and-error" strategy. The need for a more generic approach is obvious [91, 92]. A promising attempt to systematically optimize MS detection for IC, focusing on ESI separation conditions, was made using response surface methodology (RSM) [93].

Finally, IC-MS has to find and to occupy its place in the tool box of geochemical analysis. Integrated analytical concepts, for example, developed by Liu *et al.* for the study of atmospheric chemical reactions [51], might offer new perspectives for environmental geochemical applications of these promising hyphenation techniques.

4.6   REFERENCES

[1] Scott, C.D. (1968) Analysis of urine for its ultraviolet-absorbing constituents by high-pressure anion-exchange chromatography. *Clinical Chemistry*, **14**, 521.

[2] Pitt, W.W., Scott, C.D., Johnson, W.F., and Jones, G. Jr. (1970) A benchtop, automated, high-resolution analyzer for ultraviolet absorbing constituents of body fluids. *Clinical Chemistry*, **16**, 657.

[3] Jolley, R.L. and Freeman, M.L. (1968) Automated carbohydrate analysis of physiologic fluids. *Clinical Chemistry*, **14**, 538.

[4] Scott, C.D., Jolley, R.L., Pitt, W.W. Jr., and Johnson, W.F. (1970) Prototype systems for the automated, high resolution analyses of UV-absorbing constituents and carbohydrates in body fluids. *American Journal of Clinical Pathology*, **53**, 701.

[5] Katz, S., Pitt, W.W. Jr., and Scott, C.D. (1972) The determination of stable organic compounds in waste effluents at microgram per liter levels by automatic high-resolution ion exchange chromatography. *Water Research*, **6**, 1029.

[6] Turkelson, V.T. and Richards, M. (1978) Separation of the citric acid cycle acids by liquid chromatography. *Analytical Chemistry*, **50**, 1420.

[7] Rich, W.E., Smith, F., McNeill, L., and Sidebottom, T. (1979) Ion exclusion coupled to ion chromatography: Instrumentation and application, in *Ion Chromatographic Analysis of Environmental Pollutants, Vol. 2*, Ann Arbor Science, Ann Arbor, MI, p. 17.

[8] Rich, W.E., Johnson, E.L., Lois, L. *et al.* (1981) Organic acids by ion chromatography, in *Liquid Chromatography in Clinical Analysis*, The Humana Press.

[9] Weiß, J. and Reinhard, S. (1995) Aliphatische Carbonsäuren – Ein schwieriges chromatographisches Problem? *GIT Spezial*, **2**, 105.

[10] Buchanan, D.N. and Thoene, J.G. (1982) Dual-column high-performance liquid chromatographic urinary organic acid profiling. *Analytical Biochemistry*, **124**, 108.

[11] Bouyoucos, S.A. (1982) Determination of organic anions by ion chromatography using a hollow fiber suppressor. *Journal of Chromatography A*, **242**, 170.

[12] Robertson, W.G., Scurr, D.S., Smith, A., and Orwell, R.L. (1982) The determination of oxalate in urine and urinary calculi by a new ion-chromatographic technique. *Clinica Chimica Acta*, **126**, 91.

[13] Rocklin, R.D., Slingsby, R.W., and Pohl, C.A. (1986) Separation and detection of carboxylic acids by ion chromatography. *Journal of Liquid Chromatography A*, **9**, 757.

[14] Oshima, R., Kurosu, Y., and Kumanotani, J. (1979) Separation of acidic and neutral saccharides by high-performance gel chromatography on cation-exchange resins in the H+-form with acidic eluents. *Journal of Chromatography A*, **179**, 376.

[15] Ladish, M.R., Heubner, A.L., and Tsao, G.T. (1978) High speed liquid chromatography of cellodextrins and other saccharide mixtures using water as eluent. *Journal of Chromatography A*, **147**, 185.

[16] Linden, J.C. and Lawhead, C.L. (1975) Liquid chromatography of saccharides. *Journal of Chromatography A*, **105**, 125.

[17] Yang, R.T., Milligan, L.P., and Mathison, G.W. (1981) Improved sugar separation by high performance liquid chromatography using porous microparticle carbohydrate columns. *Journal of Chromatography A*, **209**, 316.

[18] Rocklin, R.D. and Pohl, C.A. (1983) Determination of carbohydrates by anion exchange chromatography with pulsed amperometric detection. *Journal of Liquid Chromatography*, **6**, 1577.

[19] Hughes, S., Meschi, P.L., and Johnson, D.C. (1981) Amperometric detection of simple alcohols in aqueous solutions by application of a triple-pulse potential waveform at platinum electrodes. *Analytica Chimica Acta*, **132**, 1.

[20] Hughes, S. and Johnson, D.C. (1983) Triple-pulse amperometric detection of carbohydrates after chromatographic separation. *Analytica Chimica Acta*, **149**, 1.

[21] Pacholec, F., Eaton, D.R., and Rossi, D.T. (1986) Characterization of mixtures of organic acids by ion-exclusion partition chromatography–mass spectrometry. *Analytical Chemistry*, **58**, 2581.

[22] Simpson, R.C., Fenselau, C.C., Hardy, M.R. *et al.* (1990) Adaptation of a thermospray LC–MS interface for use with high-performance anion exchange chromatography of carbohydrates. *Analytical Chemistry*, **62**, 248.

[23] Niessen, W.M.A., van der Hoeven, R.A.M., van der Greef, J. *et al.* (1992) High-performance anion-exchange chromatography/thermospray mass spectrometry in the analysis of oligosaccharides. *Rapid Communications in Mass Spectrometry*, **6**, 474.

[24] van der Hoeven, R.A.M., Niessen, W.M.A., Schols, H.A. *et al.* (1992) Characterization of sugar oligomers by on-line high-performance anion-exchange

chromatography–thermospray mass spectrometry. *Journal of Chromatography A*, **627**, 63.

[25] Niessen, W.M.A., van der Hoeven, R.A.M., van der Greef, J. *et al.* (1993) Recent progress in high-performance anion-exchange chromatography–thermospray mass spectrometry of oligosaccharides. *Journal of Chromatography A*, **647**, 319.

[26] Schols, H.A., Mutter, M., Voragen, A.G.J. *et al.* (1994) The use of combined high-performance anion exchange chromatography–thermospray-mass spectrometry in the structural analysis of pectic oligosaccharides. *Carbohydrate Research*, **261**, 335.

[27] Alexander, J.N. IV and Quinn, C.J. (1993) Organic acid analysis by ion chromatography–particle beam mass spectrometry. *Journal of Chromatography A*, **647**, 95.

[28] Kim, I.S., Sasinos, F.I., Stephens, R.D., and Brown, M.A. (1990) Analysis for daminozide in apple juice by anion-exchange chromatography–particle beam mass spectrometry. *Journal of Agricultural and Food Chemistry*, **38**, 1223.

[29] Brown, M.A., Kim, I.S., Sasinos, F.I., and Stephens, R.D. (1990) Analysis of target and nontarget pollutants in aqueous and hazardous waste samples by liquid chromatography/particle beam mass spectrometry, in *ACS Symposium Series*, ACS, Washington, DC, p. 198.

[30] Conboy, J.J. and Henion, J. (1992) High-performance anion-exchange chromatography coupled with mass spectrometry for the determination of carbohydrates. *Biological Mass Spectrometry*, **21**, 397.

[31] Xiang, X., Ko, C.Y., and Guh, H.Y. (1996) Ion-exchange chromatography/electrospray mass spectrometry for the identification of organic and inorganic species in topiramate tablets. *Analytical Chemistry*, **68**, 3726.

[32] Johnson, S.K., Houk, L.L., Feng, J. *et al.* (1997) Determination of small carboxylic acids by ion exclusion chromatography with electrospray mass spectrometry. *Analytica Chimica Acta*, **341**, 205.

[33] Widiastuti, R., Haddad, P.R., and Jackson, P.E. (1992) Approaches to gradient elution in ion-exclusion chromatography of carboxylic acids. *Journal of Chromatography A*, **602**, 43.

[34] Fischer, K., Bipp, H.-P., Bieniek, D., and Kettrup, A. (1995) Determination of monomeric sugar and carboxylic acids by ion-exclusion chromatography. *Journal of Chromatography A*, **706**, 361.

[35] Fischer, K., Kotalik, J., and Kettrup, A. (1999) Chromatographic properties of the ion-exclusion column IonPac ICE-AS 6 and application in environmental analysis. Part I: Chromatographic properties. *Journal of Chromatographic Science*, **37**, 477.

[36] Meyer, A., Höffler, S., and Fischer, K. (2007) Anion-exchange chromatography–electrospray ionization mass spectrometry method development for the environmental analysis of aliphatic polyhydroxy carboxylic acids. *Journal of Chromatography A*, **1170**, 62.

[37] Shellie, R.A., Ng, B.K., Dicinoski, G.W. *et al.* (2008) Prediction of analyte retention for ion chromatography separations performed using elution profiles comprising multiple isocratic and gradient steps. *Analytical Chemistry*, **80**, 2474.

[38] Zakaria, P., Dicinoski, G.W., Ng, B.K. *et al.* (2009) Application of retention modelling to the simulation of separation of organic anions in suppressed ion chromatography. *Journal of Chromatography A*, **1216**, 660.

[39] Bylund, D., Norström, S.H., Essén, S.A., and Lundström, U.S. (2007) Analysis of low molecular mass organic acids in natural waters by ion exclusion chromatography tandem mass spectrometry. *Journal of Chromatography A*, **1176**, 89.

[40] Pól, J., Hohnová, B., Jussila, M., and Hyötyläinen, T. (2006) Comprehensive two-dimensional liquid chromatography–time-of-flight mass spectrometry in the analysis of acidic compounds in atmospheric aerosols. *Journal of Chromatography A*, **1130**, 64.

[41] Ali, T., Bylund, D., Essén, S.A., and Lundström, U.S. (2011) Liquid extraction of low molecular mass organic acids and hydroxamate siderophores from boreal forest soil. *Soil Biology and Biochemistry*, **43**, 2417.

[42] Chen, Z., Kim, K.-R., Owens, G., and Naidu, R. (2008) Determination of carboxylic acids from plant root exudates by ion exclusion chromatography with ESI-MS. *Chromatographia*, **67**, 113.

[43] Erro, J., Zamarreno, A.M., Yvin, J.-C., and Garcia-Mina, J.M. (2009) Determination of organic acids in tissues and exudates of maize, lupin and chickpea by high-performance liquid chromatography–tandem mass spectrometry. *Journal of Agricultural and Food Chemistry*, **57**, 4004.

[44] Rellán-Alvarez, R., Lopez-Gomollón, S., Abadia, J., and Alvarez-Fernández, A. (2011) Development of a new high-performance liquid chromatography–electrospray ionization time-of-flight mass spectrometry method for the determination of low molecular mass organic acids in plant tissue extracts. *Journal of Agricultural and Food Chemistry*, **59**, 6864.

[45] Jimenez, S., Ollat, N., Deborde, C. *et al.* (2011) Metabolic response in roots of *Prunus* rootstocks submitted to iron chlorosis. *Journal of Plant Physiology*, **168**, 415.

[46] Sagardoy, R., Morales, F., Rellán-Alvarez, R. *et al.* (2011) Carboxylate metabolism in sugar beet plants grown with excess Zn. *Journal of Plant Physiology*, **168**, 730.

[47] Navascues, J., Pérez-Rontomé, C., Sánchez, D.H. *et al.* (2012) Oxidative stress is a consequence, not a cause, of aluminum toxicity in the forage legume *Lotus corniculatus*. *New Phytologist*, **193**, 625.

[48] Fisseha, R., Dommen, J., Sax, M. *et al.* (2004) Identification of organic acids in secondary organic aerosol and the corresponding gas phase from chamber experiments. *Analytical Chemistry*, **76**, 6535.

[49] Fisseha, R., Dommen, J., Gaeggeler, K. *et al.* (2006) Online gas and aerosol measurement of water soluble carboxylic acids in Zurich. *Journal of Geophysical Research*, **111**, D12316.

[50] Gaeggeler, K., Prevot, A.S.H., Dommen, J. *et al.* (2008) Residential wood burning in an Alpine valley as a source for oxygenated organic compounds, hydrocarbons, and organic acids. *Atmospheric Environment*, **42**, 8278.

[51] Liu, Y., Monod, A., Tritscher, T. *et al.* (2012) Aqueous phase processing of secondary organic aerosol from isoprene photooxidation. *Atmospheric Chemistry and Physics*, **12**, 5879.

[52] Brent, L.C., Reiner, J.L., Dickerson, R.R., and Sander, L.C. (2014) Method for characterization of low molecular weight organic acids in atmospheric aerosol using ion chromatography mass spectrometry. *Analytical Chemistry*, **86**, 7328.

[53] Berggren, M., Laudon, H., Haei, M. *et al.* (2010) Efficient aquatic bacterial metabolism of dissolved low-molecular-weight compounds from terrestrial sources. *ISME Journal*, **4**, 408.

[54] Berggren, M., Ström, L., Laudon, H. *et al.* (2010) Lake secondary production fueled by rapid transfer of low molecular weight organic carbon from terrestrial sources to aquatic consumers. *Ecology Letters*, **13**, 870.

[55] Miller, M. and Schnute, W. (2012), Direct determination of small organic acids in sea water by IC–MS, *Thermo Scientific Application Note 1000*.

[56] Wang, L., Henday, S., Ogren, B., Schnute, W., Hazlebeck, D., Roberts, A., Zhang, L., and Corpuz, R. (2009) Determination of 32 low molecular mass organic acids in biomass using IC/MS, *Dionex (Thermo Scientific) Application Note, LPN 2262-01*.

[57] Käkölä, J.M., Alén, R.J., Isoaho, J.P., and Matilainen, R.B. (2008) Determination of low-molecular-mass aliphatic carboxylic acids and inorganic anions from kraft black liquors by ion chromatography. *Journal of Chromatography A*, **1190**, 150.

[58] Rosenberg, E. (2003) The potential of organic (electrospray- and atmospheric pressure chemical ionisation) mass spectrometric techniques coupled to liquid-phase separation for speciation analysis. *Journal of Chromatography A*, **1000**, 841.

[59] Weiss, J. and Jensen, D. (2003) Modern stationary phases for ion chromatography. *Analytical and Bioanalytical Chemistry*, **375**, 81.

[60] Engling, G., Carrico, C.M., Kreidenweis, S.M. *et al.* (2006) Determination of levoglucosan in biomass combustion aerosol by high-performance anion-exchange chromatography with pulsed amperometric detection. *Atmospheric Environment*, **40**, 299.

[61] Iinuma, Y., Engling, G., Puxbaum, H., and Herrmann, H. (2009) A highly resolved anion-exchange chromatographic method for determination of saccharidic tracers for biomass combustion and primary bio-particles in atmospheric aerosol. *Atmospheric Environment*, **43**, 1367.

[62] Anon (2004), Analysis of carbohydrates by high-performance anion-exchange chromatography with pulsed amperometric detection (HPAE-PAD), *Technical Note 20, Dionex (Thermo Scientific), Sunnyvale, CA.*

[63] Corradini, C., Cavazza, A., and Bignardi, C. (2012) High-performance anion-exchange chromatography coupled with pulsed electrochemical detection as a powerful tool to evaluate carbohydrates of food interest: principles and application. *International Journal of Carbohydrate Chemistry*, article ID: 487564, doi:10.1155/2012/487564.

[64] Simoneit, B.R.T. (2002) Biomass burning – a review of organic tracers for smoke from incomplete combustion. *Applied Geochemistry*, **17**, 129.

[65] Oja, V. and Suuberg, E.M. (1999) Vapor pressures and enthalpies of sublimation of D-glucose, D-xylose, cellobiose, and levoglucosan. *Journal of Chemical and Engineering Data*, **44**, 26.

[66] Saarikoski, S., Timonen, H., Saarnio, K. *et al.* (2008) Sources of organic carbon in fine particulate matter in Northern European urban air. *Atmospheric Chemistry and Physics*, **8**, 6281.

[67] Piazzalunga, A., Fermo, P., Bernadoni, V. *et al.* (2010) A simplified method for levoglucosan quantification in wintertime atmospheric particulate matter by high performance anion-exchange chromatography coupled with pulsed amperometric detection. *International Journal of Environmental Analytical Chemistry*, **90**, 934.

[68] Aiken, A., DeFoy, B., Wiedinmeyer, C. *et al.* (2011) Mexico city aerosol analysis during MILAGRO using high resolution aerosol mass spectrometry at the urban supersite (T0) – Part 2: Analysis of the biomass burning contribution and the non-fossil carbon fraction. *Atmospheric Chemistry and Physics*, **10**, 5315.

[69] Saarnio, K., Teinilä, K., Aurela, M. *et al.* (2010) High-performance anion-exchange chromatography–mass spectrometry method for determination of levoglucosan, mannosan, and galactosan in atmospheric fine particulate matter. *Analytical and Bioanalytical Chemistry*, **398**, 2253.

[70] Saarnio, K., Niemi, J.V., Saarikoski, S. *et al.* (2012) Using monosaccharide anhydrides to estimate the impact of wood combustion on fine particles in the Helsinki Metropolitan Area. *Boreal Environment Research*, **17**, 163.

[71] Saarnio, K., Teinilä, K., Saarikoski, S. *et al.* (2013) Online determination of levoglucosan in ambient aerosols with particle-into-liquid sampler – high performance anion-exchange chromatography–mass spectrometry (PILS-HPAEC–MS). *Atmospheric Measurement Techniques*, **6**, 2839.

[72] Piot, C., Jaffrezo, J.-L., Cozic, J. *et al.* (2012) Quantification of levoglucosan and its isomers by high performance liquid chromatography–electrospray ionization tandem mass spectrometry and its applications to atmospheric and soil samples. *Atmospheric Measurement Techniques*, **5**, 141.

[73] Hornak, K. and Pernthaler, J. (2014) A novel ion-exclusion chromatography–mass spectrometry method to measure concentrations and cycling rates of carbohydrates and amino sugars in freshwaters. *Journal of Chromatography A*, **1365**, 115.

[74] Kirchgeorg, T., Schüpbach, S., Kehrwald, N. *et al.* (2014) Method for the determination of specific molecular markers of biomass burning in lake sediments. *Organic Geochemistry*, **71**, 1.

[75] Guignard, C., Jouve, L., Bogéat-Triboulot, M.B. *et al.* (2005) Analysis of carbohydrates in plants by high-performance anion-exchange chromatography coupled with electrospray mass spectrometry. *Journal of Chromatography A*, **1085**, 137.

[76] Coulier, L., Zha, Y., Bas, R., and Punt, P.J. (2013) Analysis of oligosaccharides in lignocellulosic biomass hydrolysates by high-performance anion-exchange chromatography coupled with mass spectrometry (HPAEC–MS). *Bioresource Technology*, **133**, 221.

[77] Caseiro, A., Marr, I.L., Claeys, M. *et al.* (2007) Determination of saccharides in atmospheric aerosol using anion-exchange high-performance liquid chromatography and pulsed-amperometric detection. *Journal of Chromatography A*, **1171**, 37.

[78] Krummen, H., Hilkert, A.W., Juchelka, D. *et al.* (2004) A new concept for isotope ratio monitoring liquid chromatography/mass spectrometry. *Rapid Communications in Mass Spectrometry*, **18**, 2260.

[79] Boschker, H.T.S., Moerdijk-Poortvliet, T.C.W., van Breugel, P. *et al.* (2008) A versatile method for stable carbon isotope analysis of carbohydrates by high-performance liquid chromatography/isotope ratio mass spectrometry. *Rapid Communications in Mass Spectrometry*, **22**, 3902.

[80] Bodé, S., Denef, K., and Boeckx, P. (2009) Development and evaluation of a high performance liquid chromatography isotope ratio mass spectrometry methodology for $\delta^{13}C$ analysis of amino sugars in soil. *Rapid Communications in Mass Spectrometry*, **23**, 2519.

[81] Morrison, D.J., Taylor, K., and Preston, T. (2010) Strong anion-exchange liquid chromatography coupled with isotope ratio mass spectrometry using a liquiface interface. *Rapid Communications in Mass Spectrometry*, **24**, 1755.

[82] Indorf, C., Stamm, F., Dyckmans, J., and Joergensen, R.G. (2012) Determination of saprotrophic fungi turnover in different substrates by glucosamine-specific $\delta^{13}C$ liquid chromatography/isotope ratio mass spectrometry. *Fungal Ecology*, **5**, 694.

[83] Sacher, F., Lenz, S., and Brauch, H.J. (1997) Analysis of primary and secondary aliphatic amines in waste water and surface water by gas chromatography–mass spectrometry after derivatization with 2,4-dinitrofluorobenzene or benzenesulfonyl chloride. *Journal of Chromatography A*, **764**, 85.

[84] Verriele, M., Plaisance, H., Depelchin, L. *et al.* (2012) Determination of 14 amines in air samples using midget impingers sampling followed by analysis with ion chromatography in tandem with mass spectrometry. *Journal of Environmental Monitoring*, **14**, 402.

[85] Saccani, G., Tanzi, E., Pastore, P. *et al.* (2005) Determination of biogenic amines in fresh and processed meat by suppressed ion chromatography–mass spectrometry. *Journal of Chromatography A*, **1082**, 43.

[86] Hermans, C., Jonkers, A.C.A., and de Bokx, P.K. (2010) Determination of amines in the presence of excess ammonia by ion chromatography. *Journal of Chromatographic Science*, **48**, 544.

[87] Abaye, D.A., Morrison, D.J., and Preston, T. (2010) Strong anion exchange liquid chromatographic separation of protein amino acids for natural 13C abundance determination by isotope ratio mass spectrometry. *Rapid Communications in Mass Spectrometry*, **25**, 429.

[88] Wood, J. and Hoefler, F. (2011) Mass sensitivity of capillary IC systems explained, *Technical Note 90, LPN 2649, Dionex (Thermo Scientific), Sunnyvale, CA.*

[89] Abian, J., Oosterkamp, A.J., and Gelpi, E. (1999) Comparison of conventional, narrow-bore and capillary liquid chromatography/mass spectrometry for electrospray ionization mass spectrometry: practical considerations. *Journal of Mass Spectrometry*, **34**, 244.

[90] Eghbali, H., Bruggink, C., Agroskin, Y. *et al.* (2012) Performance evaluation of ion-exchange chromatography in capillary format. *Journal of Separation Science*, **35**, 3461.

[91] Antignac, J.-P., de Wasch, K., Monteau, F. *et al.* (2005) The ion suppression phenomenon in liquid chromatography–mass spectrometry and its consequences in the field of residue analysis. *Analytica Chimica Acta*, **529**, 129.

[92] Clifford, M.N., Lopez, V., Poquet, L. *et al.* (2007) A systematic study of carboxylic acids in negative ion mode electrospray ionisation mass spectrometry providing a structural model for ion suppression. *Rapid Communications in Mass Spectrometry*, **21**, 2014.

[93] Wang, J. and Schnute, W.C. (2009) Optimizing mass spectrometric detection for ion chromatographic analysis. I. Common anions and selected organic acids. *Rapid Communications in Mass Spectrometry*, **23**, 3439.

5

ANALYSIS OF OXYHALIDES AND HALOACETIC ACIDS IN DRINKING WATER USING IC-MS AND IC-ICP-MS

KOJI KOSAKA

Department of Environmental Health, National Institute of Public Health, 2-3-6 Minami, Wako Saitama 351-0197, Japan

5.1 INTRODUCTION

Oxyhalides are soluble anions in water, and it is difficult to remove them during conventional water purification process. Some oxyhalides are known to be contaminants in drinking water. Bromate (molecular formula: BrO_3^-) is a disinfection by-product upon ozonation [1]. Bromate is classified into Class B2 (possibly carcinogenic to humans) by the International Agency for Research on Cancer [2]. Values of guidelines and standards for oxyhalides, including bromate, for several countries and an organization are listed in Table 5.1 [2–10]. In the United States, maximum contaminant level (MCL) of bromate is set at $10\,\mu g\,l^{-1}$ [11]. Provisional guideline value for bromate in World Health Organization's (WHO's) Guidelines for Drinking-Water Quality, [2] its interim maximum acceptable concentration (MAC) in drinking water in Canada [6], and its standard value in drinking water quality in Japan [10] are also $10\,\mu g\,l^{-1}$. The guideline value for bromate in Australia is $20\,\mu g\,l^{-1}$ [9]. Chlorate (molecular formula: ClO_3^-) is a contaminant in sodium hypochlorite solution used as a disinfectant for drinking water [2]. The primary concern about chlorate is oxidative damage to red blood cells [2]. In the United States, chlorate in drinking water is not regulated, but is included in the contaminant candidate list 3 (CCL3) [12]. A health reference

Application of IC-MS and IC-ICP-MS in Environmental Research, First Edition.
Edited by Rajmund Michalski.

TABLE 5.1 Values of Guidelines and Standards for Oxyhalides and HAAs

Country/ Organization	Bromate ($\mu g \, l^{-1}$)	Chlorate ($\mu g \, l^{-1}$)	Perchlorate ($\mu g \, l^{-1}$)	HAAs ($\mu g \, l^{-1}$)	References
WHO[a]	10	700	—	MCAA: 20, DCAA: 50, TCAA: 200	[2]
US EPA[b]	10	210[c]	15[d]	HAA5: 60	[3–5]
Canada[a]	10	1000	6[e]	HAA5: 80[f]	[6–8]
Australia[a]	20	—	—	MCAA: 150, DCAA: 100, TCAA: 100	[9]
Japan[b]	10	600	25[g]	MCAA: 20, DCAA: 30, TCAA: 30	[10]

[a]Guideline value.
[b]Standard value.
[c]Health reference level.
[d]Health advisory level.
[e]Guideline value recommended by Health Canada.
[f]As low as reasonably achievable.
[g]Index value.

level (HRL) of chlorate was calculated to be $210 \, \mu g \, l^{-1}$ by the United States Environmental Protection Agency (US EPA) [3]. Also, its provisional guideline value in WHO's Guidelines for Drinking-Water Quality [2], MAC in Canada [6], and standard value in Japan [10] are 700, 1000, and $600 \, \mu g \, l^{-1}$, respectively. Perchlorate (molecular formula: ClO_4^-) is used for many purposes in the form of perchlorate compounds [13, 14]. A known effect of perchlorate is interference of iodide uptake through the thyroid gland [15]. Although perchlorate is not regulated in the United States, as with chlorate, perchlorate is also classified in CCL3 [12]. An interim health advisory level for perchlorate in drinking water is set at $15 \, \mu g \, l^{-1}$ [4]. In Canada, a guideline value of $6 \, \mu g \, l^{-1}$ for perchlorate in drinking water is recommended by Health Canada [7]. The Joint Food and Agriculture Organization of the United Nations/WHO Expert Committee on Food Additives [16] has released $0.01 \, mg \, kg^{-1}$ as a provisional maximum tolerable daily intake of perchlorate. In Japan, perchlorate is classified as an item for further study with a $25 \, \mu g \, l^{-1}$ index value [10].

Haloacetic acids (HAAs) are one of the major groups of disinfection by-products during chlorination (molecular formula: CX_3COOH [X = hydrogen or halogen; more than one X is halogen]) [17–19]. There are nine chloro- and bromo-HAAs, or HAA9: monochloroacetic acid (MCAA), dichloroacetic acid (DCAA), trichloroacetic acid (TCAA), monobromoacetic acid (MBAA), dibromoacetic acid (DBAA), tribromoacetic acid (TBAA), bromochloroacetic acid (BCAA), dibromochloroacetic acid (DBCAA), and bromodichloroacetic acid (BDCAA). In the United States, the total concentrations of five HAAs (HAA5; MCAA, DCAA, TCAA, MBAA, and DBAA) among HAA9 were regulated with $60 \, \mu g \, l^{-1}$ of MCL under Stage 2 Disinfectants and the Disinfection By-Products Rule (Table 5.1) [5]. In Canada, a MAC for HAA5 in drinking water is $80 \, \mu g \, l^{-1}$ [6]. It is recommended that treatment plants strive to maintain HAA levels as low as reasonably achievable (ALARA) without compromising disinfection [8]. In the Guideline for Drinking-Water Quality [2], guideline values of MCAA, DCAA, and TCAA are set at 20, 50, and 200 µg l^{-1}, respectively (provisional guideline value for DCAA). Guideline values of these three HAAs in Australia are 150, 100, and 100 µg l^{-1}, respectively [9], and their standard values in Japan are 20, 30, and 30 µg l^{-1}, respectively [10]. In Japan, the remaining six HAAs of the HAA9 are also classified as items for further study without index values [10].

Oxyhalides were initially analyzed using ion chromatography with conductivity detection (IC-CD) [20, 21]. To enhance sensitivities for oxyhalides, inline sample concentration and two-dimensional IC were also applied to IC-CD [22–24]. As for bromate analysis, IC with postcolumn reaction (PCR) was also applied [25, 26]. On the other hand, IC with mass spectrometry (MS) or tandem mass spectrometry (MS/MS) has been applied for oxyhalides analysis because IC-MS(/MS) is more sensitive and accurate [27–29]. In addition, IC with inductively coupled plasma (ICP)-MS is applied for oxyhalide analysis [30–32]. In the case of HAAs, they have been analyzed using gas chromatography (GC) with electron capture detection (ECD) and MS after derivatization followed by liquid–liquid extraction [33, 34]. Similar to the case of oxyhalides, IC-MS(/MS) and IC-ICP-MS have also been investigated for HAA analysis [32]. For HAA analysis, sensitivities of IC-MS(/MS) and IC-ICP-MS are similar or slightly lower compared to GC-ECD and GC-MS. However, GC-ECD and GC-MS are time-consuming methods, so IC-MS(/MS) and IC-ICP-MS have the advantage of simplicity. As an example of comparisons of the methodology, detection limits (DLs) of oxyhalides and HAAs in EPA Methods are listed in Table 5.2 [20–30, 33]

In this chapter, the analytical methods of oxyhalides and HAAs using IC-MS(/MS) and IC-ICP-MS are introduced. Instead of IC, liquid chromatography (LC) is occasionally applied, and thus, analyses of oxyhalides and HAAs using LC-MS(/MS) are also discussed. Among the oxyhalides, bromate, chlorate, and perchlorate are the main focus. Moreover, occurrence of oxyhalides and HAAs in drinking water is introduced. Reviews of the application of IC-MS(/MS) and IC-ICP-MS for the analyses of ionic compounds, including oxyhalides and HAAs, were reported by several researchers [31, 32]. Analytical conditions such as the analytical column, mobile phase, and DLs of target compounds are listed in these reviews. Compared to these reviews, practical characteristics of the analyses and occurrence of oxyhalides and HAAs are the focus of this chapter.

5.2   SOURCE OF OXYHALIDES AND HAAs

Bromate compounds (e.g., potassium bromate) are used in permanent-wave neutralizing solutions and dyeing of textiles using sulfur dyes and as an oxidizer to mature flour during milling [2]. Reports of bromate detection in environmental water were limited, but bromate contamination in environmental water has occasionally been reported. For example, due to its status as a former industrial site, groundwater in the Hertfordshire Chalk in the United Kingdom was contaminated by bromate [35]. In the closed areas of the contamination, usage of public water supply boreholes was

TABLE 5.2 Detection Limit of Oxyhalides and HAAs in EPA Methods

Number	Methodology	Bromate ($\mu g\,l^{-1}$)[a]	Chlorate ($\mu g\,l^{-1}$)[a,b]	Perchlorate ($\mu g\,l^{-1}$)[a,b]	HAAs ($\mu g\,l^{-1}$)[a,b]	References
300.1	IC-CD	1.32–1.44	1.31–2.55	—	—	[20]
302.0	Two-dimensional IC-CD	0.12 (0.18)	—	—	—	[24]
314.0	IC-CD	—	—	0.53	—	[21]
314.1	IC-CD (inline column concentration)	—	—	0.03 (0.13–0.14)	—	[22]
314.2	Two-dimensional IC-CD	—	—	0.012–0.018 (0.038–0.060)	—	[23]
317.0	IC-PCR	0.12	—	—	—	[25]
321.8	IC-ICP-MS	0.3	—	—	—	[30]
326.0	IC-PCR	1.2[c], 0.17	1.7[c]	—	—	[26]
331.0	Nonsuppressed IC-MS/(MS)	—	—	MS: 0.008 (0.056), MS/MS: 0.005 (0.022)	—	[27]
332.0	Suppressed IC-MS	—	—	0.02 (0.10)	—	[28]
552.3	GC-ECD[d]	—	—	—	0.012–0.17[e]	[33]
557.0	Suppressed IC-MS/MS	0.020 (0.042)	—	—	0.015–0.2[e] (0.062–0.58)[e]	[29]

[a]Detection limit or method detection limit.
[b]Value in parenthesis is the lowest concentration minimum reporting level.
[c]Value using IC-CD.
[d]Derivatization followed by liquid–liquid extraction using methyl *tert*-butyl ether.
[e]Range among HAA9.

restricted. Bromate is formed from bromide during ozonation [1]. The degree of bromate formation during ozonation is dependent on ozonation conditions (e.g., ozone dose, pH, bromide concentration, and total organic carbon [TOC]). Bromide is naturally occurring in water. Its levels are higher when the source waters are affected by seawater or some types of wastewater. In the United States, mean and median of bromide in influent water at 449 large water purification plants were 68 and 36 $\mu g\,l^{-1}$, respectively [17], which were calculated using the Information Collection Rule (ICR) data. The ICR data of each water purification plant was mean concentration of bromide of all reported data during the 12 months of the ICR collection period [17]. In Japan, bromide levels in source water at 85 water purification plants were in the range of <50–445 $\mu g\,l^{-1}$ [36]. In 53 of the 85 water purification plants, bromide concentrations were <50 $\mu g\,l^{-1}$. Bromate also contains sodium hypochlorite solution as an impurity [37, 38]. Bromide is an impurity of sodium chloride, a raw material of sodium hypochlorite solution. Bromate is produced from bromide during the electrolysis of sodium chloride solution.

Chlorate compounds (e.g., potassium and sodium bromate) are used as herbicides, explosives, and oxidizers [2, 3]. The primary source of chlorate in drinking water is considered to be impurity of sodium hypochlorite solution [3, 38]. Chlorate is formed during electrolysis of sodium chloride solution [39]. Chlorate formation is higher with platinum and lead dioxide anodes when electrolysis of sodium chloride solution is conducted with six types of anodes and a titanium cathode (Figure 5.1) [39]. Also, chlorate concentration in sodium hypochlorite solution increases during storage because chlorate is formed by the decomposition of hypochlorite [3]. In addition, chlorate is known to be a disinfection by-product of chlorine dioxide [2]. Chlorite (molecular formula: ClO_2^-), another oxyhalide, is also a disinfection by-product of

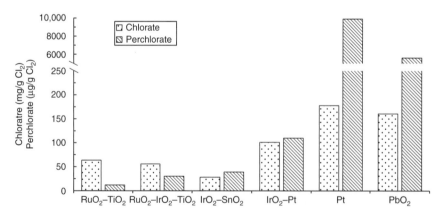

Figure 5.1 Generation of chlorate and perchlorate during electrolysis of sodium chloride solution using six types of anodes (ruthenium dioxide, RuO_2; titanium dioxide, TiO_2; iridium dioxide, IrO_2; tin dioxide, SnO_2; platinum, Pt; lead dioxide, PbO_2) (sodium chloride, 30 g l^{-1}; current, 2 A; immersed surface area of electrode plate, about 20 cm^2; reaction time, 120 min; cathode, Ti). Drawn by the author using data in Ref. [39].

chlorine dioxide [2]. Moreover, chlorate is naturally occurring in water due to its formation in the atmosphere [40, 41].

In the case of perchlorate, perchlorate compounds (e.g., ammonium, sodium, and potassium perchlorate and perchloric acid) were used as solid propellants, explosives, fireworks, etching reagents, etc. [13, 14, 42]. Perchlorate-producing facilities and usage of perchlorate-containing products were reported to be sources of perchlorate contamination in water [13, 14]. Perchlorate is also known to be naturally occurring, including as an impurity of the nitrate deposits of the Atacama Desert in Chile [40]. The primary source of perchlorate contamination in drinking water is considered to be contamination in source water. However, similar to chlorate, perchlorate is a contaminant in sodium hypochlorite solution, and disinfection using sodium hypochlorite solution occasionally affects perchlorate levels in drinking water. Perchlorate is formed from chlorate during the electrolysis of sodium chloride solution. It has been reported that perchlorate is formed in high concentration when platinum and lead dioxide are used as anode materials (Figure 5.1) [39]. Perchlorate is also formed at high concentrations with a boron-doped diamond anode during electrolysis [43].

HAAs are one of the major disinfection by-products during chlorination [18, 19]. Hua and Reckhow [18] reported that percentages of unknown total organic halide (TOX) to TOX after chlorination of six sources of water at water purification plants ranged from 55% to 66% (mean 60%) when disinfection by-products analyzed were four THMs (THM4), HAA9, three dihaloacetonitriles, two haloketones, and chloropicrin. Chlorination conditions were as follows: reaction time, 48 h; chlorine concentration after 48 h, 0.5 mg Cl_2 l^{-1}; pH 7; temperature, 20 °C. Among the disinfection by-products analyzed, THM4 was the largest group, followed by HAA9. For example, percentages of THM4 and HAA9 to TOX after chlorination of Newport News water were 27.6% and 15.4%, respectively [18].

The formation of disinfection by-products, including HAAs, is affected by chlorination conditions (e.g., chlorine dose, reaction time, TOC, and pH) [19]. Formation of bromo-HAAs increased when the bromide level in water was higher. It was reported that formation of chloro- and bromo-disinfection by-products by chloramination was lower than that by chlorination [18, 19]. However, in the case of iodo-disinfection by-products, including iodo-HAAs, their concentrations were higher after chloramination [19]. Iodo-HAAs were reported to be more cytotoxic and genotoxic in mammalian cells than chloro- and bromo-HAAs [44]. The iodoacids, including iodo-HAAs, were detected from samples of most water purification plants when occurrence of the five iodoacids in chloraminated and chlorinated drinking water samples from 23 cities in the United States and Canada was investigated. [44].

Echigo et al. [45] investigated formation potentials of the sum of three HAAs (MCAA, DCAA, and TCAA) from aliphatic compounds, amino acids, and aromatic compounds during chlorination (target compound, 3 mg C l^{-1}; chlorine dose, 30 mg Cl_2 l^{-1}; pH 7; reaction time, 24 h). In the case of nonaromatic compounds, DCAA was the primary HAA formed. On the other hand, in the case of aromatic compounds, TCAA was the primary HAA formed. The formation potentials of the sum of the three HAAs ranged from 0.003 to around 60 μmol mg C. The compounds with high formation potentials were amino acids such as asparagic acid

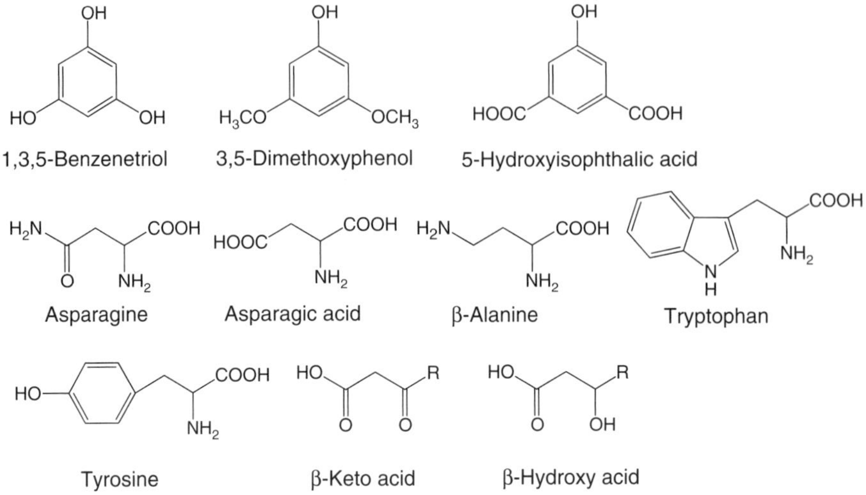

1,3,5-Benzenetriol   3,5-Dimethoxyphenol   5-Hydroxyisophthalic acid

Asparagine   Asparagic acid   β-Alanine   Tryptophan

Tyrosine   β-Keto acid   β-Hydroxy acid

Figure 5.2   High-yield HAA precursors during chlorination (Echigo *et al.*, 2007). Source: Reprint from Ref. [47] with permission of Jpn. soc. Civil Eng.

and β-alanine, α-keto acids, α-hydroxyl acids, and phenolic compounds such as e.g., 1,3,5-benzenetriol (Figure 5.2) [47].

5.3   ANALYSIS OF OXYHALIDES AND HAAs

5.3.1   Suppressed IC-MS

Suppressed IC-MS(/MS) is a common IC-MS(/MS) system and is applied for analysis of oxyhalides and HAAs by many researchers [31, 32]. Simultaneous analyses of oxyhalides and HAAs were frequently investigated, since both are categorized as disinfection by-products, including impurities of disinfectants. In suppressed IC-MS(/MS), in two ways organic solvents (e.g., acetonitrile and methanol) are added to the mobile phase to enhance sensitivities of the target compounds. One is mixing of organic solvents with the mobile phase as postcolumn solution [28] and the other one is the use of organic solvents as a mobile phase [46]. Aqueous solution of nonvolatile salt (e.g., sodium hydroxide, potassium hydroxide, or sodium carbonate) can be used as the mobile phase in suppressed IC-MS(/MS) because sodium and potassium ions are exchanged to hydrogen ion by a suppressor.

Pairs of precursor and product ions for oxyhalides by MS/MS were $[M-H]^-$ and $[M-H-O]^-$ [47]. For example, those of bromate, chlorate, and perchlorate were $^{79}BrO_3-$ $^{81}BrO_3-$ and $^{79}BrO_2-$ $81BrO_2-$, $^{35}ClO_3-$ and $^{35}ClO_2-$, and $^{35}ClO_4-$ and $^{35}ClO_2-$, respectively. In the case of HAAs, pairs of precursor and product ions by MS/MS were dependent upon HAAs [47]. Among the HAA9, the precursor ion of TBAA was $[M-COOH]^-$ and those of the remaining 8 HAAs were $[M-H]^-$. The product ions of MCAA, MBAA, and TBAA usually selected were halide ions $([Cl]^-, [Br]^-,$ and $[Br]^-$, respectively), and those of the remaining six HAAs were

[M−COOH]⁻. Among halides, bromide and iodide were occasionally analyzed together with oxyhalides and HAAs. In that case, the same m/z of precursor and product ions were selected (c.g., m/z of 79 (81) and 79 (81) for bromide) [48].

Common anions (e.g., chloride, sulfate, and nitrate) are usually in the range of several to several tens of milligrams per liter in environmental water. When such common anions coeluted with target compounds, sensitivities of target compounds of MS(/MS) decrease due to ion suppression. In addition, the m/z of hydrogen sulfate (98) interferes with the m/z of perchlorate [28]. Thus, chromatographic separation of such common anions and target compounds is important in IC-MS(/MS). On the other hand, when separation is difficult, techniques to mitigate or correct the reduction in sensitivities of target compounds by MS(/MS) are applied. For example, chloride, sulfate, and carbonate ions in sample water are removed by pretreatment cartridges of silver, barium, and hydrogen, respectively [47–49]. Also, isotope-labeled compounds are occasionally used as internal standards to correct the change of the sensitivities of target compounds. Particularly, ^{18}O-perchlorate is frequently used for perchlorate analysis [28]. Some of the isotope-labeled compounds (e.g., ^{18}O-perchlorate and ^{18}O-bromate) are commercially available. On the other hand, in a study the isotope-labeled compound was synthesized by researchers and used [41].

Among oxyhalides, analysis of perchlorate using suppressed IC-MS(/MS) has been investigated under different analytical conditions (e.g., analytical column and graduation conditions). For example, in EPA Method 332.0 [28], two combinations of analytical column and mobile phase were shown. That is, the combinations were an IonPac AS16 column (Dionex) and aqueous solution containing 75 mM (or 65 mM) potassium hydroxide, and an A Supp 5–100 column (Metrohm) and mixture of aqueous solution containing 30 mM sodium hydroxide and 30% methanol. In the former case, reagent water (or mixture of acetonitrile and water [50:50 v/v]) was used as postcolumn solution. The DL and the lowest concentration minimum reporting level (LCMRL) of perchlorate were 0.02 and 0.10 µg l^{-1}, respectively (Table 5.2) [28]. Barron and Paull [47] reported simultaneous analysis of four oxyhalides (iodate, bromate, chlorate, and perchlorate) and 10 HAAs (HAA9 and trifluoroacetic acid) using suppressed IC-MS/MS. The analytical column was an IonPac AS16 column (Dionex), and the mobile phase was aqueous solution containing sodium hydroxide. Postcolumn solution was methanol. The elution of oxyhalides was earlier in the order of iodate, bromate, chlorate, and perchlorate. The elution of HAAs was earlier in order of mono-HAAs, di-HAAs, and tri-HAAs. Also, the elution of chloro-HAAs was earlier than bromo-HAAs when the number of halogens of HAAs was the same. In that method [47], HAAs were preconcentrated using solid-phase extraction with subsequent common anion removal.

Simultaneous analyses of 14 target compounds (chlorite, chlorate, bromate, perchlorate, HAA9, and bromide) were also reported by Asami et al. [48] (Figure 5.3). The quantification limits (QLs) in ultrapure water were 0.05–0.5 µg l^{-1}. Both of the two quenching agents of residual free chlorine (sodium ascorbate and ammonium chloride) decreased the sensitivities of the target compounds, although the target compounds affected were different. The ion suppression derived from ammonium chloride was improved by using barium/silver/hydrogen pretreatment cartridges, although

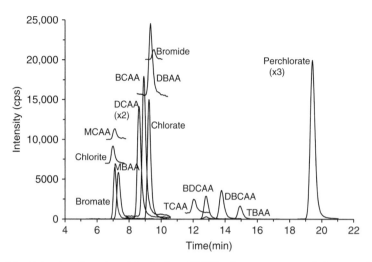

Figure 5.3 Selective Reaction Monitoring (SRM) chromatograms of four oxyhalides, HAA9, and bromide in ultrapure water using suppressed IC-MS/MS (target compounds, $1\,\mu g\,l^{-1}$; analytical column, IonPac AS20; mobile phase, aqueous solution of potassium hydroxide; postcolumn solution, mixture of acetonitrile and water [90:10 v/v]). Source: Reprint from Ref. [48] with permission of Jpn. Soc. Environ. Chem.

chlorite and bromide were partially removed by the cartridge (Figure 5.4) [48]. Thus, it was concluded that the pretreatment cartridges or more than fivefold dilution was recommended for pretreatment procedures, and its selection was dependent upon target compounds [48]. Preferable temperatures in ionization source were dependent upon target compounds, and sensitivities of some target compounds decreased at high temperatures (e.g., TCAA). The target compounds were separated into three periods during analysis, and the temperatures in the ionization source were changed by the periods [48]. In EPA Method 557.0 [29], bromate, HAA9, and dalapon were simultaneously analyzed using suppressed IC-MS/MS. The analytical column was an IonPac AS24 column (Dionex), the mobile phase was aqueous solution of hydroxide, and postcolumn solution was acetonitrile. The column temperature was set at 15 °C. Four isotope-labeled HAAs (MCAA[2-^{13}C], MBAA[1-^{13}C], DCAA[2-^{13}C], and TCAA[2-^{13}C]) were used for HAA9 analyses. MBAA[1-^{13}C] and DCAA[2-^{13}C] were also used for internal standards of bromate and dalapon, respectively. In that method [29], the target compounds were separated with common anions. In addition, the flow channel was tentatively changed so that mobile phase was not introduced into a mass selective detector during elution of the common anions. Thus, the 11 target compounds could be analyzed without sample pretreatment, and their DLs and LCMRLs by suppressed IC-MS/MS were from 0.015 to 0.20 and from 0.042 to $0.41\,\mu g\,l^{-1}$, respectively (Table 5.2) [29]. Using the same column (IonPac AS24 column), Li *et al.* [50] reported simultaneous analyses of three oxyhalides (chlorite, bromate, and chlorate) and HAA9 (Figure 5.5). Analyses of the 12 target compounds in

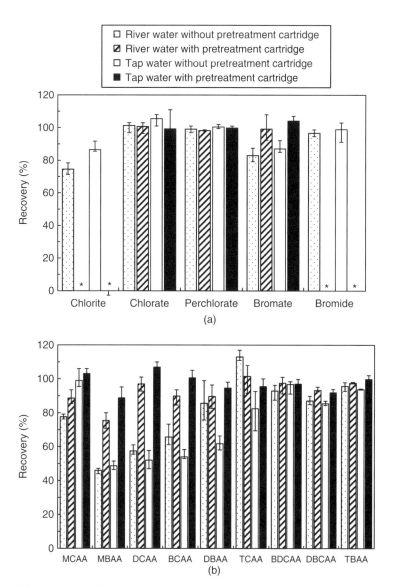

Figure 5.4 Recovery of (a) oxyhalide and bromide and (b) HAA9 in river and tap waters with and without pretreatment cartridge (*recovery of chlorite and bromide with pretreatment cartridge was excluded; error bar, maximum and minimum values [$n = 3$]) (target compound in river and tap waters, 5 and $4\,\mu g\,l^{-1}$, respectively; pretreatment cartridge, barium/silver/hydrogen cartridge; dose of ammonium chloride in tap water, $10\,mg\,l^{-1}$). Source: Reprint from Ref. [48] with permission of Jpn. Soc. Environ. Chem.

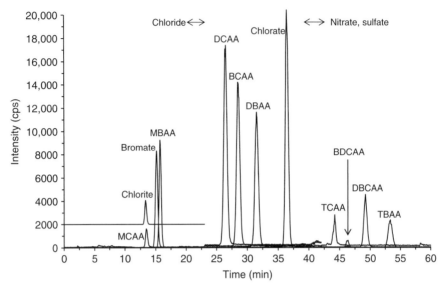

Figure 5.5 SRM chromatograms of three oxyhalides and HAA9 in ultrapure water using suppressed IC-MS/MS (target compounds, $2\,\mu g\,l^{-1}$; analytical column, IonPac AS24; mobile phase, aqueous solution of potassium hydroxide; postcolumn solution, acetonitrile). Source: Reprint from Ref. [50] with permission of Jpn. Industrial Water Assoc.

drinking water were not affected by common anions and ethylenediamine, a quenching agent of residual-free chlorine. Moreover, bromate and six HAAs (HAA5 and monoiodoacetic acid) were also analyzed without sample pretreatment by suppressed IC-MS/MS [46]. The analytical column was Metrosep A Supp plus Metrostep A Supp 5 columns (Metrohm), and the mobile phase was (A) aqueous solution containing 3.2 mM sodium carbonate and (B) aqueous solution containing 1.0 mM sodium hydrogen carbonate. The method detection limits (MDLs) and LCMRLs were from 0.009 to 0.022 and from 0.1 to $0.368\,\mu g\,l^{-1}$, respectively.

5.3.2   Nonsuppressed IC-MS and LC-MS

Anions are occasionally separated using an ion-exchange column without suppressor and are detected using MS(/MS). Similar to the case of suppressed IC-MS(/MS), this analytical system is also categorized into IC-MS(/MS) and called nonsuppressed IC-MS(/MS). It is of note that nonsuppressed IC-MS(/MS) is occasionally categorized into LC-MS(/MS).

Solution of volatile salt is mostly used as mobile phase when nonsuppressed IC-MS(/MS) is applied. Perchlorate was analyzed using nonsuppressed IC-MS/MS [49]. In that study, the analytical column was an IonPac AG16 column (Dionex) and the mobile phase was (A) water and (B) aqueous solution containing 50 mM ammonium hydroxide. The MDL was $0.05\,\mu g\,l^{-1}$. Barium cartridge was used to remove sulfate in sample water. In EPA Method 331.0 [27], perchlorate was also analyzed using nonsuppressed IC-MS(/MS). Analytical column was an IonPac AS21 (Dionex), a hydroxide-selective column, and mobile phase was aqueous solution of methylamine. The DLs and LCMRLs of perchlorate were 0.008 and 0.056 μg l^{-1}, respectively, using IC-MS, and 0.005 and 0.022 μg l^{-1}, respectively, using IC-MS/MS (Table 5.2) [27]. In EPA Method 331.0 [27], it was described that the LC system must be compatible with high alkali solution (approximately pH 12) because methylamine solution was used as eluent. Thus, it was suggested that some components should be replaced with a material suitable for high pH. In another study [51], perchlorate was also determined using nonsuppressed IC-MS with the same column (i.e., IonPac AS21 column) and a different mobile phase. That is, the mobile phase was (A) aqueous solution containing 73 mM ammonium carbonate and 20 mM ammonium hydroxide and (B) acetonitrile. In that method [51], perchlorate was eluted by not only hydroxide (pH) but also carbonate because pH of the ammonium solution was lower than that of methylamine solution. The retention time of perchlorate was around 11.6 min. As for the mobile phase using carbonate solution, it was also reported that perchlorate was analyzed using nonsuppressed IC-MS/MS with an IC-Pak Anion HR column (Waters) [52]. The mobile phase was (A) aqueous solution of 25 mM ammonium bicarbonate and (B) mixture of solution of ammonium hydroxide (pH 10) and acetonitrile (50%). Retention time of perchlorate was 8–9 min, and its QL was 0.05 μg l^{-1}.

Anions have also been analyzed using LC-MS(/MS) with reversed-phase columns. In general, separation of oxyhalides and HAAs with common anions using reversed-phase columns is more difficult than those using IC because oxyhalides and HAAs are ionic and hydrophilic. The mobile phase is usually aqueous solution containing acid to reduce the dissociation and to improve their retention and separation [53]. Four oxyhalides (bromate, chlorate, iodate, and perchlorates) in natural waters were analyzed using LC-MS/MS with a Synergi Max-RP C12 column (Phenomenex) [54]. The mobile phase was (A) aqueous solution of formic acid and (B) methanol. Their MDLs and minimum reporting levels (MRLs) ranged from 0.0219 to 0.694 and from 0.05 to 0.1 μg l^{-1}, respectively. Ion suppression was observed by the coelution of common anions [54]. Hydronium and barium cartridges were used as pretreatment cartridges to remove carbonate and sulfate, respectively, in the samples. Also, perchlorate was analyzed using a Gemini C18 column (Phenomenex) [55] and a KP-RPPX column (K' [Prime] Technology) [56]. For bromate analysis, application of ultra-HPLC (UHPLC)–MS/MS was also reported [57]. The analytical column was an Acquity BEH C18 column (Waters), and the mobile phase was aqueous solution containing 0.1% formic acid. The QL of bromate was 0.04 μg l^{-1}. The retention time was very short (0.4 min), but the peak was not affected by matrix in drinking water. One reason was considered to be the selective reaction monitoring (SRM) mode and configuration of the ionization source [57].

HAAs were also analyzed using LC-MS/MS and UHPLC-MS/MS. HAA9 were simultaneously analyzed using UHPLC-MS/MS [53]. The mobile phase (A) was acetonitrile. From the points of separation and sensitivity of the HAAs, mobile phase (B) was compared between aqueous solution containing acetic acid and formic acid, and the percentage of these acids was also investigated to be 0.05% to 0.3% (v/v). Thus, aqueous solution containing 0.1% acetic acid was selected as the mobile phase (B). The separation column used was an Acquity BEH C8 column (Waters). The retention times of HAA9 were less than 7.5 min. The DLs and QLs, excluding TBAA, ranged from 0.16 to 1.44 and from 0.56 to 4.80 µg l^{-1}, respectively. The DL and QL of TBAA were 8.87 and 24.29 µg l^{-1}, respectively. The recoveries of HAA9 in drinking water were sufficient (80.1–108%). HAA9 analyses using UHPLC-MS/MS were also conducted with a HSS T3 column (Waters) [58]. Moreover, HAA analysis using ion-paring LC-MS/MS has been investigated, although the details are not discussed in this chapter [32].

Mixed-mode columns, having both reversed-phase and anion-exchange retention properties, are also used for oxyhalides and HAAs analysis. Bromate was analyzed using LC-MS/MS with an Acclaim Trinity P1 column (Thermo Fisher Scientific) (Figure 5.6) [59]. The mobile phase was (A) aqueous solution (pH 5) containing 20 mM ammonium acetate and 0.05% v/v acetic acid and (B) mixture of aqueous solution (pH 5) containing 200 mM ammonium acetate and 0.5% v/v acetic acid and acetonitrile. The QL of bromate was 0.2 µg l^{-1}. In that study [59], the commercially available ^{18}O-bromate was used as the internal standard. It was reported that the ^{18}O-bromate contained native bromate as impurity. Thus, it was important to consider the dose of ^{18}O-bromate added to the sample. The mixed-mode column was

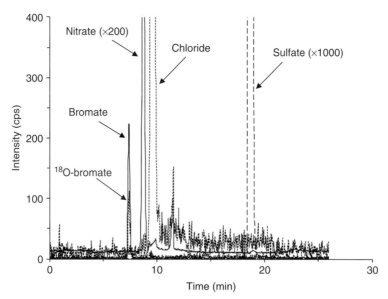

Figure 5.6 SRM chromatograms of bromate and common anions in drinking water using LC-MS/MS (analytical column, Acclaim Trinity P1; mobile phase, [A] aqueous solution [pH 5] containing 20 mM ammonium acetate and 0.05% v/v acetic acid and [B] mixture of aqueous solution [pH 5] containing 200 mM ammonium acetate and 0.5% v/v acetic acid and acetonitrile). Source: Reprint from Ref. [59] with permission of Jpn. Soc. Anal. Chem.

also applied for HAA analysis. Three HAAs (MCAA, DCAA, and TCAA) were chromatographically separated with common anions by an Acclaim HAA column (Thermo Fisher Scientific), and the three HAAs in tap water were analyzed using LC-MS/MS without sample pretreatment [60].

Recently, hydrophilic interaction liquid chromatography (HILIC) has been applied for the separation of polar, weakly acidic or basic compounds [32, 61]. Bromate and perchlorate were analyzed using HILIC LC-MS/MS, and the separation columns were a X-Bridge HILIC column (Waters) and an Inertsil HILIC column (GL Science), respectively [62, 63]. Also, 10 HAAs (HAA9 and monoiodoacetic acid) and 2-bromobutylic acid were analyzed using HILIC UHPLC-MS/MS [64]. The analytical column was an Acquity UPLC BEH HILIC column (Waters) and the mobile phase was (A) aqueous solution (pH 4.1) containing 5 mM formic acid, 10 mM ammonium formate, and (B) acetonitrile. The retention times of the compounds were earlier, in order of tri-HAAs, di-HAAs, and mono-HAAs. This tendency of the elution order was opposite of those using ion-exchange and reversed-phase columns. The DLs and QLs of the compounds without preconcentration were in the range of 0.08–2.7 and 0.39–9.8 μg l^{-1}, respectively [64]. These values were lower than those obtained using LC-MS/MS with a BetaMax acid column (Thermo Hypersil-Keystone), a C12 column. However, for HILIC UHPLC-MS/MS, dissolution of the sample in 90% acetonitrile was required to prepare the composition of the sample; that is, the sample was diluted before analysis [64].

5.3.3 IC-ICP-MS

Similar to IC-MS(/MS), IC-ICP-MS has been investigated for the analysis of oxyhalides and HAAs [31, 32, 65]. The values of m/z of compounds with chlorine, bromine, and iodine are 35 (^{35}Cl), 79 (^{79}Br), and 127 (^{127}I), respectively. Sensitivities of ^{79}Br and ^{127}I are higher than that of ^{35}Cl. Sodium salt in the mobile phase was not desirable in IC-ICP-MS because continuous introduction of sodium changed plasma conditions and caused clogging of cone orifices [66]. The sensitivities of the target compounds also decreased by sodium and potassium [68]. Thus, such nonvolatile salts were eliminated at a suppressor when suppressed IC-ICP-MS is applied. On the other hand, volatile salts such as ammonium are preferred for nonsuppressed IC-ICP-MS.

Bromate analysis is a popular application of IC-ICP-MS [65]. Several combinations of analytical columns and mobile phases were investigated. In ICP-MS, bromine is monitored for bromate analysis. In EPA Method 321.8 [30], the analytical column was a PA100 column (Dionex) and the mobile phase was aqueous solution containing 5 mM nitric acid and 25 mM ammonium nitrate. The DL of bromate was 0.3 μg l^{-1} (Table 5.2) [30]. Chromatographic separation of bromate with other bromine-containing compounds is needed. Also, bromine analysis is influenced by argon (Ar) species, carrier gas of ICP-MS (i.e., ^{40}Ar^{21}H$^+$ for ^{81}Br and ^{40}Ar^{38}Ar^{1}H$^+$ for ^{79}Br) [66]. The interference of ^{79}Br is substantially lower than that of ^{81}Br, so ^{79}Br is usually monitored in ICP-MS [66]. As for the coeluent polyatomic ion, ^{40}Ar^{39}K$^+$ is also known [30].

Pantsar-Kallio and Mannien [68] reported determination of five oxyhalides (chlorite, chlorate, perchlorate, bromate, and iodate) and three halides (chloride, bromide, and iodide) using nonsuppressed IC-ICP-MS. The analytical column was an IC-Pak A column (Waters). Carbonate solution was initially used as the mobile phase. However, sensitivities decreased after long aspiration of carbonate solution because of carbon accumulation in the skimmer cone. Thus, aqueous solution of potassium nitrate and nitric acid was finally used as the mobile phase. The DLs of target compounds were ranged from 0.1 to $5\,\mu g\,l^{-1}$, except for chloride. The DL of chloride was $500\,\mu g\,l^{-1}$. As for HAAs, Liu et al. [67] investigated analysis of HAA9 using suppressed IC-ICP-MS. The analytical column was an IonPac AS16 column (Dionex) and mobile phase was aqueous solution of sodium hydroxide. For chloro-HAAs, sensitivity of m/z 51 (^{35}ClO) was higher than that of m/z 35 (^{35}Cl), and thus, ^{35}ClO was monitored for determination. The DLs of chloro- and bromo-HAAs were in the range of 15.6–23.6 and 0.45–$0.99\,\mu g\,l^{-1}$, respectively. Determination of chloro-HAAs was affected by chloride, and a silver cartridge was used as a pretreatment cartridge for chloride removal. Also, bromo- and iodo-compounds (halides, oxyhalides, and HAAs) were analyzed by nonsuppressed IC-ICP-MS [69]. The analytical column was an IonPac AS11-HC column (Dionex) and the mobile phase was (A) distilled ionized water and (B) aqueous solution containing 200 mM ammonium nitrate. In that study [69], germanium dioxide was used as an internal standard. The MDLs of bromo- and iodo-compounds in Missouri River water ranged from 1.36 to 3.28 and from 0.33 to $0.72\,\mu g\,l^{-1}$, respectively.

5.4   APPLICATION FOR MONITORING OF OXYHALIDES AND HAAs IN DRINKING WATER

As discussed in the previous section, many studies on analyses of oxyhalides and HAAs in water using IC-MS(/MS) and IC-ICP-MS have been conducted. The main objectives of these studies were the development of the analytical methods. Studies on the monitoring of oxyhalides and HAAs using IC-MS(/MS) and IC-ICP-MS seem to be limited but gradually increase. In this section, such monitoring studies are introduced.

5.4.1   Oxyhalides

IC-ICP-MS and IC-MS/MS of bromate analysis are adopted in EPA Methods (Table 5.2) [29, 30]. However, IC-PCR is a popular method for bromate analysis in drinking water after its development. One reason considered is that IC-MS/MS was recently developed for bromate analysis in drinking water and also that ICP-MS and MS(/MS) were used for analysis of many compounds (e.g., metals and organic micropollutants), but the IC-PCR instrument was dedicated for bromate analysis. For example, in surveys throughout the United States [17], bromate concentrations in finished water at water purification plants were investigated using IC-PCR. Bromate concentrations in finished water at water purification plants applying ozone were much higher than those applying chlorine dioxide. Means of bromate concentrations in finished water at 30 water purification plants applying ozone ranged from 0 to $7.2\,\mu g\,l^{-1}$ (MRL, 0.02 or $5\,\mu g\,l^{-1}$). The effects of bromate levels by ozonation and sodium hypochlorite solution were reported by [37]. Bromate concentration in raw water was $<0.05\,\mu g\,l^{-1}$ and increased from 0.05 to $0.58\,\mu g\,l^{-1}$ after ozonation at a water purification plant applying ozonation. The bromate level further increased to $3.3\,\mu g\,l^{-1}$ in clear well effluent after disinfection using sodium hypochlorite solution. In that study, bromate concentration in raw water, filter effluent after ozonation, and plant effluent after the addition of sodium hypochlorite solution at another water purification plant were 0.06, 2.96, and $4.19\,\mu g\,l^{-1}$, respectively [37]. Also, bromate concentration in drinking water was $168\,\mu g\,l^{-1}$ due to the use of sodium hypochlorite solution with high bromate level ($668\,mg\,l^{-1}$) [38]. On the other hand, LC-MS/MS was applied for simultaneous analyses of four oxyhalides, including bromate in bottled and natural waters in the United States [54]. Concentrations of bromate, chlorate, iodate, and perchlorate in 21 bottled waters were in the range of <0.1–76, <0.1–5.8, <0.1–25, and <0.05–$0.11\,\mu g\,l^{-1}$, respectively. Those in 11 natural waters were in the range of <0.1–4.6, <0.1–270, <0.1–6.2, and <0.05–$6.8\,\mu g\,l^{-1}$, respectively. Bromate concentrations in the bottled waters were higher when ozonation was applied. Chlorate concentrations in the bottled waters were higher when chlorine disinfectant was added.

For chlorate analysis in drinking water, IC-CD has mostly been applied [3, 17], although IC-MS/MS and LC-MS/MS are more sensitive and accurate. This is because guideline value, standard value, and HRL of chlorate in drinking water are high enough to determine by using IC-CD. In addition, chlorate concentrations in drinking water ranged from several tens to several hundreds of micrograms per liter when hypochlorite and chlorine dioxide are used as disinfectant [3, 17]. For example, under preliminary monitoring of the third unregulated contaminant monitoring rule, the median and maximum concentrations of chlorate in 6561 samples of approximately 600 utilities were 34 and $6200\,\mu g\,l^{-1}$, respectively [3]. The chlorate concentrations of more than 10% of the samples, and more than 30% of the utilities, were larger than the HRL by the US EPA ($210\,\mu g\,l^{-1}$) [3]. However, occurrence of chlorate and perchlorate in water purification plants throughout Japan was investigated using IC-MS/MS (Table 5.3) [70]. Chlorate and perchlorate were discharged into the Tone River and its tributary in the upper Tone River basin via industrial effluents [70]. Perchlorate was also discharged into a tributary of the Tone River in the middle basin via industrial effluent [71]. Thus, their levels in the Tone River basin were higher than in other regions (Table 5.3) [70]. In addition, due to disinfection using sodium hypochlorite solution, chlorate and perchlorate concentrations increased in finished waters. Also, IC-MS/MS and LC-MS/MS are occasionally applied for analysis of chlorate in environmental, natural, and bottled waters [41, 54, 72]. Chlorate concentrations in natural and bottled waters were $<10\,\mu g\,l^{-1}$ in the cases where they were not disinfected by chlorine disinfectants [72]. Also, concentrations of chlorate and perchlorate in 31 unconfined aquifers, 16 confined aquifers, and 3 springs in Tokyo were investigated (Figure 5.7) [73]. Usage of these groundwater sources was dependent upon groundwater (e.g., domestic use, bathing, drinking water).

TABLE 5.3   Concentrations of Chlorate and Perchlorate in Source and Finished Water at Water Purification Plants in Japan [70]

Sample[a]	Chlorate			Perchlorate		
	Detection Rate	Median (μg l^{-1})	Range (μg l^{-1})	Detection Rate	Median (μg l^{-1})	Range (μg l^{-1})
Tone River basin						
Source water (upper basin, SW)[b]	6/6	2.3	1.6–2.7	6/6	0.15	0.09–0.82
Finished water (upper basin, SW)[b]	6/6	67	1.8–92	6/6	0.57	0.12–0.86
Source water (middle basin, SW)[b]	19/19	21	6.4–26	19/19	10	0.15–15
Finished water (middle basin, SW)[b]	22/22	53	18–290	22/22	8.4	0.22–14
Source water (lower basin, SW)[b]	19/19	11	1.3–16	19/19	1.1	0.15–2.7
Finished water (lower basin, SW)[b]	17/17	120	10–240	17/17	1.1	0.23–1.8
Source water (upper basin, GW)[c]	21/21	21	5.6–77	21/21	1.1	0.93–32
Finished water (upper basin, GW)[c]	5/5	89	62–430	5/5	0.45	0.08–24
Other region						
Source water (SW)[b]	35/42	1.0	<0.05–44	30/42	0.08	<0.05–2.5
Finished water (SW)[b,d]	42/42	59	0.87–2900	36/42	0.11	<0.05–6.1
Source water (GW)[c,e]	17/20	0.36	<0.05–12	15/20	0.09	<0.05–1.2
Finished water (GW)[c–e]	20/20	92	0.17–990	20/20	0.16	0.06–1.2

[a]Sampling was conducted in September 2006 to February 2007.
[b]Source water type was surface water (SW).
[c]Source water was groundwater (GW).
[d]Some of finished water samples were tap water.
[e]Private water supply.

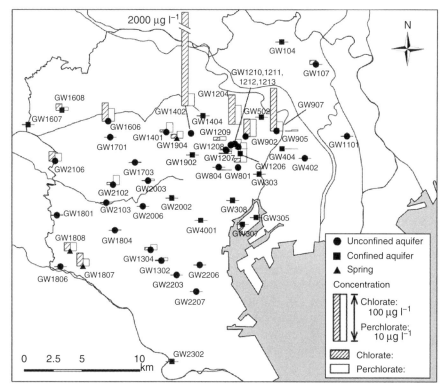

Figure 5.7   Concentrations of chlorate and perchlorate in groundwater in Tokyo.   Source: Reprint from Ref. [73] with permission of Jpn. Soc. Civil Eng.

Chlorate concentrations in these three types of groundwater were from <0.05 to 2000, from <0.05 to 14, and from 9.3 to 28 μg l^{-1}, respectively [73]. Perchlorate concentrations were from <0.01 to 4.1, from <0.01 to 4.2, and from 1.4 to 1.8 μg l^{-1}, respectively. Thus, contamination by chlorate and perchlorate was observed in some groundwater sources. The chlorate concentrations in one unconfined aquifer and two confined aquifers were not determined in that study [73]. Rao *et al.* [41] reported on chlorate concentrations in environmental waters, soils, and plants whose perchlorate concentrations were previously measured. The results indicated that similarly to perchlorate, chlorate was produced in the atmosphere and deposited on the land surface. It was also implied that chlorate has relatively low stability in the subsurface or biosphere compared to perchlorate [41].

For perchlorate, IC-MS/MS and LC-MS/MS have commonly been applied for its monitoring in drinking water in countries in different continents (e.g., North America, Asia, and Europe) [14, 74]. Of course, monitoring of perchlorate has also been conducted using IC-CD. Perchlorate concentrations in drinking water were generally <1.0 μg l^{-1}, but those in some drinking water sources were observed at ≥1.0 μg l^{-1} in many countries [14]. In addition, relatively large contamination by perchlorate was reported in some countries [14]. In Japan, perchlorate contamination was found in the Tone River basin, as described earlier (Table 5.3) [70]. After the discovery of the contamination in the basin, perchlorate discharges has been reduced at the main industrial sources, and its level in the basin decreased [14]. Perchlorate concentration was also investigated in natural and bottled waters [54, 72, 75]. It was reported that perchlorate concentrations in natural and bottled waters were generally up to sub-microgram levels per liter, but those in only some natural and bottled waters were several micrograms per liter [14].

In addition, concentrations of oxyhalides in hypochlorite solutions were investigated using IC-MS/MS and LC-MS/MS. Asami *et al.* [38] reported concentrations of chlorate and perchlorate in 38 hypochlorite solutions (sodium hypochlorite, 32; on-site generation, 6) used at water purification plants by IC-MS/MS. Concentrations of chlorate and perchlorate in on-site generation were lower than those in sodium hypochlorite solution. However, concentrations of chlorate and perchlorate per free available chlorine were not so different between on-site generation and sodium hypochlorite solution when free available chlorine in sodium hypochlorite solution did not decrease. Both concentrations of chlorate and perchlorate were higher when the measured free available chlorine was lower due to decomposition of hypochlorite (Figure 5.8) [38]. Stanford *et al.* [76] also determined concentrations of bromate, chlorate, and perchlorate in the types of hypochlorite solutions (sodium hypochlorite, 5; on-site generation, 12; calcium hypochlorite, 2) using LC-MS/MS. Their concentrations per free available chlorine among the types of hypochlorite solutions were not so different. That is, bromate concentrations per free available chlorine in the three types of hypochlorite solutions were in the range of 60–280,

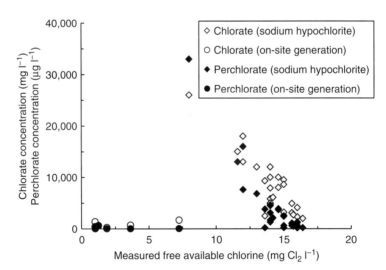

Figure 5.8 Relationships between measured available free chlorine and concentrations of chlorate and perchlorate in hypochlorite solutions. Source: Reprint from Ref. [38] with permission of copyright holders, Int. Water Assoc. Publishing.

20–840, and 80–90 μg g^{-1} Cl$_2$, respectively. Chlorate concentrations per free available chlorine were in the range of 15–220, 14–270, and 12–14 mg g^{-1} Cl$_2$, respectively. Perchlorate concentrations per free available chlorine were in the range of 1.8–160, 0.6–520, and 0.8–0.9 μg g^{-1} Cl$_2$, respectively [76].

5.4.2 HAAs

Occurrence of HAAs in drinking water has been reported using GC-ECD and GC-MS in many studies. For example, occurrence of disinfection by-products, including HAA5, in drinking water throughout the United States was summarized by the US EPA [17]. Concentrations of HAA5 in means of distribution system samples at 213 water purification plants using surface water as source water ranged from <DL to 116 μg l^{-1} (median, 24.38 μg l^{-1}). Those at 82 water purification plants using groundwater as source water ranged from <DL to 71 μg l^{-1} (median, 2.24 μg l^{-1}). Means of distribution system samples were the calculated mean of four distribution system samples [17]. Also, Weinberg *et al.* [77] investigated concentrations of disinfection by-products of 12 water purification plants and their distribution systems in EPA regions 3–7 and 9. Frequency of the investigation was four or five at each of water purification plant. The mean concentrations of HAA5 and HAA9 in the distribution system of each water purification plant ranged from 12 to 65 and from 16 to 79 μg l^{-1}, respectively [77].

On the other hand, recently, in the United Kingdom, concentrations of HAA9 in final drinking water were investigated at 20 sites (A–T sites) of England and Wales in 2010 and 2011 using EPA Method 557.0, an analytical method using IC-MS/MS [78]. Frequency of sampling at 11 sites was 4, and that of the remaining 9 sites was 2 or 3. The concentrations of HAA9 in final drinking water at 20 sites during the four surveys ranged from <DL to 21.4 μg l^{-1} (mean, 9.8 μg l^{-1}), from 2.4 to 27.8 μg l^{-1} (mean, 13.75 μg l^{-1}), from 0.3 to 48.6 μg l^{-1} (mean, 19.54 μg l^{-1}), and from 5.3 to 33.9 μg l^{-1} (mean, 16.05 μg l^{-1}), respectively [78]. Also, mean concentrations of HAA9 in final drinking water at each of 20 sites are shown in Figure 5.9 [78]. Among HAA9, contributions of concentrations of DCAA and TCAA to HAA9 concentrations were generally higher. The contributions of BCAA and DBAA were also relatively high at some sites [78]. In Japan, concentrations of MCAA, DCAA, and TCAA in drinking water are regulatory monitored by water utilities. In fiscal 2013, their analytical methods using LC-MS(/MS) in drinking water were newly adopted as one of the standard methods of the three HAAs in Japan. From the results of an external quality control survey in fiscal 2013 by Ministry of Health, Labour and Welfare (MHLW) [79], it was presumed that drinking water quality inspection of the three HAAs was conducted using LC-MS(/MS) at 22 water utilities, 8 local institutes for health, and 21 registration inspection institutes.

SUMMARY

IC-ICP-MS and particularly IC-MS(/MS) are promising methods for determining oxyhalides and HAAs in drinking water. These methods are affected by common

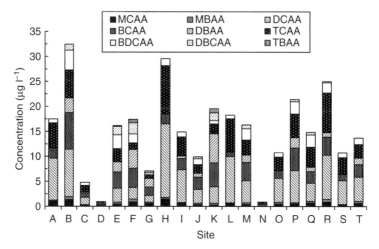

Figure 5.9   Mean concentration of HAA9 at two to four sampling events in 20 final drinking water samples in England and Wales. Drawn by the author using data in Ref. [78].

anions, and thus, column selections and optimization of gradient conditions are important. Additionally, further development of analytical columns that have higher separation capacity is desired for analysis without pretreatment. To date, from the point of routine monitoring of oxyhalides and HAAs in drinking water, except for perchlorate, application of IC-MS(/MS) and IC-ICP-MS may be limited, although some analytical methods were described as standard methods (e.g., EPA Methods). MS(/MS) and ICP-MS are useful analytical instruments for water quality inspection of water utilities, but they are expensive, and their installation may be limited. When they are installed, they are used for analyses of many contaminants, and oxyhalides and HAAs may be analyzed by other methods. Thus, rapid and simultaneous analyses are more efficient so that IC-MS(/MS) and IC-ICP-MS become more popular for routine analysis and investigative surveys of oxyhalides and HAAs in drinking water.

5.5   REFERENCES

[1] von Gunten, U. (2003) Ozonation of drinking water: Part II. Disinfection and by-product formation in presence of bromide, iodide, or chlorine. *Water Research*, **37**, 1469.

[2] WHO (2011) *Guidelines for Drinking-Water Quality*, 4th edn, WHO, Geneva.

[3] Alfredo, K., Adams, C., Eaton, A. *et al.* (2014) *The Potential Regulatory Implications of Chlorate*, American Water Works Association, Washington, DC.

[4] US EPA (2008) *Interim Drinking Water Health Advisory for Perchlorate* (EPA 822-R-08-025).

[5] US EPA (2006) *National Primary Drinking Water Regulations: Stage 2 Disinfectants and Disinfection Byproducts Rule*, Federal Register Notice.

[6] Health Canada, *Guidelines for Canadian Drinking Water Quality Summary Table*, http://www.hc-sc.gc.ca/ewh-semt/alt_formats/pdf/pubs/water-eau/sum_guide-res_recom/sum_guide-res_recom-eng.pdf (accessed January 2016).

[7] Srinivasan, A. and Viraraghavan, T. (2009) Perchlorate: health effects and technologies for its removal from water resources. *International Journal of Environmental Research and Public Health*, **6**, 1418.

[8] Health Canada (2008) *Guidelines for Canadian Drinking Water Quality: Guideline Technical Document Haloacetic Acids (H128-1/08-548E)*, Health Canada, Ottawa.

[9] National Health and Medical Research Council (NHMR) and National Resource Management Ministerial Council (NRMMC) (2011) *Australian Drinking Water Guidelines Paper 6: National Water Quality Management Strategy*, NHMRC, NRMMC, Commonwealth of Australia, Canberra.

[10] MHLW, *Drinking Water Quality Standards* (in Japanese), http://www.mhlw.go.jp/stf/seisakunitsuite/bunya/topics/bukyoku/kenkou/suido/kijun/kijunchi.html (accessed January 2016).

[11] U.S. EPA, *Basic Information about Disinfection Byproducts in Drinking Water: Total Trihalomethanes, Haloacetic Acids, Bromate, and Chlorite*, http://water.epa.gov/drink/contaminants/basicinformation/disinfectionbyproducts.cfm (accessed January 2016).

[12] US EPA (2009) *Federal Register Notice: Drinking Water Contaminant Candidate List 3-Final*.

[13] ITRC Perchlorate Team, *Perchlorate: Overview of Issues, Status, and Remedial Options*, http://www.itrcweb.org/Documents/PERC-1.pdf (accessed January 2016).

[14] Kosaka, K., Asami, M., and Kunikane, S. (2013) Perchlorate: origin and occurrence in drinking water, in *Reference Module in Earth Systems and Environmental Sciences* (ed S. Elias), Elsevier, Burlington, MA, p. 1.

[15] National Research Council (2005) *Health Implications of Perchlorate Ingestion*, National Academies Press, Washington, DC.

[16] Joint Food and Agriculture Organization of the United Nations/World Health Organization Expert Committee on Food Additives. (2010). *Evaluation of Certain Contaminants in Food: Seventy-Second Report of the Joint FAO/WHO Expert Committee on Food Additives* (WHO Technical Report Series, No. 959), http://whqlibdoc.who.int/trs/WHO_TRS_959_eng.pdf (accessed January 2016).

[17] US EPA (2005) *Occurrence Assessment for the Final Stage 2 Disinfectants and Disinfection Byproducts Rule* (EPA 815-R-05-011).

[18] Hua, G. and Reckhow, D.A. (2007) Comparison of disinfection byproduct formation from chlorine and alternative disinfectants. *Water Research*, **41**, 1667.

[19] Itoh, S. and Echigo, S. (2008) *Disinfection Byproducts in Water*, Gihodo, Tokyo (in Japanese).

[20] Hautman, D.P. and Munch, D.J. (1997) *Method 300.1: Determination of Inorganic Anions in Drinking Water by Ion Chromatography*, US EPA, Cincinnati.

[21] Hautman, D.P. and Munch, D.J. (1999) *EPA Method 314.0: Determination of Perchlorate in Drinking Water Using Ion Chromatography*, US EPA, Cincinnati.

[22] Wagner, H.P., Pepich, B.V., Pohl, C. *et al.* (2005) *EPA Method 314.1: Determination of Perchlorate in Drinking Water Using Inline Column Concentration/Matrix Elimination Ion Chromatography with Suppressed Conductivity Detection*, US EPA, Cincinnati.

[23] Wagner, H.P., Pepich, B.V., Later, D. *et al.* (2008) *Method 314.2: Determination of Perchlorate in Drinking Water Using Two-Dimensional Ion Chromatography with Suppressed Conductivity Detection (815-B-08-001)*, US EPA, Cincinnati.

[24] Wagner, H.P., Pepich, B.V., Pohl, C. *et al.* (2009) *Method 302.0: Determination of Bromate in Drinking Water Using Two-Dimensional Ion Chromatography with Suppressed Conductivity Detection (815-B-09-014)*, US EPA, Cincinnati.

[25] Wagner, H.P., Pepich, B.V., Hautman, D.P., and Munch, D.J. (2001) *Method 317.0: Determination of Inorganic Oxyhalide Disinfection Byproducts in Drinking Water Using Ion Chromatography with the Addition of a Postcolumn Reagent for Trace Bromate Analysis*, US EPA, Cincinnati.

[26] Wagner, H.P., Pepich, B.V., Hautman, D.P. *et al.* (2002) *Method 326.0: Determination of Inorganic Oxyhalide Disinfection Byproducts in Drinking Water Using Ion Chromatography Incorporating the Addition of a Suppressor Acidified Postcolumn Reagent for Trace Bromate Analysis (815-R-03-007)*, US EPA, Cincinnati.

[27] Wendelken, S.C., Munch, D.J., Pepich, B.V. *et al.* (2005) *Method 331.0: Determination of Perchlorate in Drinking Water by Liquid Chromatography Electrospray Ionization Mass Spectrometry (815-R-05-007)*, US EPA, Cincinnati.

[28] Hedrick, E., Behymer, T., Slingsby, R., and Munch, D. (2005) *Method 332.0: Determination of Perchlorate in Drinking Water by Ion Chromatography Suppressed Conductivity and Electrospray Ionization Mass Spectrometry (EPA 600-R-05-049)*, US EPA, Cincinnati.

[29] Zaffiro, A.D., Zimmerman, M., Pepich, B.V. *et al.* (2009) *Method 557: Determination of Haloacetic Acids, Bromate, and Dalapon in Drinking Water by Ion Chromatography Electrospray Ionization Tandem Mass Spectrometry (IC–ESI-MS/MS) (EPA 815-B-09-012)*, US EPA, Cincinnati.

[30] Creed, J.T., Brockhoff, C.A., and Martin, T.D. (1997) *Method 321.8: Determination of Bromate in Drinking Waters by Ion Chromatography Inductively Coupled Plasma-Mass Spectrometry*, US EPA, Cincinnati.

[31] Michalski, R., Jabłońska, M., Szopa, S., and Łyko, A. (2011) Application of ion chromatography with ICP-MS or MS detection to the determination of selected halides and metal/metalloids species. *Critical Reviews in Analytical Chemistry*, **41**, 133.

[32] Barron, L. and Gilchrist, E. (2014) Ion chromatography–mass spectrometry: a review of recent technologies and applications in forensic and environmental explosives analysis. *Analytica Chimica Acta*, **806**, 27.

[33] Domino, M.M., Pepich, B.V., Munch, D.J. *et al.* (2003) *Method 552.3: Determination of Haloacetic Acids and Dalapon in Drinking Water by Liquid–Liquid Microextraction, Derivatization, and Gas Chromatography with Electron Capture Detection, EPA 815-B-03-002*, US EPA, Cincinnati.

[34] Japan Water Works Association (2011) *Standard Methods for the Examination of Water*, Japan Water Works Association, Tokyo (in Japanese).

[35] Fitzpatrick, C.M. (2010) *The hydrogeology of bromate contamination in the Hertfordshire chalk: double-porosity effects on catchment-scale evolution, Doctoral dissertation*, University College London, London.

[36] Simazaki, D., Asami, M., Nishimura, T., Aizawa, T., Kunikane, S., and Magara, Y.(2009) Occurrence of bromate in raw water and finished waters for drinking water supply in Japan, Proceedings of 13th Jpn./Korea Symposium on Water Environment, p. 89.

[37] Weinberg, H.S., Delcomyn, C., and Unnam, V. (2003) Bromate in chlorinated drinking waters: occurrence and implications for future regulation. *Environmental Science and Technology*, **37**, 3104.

[38] Asami, M., Kosaka, K., and Kunikane, S. (2009) Bromate, chlorate, chlorite and perchlorate in sodium hypochlorite solution used in water supply. *Journal of Water Supply: Research and Technology-Aqua*, **58**, 107.

[39] Asami, M., Kosaka, K., Simazaki, D., and Takei, K. (2014) Generation characteristics of chlorate and perchlorate in electrolysis of salt water using six anodes of different materials. *Journal of Japan Society of Water Environment*, **37**, 189 (in Japanese).

[40] Dasgupta, P.K., Martinelango, P.K., Jackson, W.A. *et al.* (2005) The origin of naturally occurring perchlorate: the role of atmospheric processes. *Environmental Science and Technology*, **39**, 1569.

[41] Rao, B., Hatzinger, P.B., Böhlke, J.K. *et al.* (2010) Natural chlorate in the environment: application of a new IC-ESI/MS/MS method with a $Cl_{18}O_3$-internal standard. *Environmental Science and Technology*, **44**, 8429.

[42] Massachusetts Dep (2005) *The Occurrence and Sources of Perchlorate in Massachusetts (Draft Report)*.

[43] Bergmann, M.E.H., Rollin, J., and Iourtchouk, T. (2009) The occurrence of perchlorate during drinking water electrolysis using BDD anodes. *Electrochimica Acta*, **54**, 2102.

[44] Richardson, S.D., Fasano, S., Ellington, J.J. *et al.* (2008) Occurrence and mammalian cell toxicity of iodinated disinfection byproducts in drinking water. *Environmental Science and Technology*, **42**, 8330.

[45] Echigo, S., Yano, Y., Jo, I., and Itoh, S. (2007) Formation characteristics of haloacetic acids from common chemical structures in dissolved organic matter during chlorination. *Environmental Engineering Research*, **44**, 265 (in Japanese).

[46] Mathew, J., McMillin, R., Gandhi, J. *et al.* (2009) Trace level haloacetic acids in drinking water by direct injection ion chromatography and single quadrupole mass spectrometry. *Journal of Chromatographic Science*, **47**, 505.

[47] Barron, L. and Paull, B. (2006) Simultaneous determination of trace oxyhalides and haloacetic acids using suppressed ion chromatography–electrospray mass spectrometry. *Talanta*, **69**, 621.

[48] Asami, M., Kosaka, K., Matsuoka, Y., and Kamoshita, M. (2007) An analytical method for haloacetic acids and oxohalides using ion chromatography coupled with tandem mass spectrometry and detection of perchlorate in environmental and drinking waters. *Journal of Environmental Chemistry*, **17**, 363 (in Japanese).

[49] Winkler, P., Minteer, M., and Willey, J. (2004) Analysis of perchlorate in water and soil by electrospray LC/MS/MS. *Analytical Chemistry*, **76**, 469.

[50] Li, K., Suzuki, T., and Sekiguchi, Y. (2008) Determination of trace-level haloacetic acids in tap water with ion chromatography coupled with tandem mass spectrometry. *Industrial Water*, **591**, 56 (in Japanese).

[51] Kamoshita, M., Kosaka, K., Asami, M., and Matsuoka, Y. (2009) Analytical method for perchlorate in water by liquid chromatography–mass spectrometry using an ion exchange column. *Analytical Sciences*, **25**, 453.

[52] Krol, J. (2011) *The Determination of Perchlorate in Water Using LC/MS/MS (Application Note 720000941EN)*, Waters Corp, Milford, MA.

[53] Meng, L., Wu, S., Ma, F. *et al.* (2010) Trace determination of nine haloacetic acids in drinking water by liquid chromatography–electrospray tandem mass spectrometry. *Journal of Chromatography A*, **1217**, 4873.

[54] Snyder, A.S., Vanderford, B.J., and Rexing, D.J. (2005) Trace analysis of bromate, chlorate, iodate, and perchlorate in natural and bottled waters. *Environmental Science and Technology*, **39**, 4586.

[55] Li, Y. and George, E.D. (2006) Reversed-phase liquid chromatography/electrospray ionization tandem mass spectrometry for analysis of perchlorate in water. *Journal of Chromatography A*, **1133**, 215.

[56] Di Rienzo, R.P., Lin, K., McKay, T.T., and Wade, R.W. (2005) Analysis of perchlorate in difficult matrices by LC/MS. *Federal Facilities Environmental Journal*, **15**, 27.

[57] Alsohaimi, I.H., Alothman, Z.A., Khan, M.R. *et al.* (2012) Determination of bromate in drinking water by ultraperformance liquid chromatography–tandem mass spectrometry. *Journal of Separation Science*, **35**, 2538.

[58] Duan, J., Li, W., Si, J., and Mulcahy, D. (2011) Rapid determination of nine haloacetic acids in water using ultra-performance liquid chromatography–tandem mass spectrometry in multiple reactions monitoring mode. *Analytical Methods*, **3**, 1667.

[59] Kosaka, K., Asami, M., Takei, K., and Akiba, M. (2011) Analysis of bromate in drinking water using liquid chromatography–tandem mass spectrometry without sample pretreatment. *Analytical Sciences*, **27**, 1091.

[60] Thermo Fisher Scientific (2013) *Quantification of Haloacetic Acids in Tap Water Using a Dedicated HAA LC Column with LC–MS/MS Detection* (Application Note 590).

[61] Jandera, P. (2011) Stationary and mobile phases in hydrophilic interaction chromatography: a review. *Analytica Chimica Acta*, **692**, 1.

[62] Arias, F., Li, L., Huggins, T.G. *et al.* (2010) Trace analysis of bromate in potato snacks using high-performance liquid chromatography–tandem mass spectrometry. *Journal of Agricultural and Food Chemistry*, **58**, 8134.

[63] Chen, L., Chen, H., Shen, M. *et al.* (2010) Analysis of perchlorate in milk powder and milk by hydrophilic interaction chromatography combined with tandem mass spectrometry. *Journal of Agricultural and Food Chemistry*, **58**, 3736.

[64] Chen, C.Y., Chang, S.N., and Wang, G.S. (2009) Determination of ten haloacetic acids in drinking water using high-performance and ultra-performance liquid chromatography-tandem mass spectrometry. *Journal of Chromatographic Science*, **47**, 67.

[65] Michalski, R. and Łyko, A. (2013) Bromate determination: state of the art. *Critical Reviews in Analytical Chemistry*, **43**, 100.

[66] Divjak, B., Novič, M., and Goessler, W. (1999) Determination of bromide, bromate, and other anions with ion chromatography and an inductively coupled plasma mass spectrometer as element-specific detector. *Journal of Chromatography A*, **862**, 39.

[67] Liu, Y., Mou, S., and Chen, D. (2004) Determination of trace-level haloacetic acids in drinking water by ion chromatography–inductively coupled plasma mass spectrometry. *Journal of Chromatography A*, **1039**, 89.

[68] Pantsar-Kallio, M. and Manninen, P.K.G. (1998) Speciation of halogenides and oxyhalogens by ion chromatography–inductively coupled plasma mass spectrometry. *Analytica Chimica Acta*, **360**, 161.

[69] Shi, H. and Adams, C. (2009) Rapid IC–ICP/MS method for simultaneous analysis of iodoacetic acids, bromoacetic acids, bromate, and other related halogenated compounds in water. *Talanta*, **79**, 523.

[70] Asami, M., Kosaka, K., Yoshida, N. *et al.* (2008) Occurrence of chlorate and perchlorate in water environment, drinking water, and hypochlorite solution. *Journal of Japan Water Works Association*, **883**, 7 (in Japanese).

[71] Kosaka, K., Asami, M., Matsuoka, Y. *et al.* (2007) Occurrence of perchlorate in drinking water sources of metropolitan areas in Japan. *Water Research*, **41**, 3474.

[72] Asami, M., Kosaka, K., and Yoshida, N. (2009) Occurrence of chlorate and perchlorate in bottled beverages in Japan. *Journal of Health Science*, **55**, 549.

[73] Kosaka, K., Kuroda, K., Murakami, M. *et al.* (2013) Occurrence of chlorate and perchlorate in groundwater in Tokyo. *Journal of Japan Society of Civil Engineers, Ser. G (Environmental Research)*, **69**, 10 (in Japanese).

[74] Brandhuber, P., Clark, S., and Morley, K. (2009) A review of perchlorate occurrence in public drinking water systems. *Journal of American Water Works Association*, **101**, 63.

[75] El Aribi, H., Le Blanc, Y.J.C., Antonsen, S., and Sakuma, T. (2006) Analysis of perchlorate in foods and beverages by ion chromatography coupled with tandem mass spectrometry (IC–ESI-MS/MS). *Analytica Chimica Acta*, **567**, 39.

[76] Stanford, B.D., Pisarenko, A.N., Snyder, S.A., and Gordon, G. (2011) Perchlorate, bromate, and chlorate in hypochlorite solutions: guidelines for utilities. *Journal of American Water Works Association*, **103**, 71.

[77] Weinberg, H.S., Krasner, S.W., Richardson, S.D., and Thruston, A.D. Jr. (2002) *The Occurrence of Disinfection By-Products (DBPs) of Health Concern in Drinking Water: Results of a Nationwide DBP Occurrence Study (EPA/600/R-02/068)*, US EPA, Athens, GA.

[78] Inspectorate, D.D.W. (2011) *Evaluation of Haloacetic Acids Concentrations in Treated Drinking Water (DWI 70/2/253)*, Cranfield University, Cranfield, UK.

[79] MHLW (2014) *Results of External Quality Control Survey for Drinking Water Quality Inspection in Fiscal Year 2013* (in Japanese).

6

ANALYSIS OF VARIOUS ANIONIC METABOLITES IN PLANT AND ANIMAL MATERIAL BY IC-MS

ADAM KONRAD JAGIELSKI AND MICHAL USAREK

Department of Metabolic Regulation, Faculty of Biology, Institute of Biochemistry, University of Warsaw, Miecznikowa 1, 02-096 Warsaw, Poland

6.1 INTRODUCTION

A vast number of key metabolites characterizing biological material are either polar or ionic. All of the metabolites of *glycolysis* and *citrate cycle* forming the backbone of cell metabolism are present in the cell as anions. They can be much better resolved *en masse* in their native form by ion chromatography than by any other method available without the need of any prior derivatization.

Due to the large number of anionic species present in biological material and low concentrations of many intermediary metabolites, mass spectrometry is the preferred method of detection in such analyses, because it is both highly sensitive and resolves anions differing in their mass, allowing separate detection of coeluting anions.

The coupling of mass spectrometer to ion chromatography system is, however, not a straightforward task, and the use of such a detector in quantitative analysis has to address some issues before a successful analysis can be performed.

Ion-exchange separations employ high salt concentrations, which are incompatible with the present-day mass spectrometers, interfering with analyzed anions ionization and causing a quick deterioration of the instrument's sensitivity due to the accumulation of salts on its surfaces. In most systems, the eluent from the ion-exchange column

Application of IC-MS and IC-ICP-MS in Environmental Research, First Edition.
Edited by Rajmund Michalski.
© 2016 John Wiley & Sons, Inc. Published 2016 by John Wiley & Sons, Inc.

flows through the eluent suppressor before entering the detector, which enables eradication of commonly used sodium or potassium cations accompanied by conversion of hydroxide anions into pure water, which is fully compatible with mass spectrometry, in spite of relatively low volatility, requiring higher temperature and nitrogen flow in the source to generate ions or flow splitting. One must also be aware that suppressors may interfere with some compounds (e.g., AMP), eliminating them partially or totally from the eluent. Whenever in doubt, the analyst should always check if this is the case.

However, the most difficult problem is accurate quantitation of compounds on an MS instrument, as ionization efficiency of any given compound may vary depending on coeluting ions (sample composition) and instrument operating conditions, such as vacuum and cleanness. Some of the difficulties are addressed later in this chapter. The chromatographic separation may also be problematic if some ions are present in large excess. Even if they seem not to overload the capacity of the column, they may significantly impair the retention times of other anionic compounds. This places some restrictions on sample preparation. Moreover, some compounds may be instable in aqueous KOH gradient.

6.2 OPTIMIZATION OF HPIC AND MS SETTINGS

6.2.1 HPIC Settings

We have previously stated that mass-resolving power of MS instruments abolishes, to some extent, the necessity of obtaining full chromatographic separation of analyzed compounds. This statement should not discourage from obtaining the best resolution possible under the circumstances we are given, analysis time being one of the limiting factors. Single-mass chromatograms, pertaining to only one metabolite of interest, can be extracted from the total ion current (TIC) spectra, and the area under the peak may be used as a measure of anion's abundance. However, coeluting ions can change the ionization efficiency of the compound of interest and thus invalidate previously obtained calibration curves, rendering our quantification inaccurate. Due to a large number of anions present in the sample, unreachable ideal would be to have only one anion entering the mass spectrometer at a time. Finding the balance between resolution and analysis time is sometimes very difficult, more so, because in case of such analyses, it is a good practice to run all separations thrice.

To optimize separation method in our laboratory, high-performance ion chromatography system (HPIC) ICS3000 from Dionex with KOH eluent generator and anion self-regenerating suppressor (ASRS) has been used throughout. Samples were separated on anion-exchange high-capacity AS11HC column in KOH gradient. Conductometric, UV/Vis, and ZQ mass detector from Waters were used. Initially, one of the methods proposed in the column user's manual, that is, 43-min gradient (for details see Figure 6.1a) was employed. Due to unsatisfactory resolution, several modifications of the method have been applied.

Addition of methanol to the mobile phase (10%, v/v) improved the separation of some peaks at the same time adversely affecting the separation of other ones.

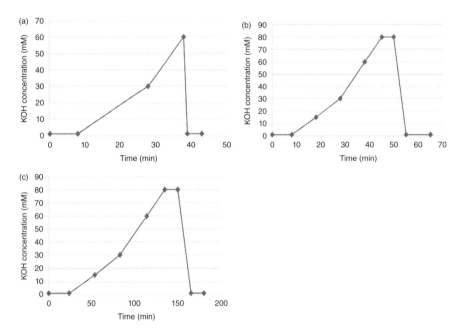

Figure 6.1   Different gradient settings used during optimization of the method: (a) default gradient for the 2 mm AS11HC column (43 min) at flow rate – 0.38 ml min^{-1}; (b) prolonged gradient used in pivotal analysis (65 min) at flow rate – 0.38 ml min^{-1}; (c) gradient adjusted to analyses performed on 2 AS11HC columns to improve resolution and column capacity (180 min) at flow rate – 0.25 ml min^{-1}.

The benefits from introducing methanol were not clear and did not overgrow disadvantages such as increased working pressure. Therefore, methanol has not been used in further analyses. The addition of methanol to the mobile phase may, however, have its merits, if we are interested in increasing the accuracy of quantitation of one of the metabolites that otherwise coelutes.

Another way of improving the separation is to increase the length and capacity of the column. Therefore, two AS11HC columns have been joined together (head–tail) and the gradient has been modified appropriately (for details, see Figure 6.1c). The resulting chromatograms of the standards mixture are presented in Figure 6.2 (chromatogram obtained on conductometric detector), Figure 6.3 (mass chromatogram), and Figure 6.4 (chromatogram from UV/Vis detector at 260 nm). Several advantages of such an approach are listed as follows:

1. The resolution increases substantially; placing compounds of interest and matrix anions further apart, increasing the accuracy of MS based quantitation and the robustness of the method.

2. The quality of signal from low-retention metabolites such as pyruvate or lactate increased, that is, peak shape. As the volumes of samples injected onto the column were quite large (i.e., up to 300 μl), the intensity of signals from

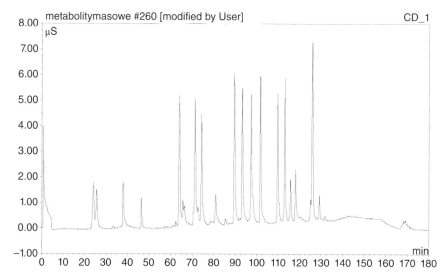

Figure 6.2　Chromatogram obtained on conductometric detector during the separation of mixture of standards utilizing 180 min KOH gradient. The standards used and the order of elution are the same as in Figure 6.3.

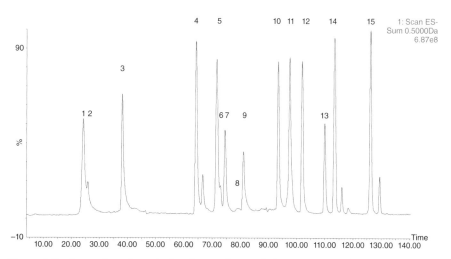

Figure 6.3　Separation of standards mixture. Summed chromatogram of masses of metabolites present in mixture. Peak annotations: 1 – lactate, 2 – acetate, 3 – pyruvate; 4 – succinate and malate; 5 – glucose-6-phosphate and 2-oxoglutarate, 6 – fructose-6-phosphate, 7 – fumarate, 8 – AMP, 9 – oxaloacetate, 10 – 3-phosphoglycerate, 11 – citrate, 12 – phosphoenolpyruvate, 13 – ADP; 14 – fructose-1,6-bisphosphate, 15 – ATP.

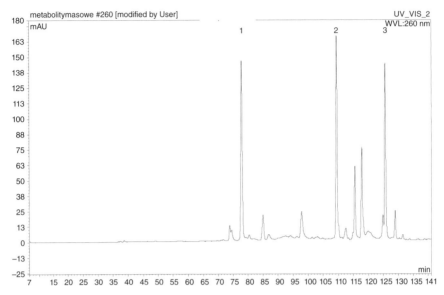

Figure 6.4 Chromatogram obtained at 260 nm during the separation of standards mixture. Peak annotation: 1 – AMP, 2 – ADP, 3 – ATP.

accompanying compounds that had little or no retention on the column was substantial and often interfered with low-retention metabolites of interest when analyzed on one column. Moreover, frequently, the content of quickly eluting metabolites such as lactate and pyruvate is quite high in biological fluids such as blood in comparison to other metabolites of interest. When working with one column only, it may be sometimes prudent to run two analyses with a bigger and smaller injection, when the early eluting compounds elute in wide bands instead of peaks.

However, the method has its drawbacks:

1. It is extremely time-consuming – it took 3 h to perform single analysis.
2. It is less cost-effective, as two columns had to be applied at the same time.

Despite these drawbacks, when analysis time is not a limiting factor, we would highly recommend this method, as in our experience it produces the most accurate results.

However, because this method is very time-consuming, usually we use only one column with the gradient slightly modified compared to the standard, Dionex-developed method (for details, see Figure 6.1b). Chromatogram of standards mixture separation recorded on ZQ mass spectrometer is presented in Figure 6.5.

The method does not allow detection and quantitation of phosphodihydroxyacetone and phosphoglyceraldehyde, as these compounds could not be detected under the

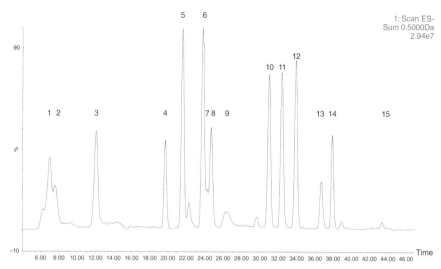

Figure 6.5 Separation of standards mixture. Summed chromatogram of masses of metabolites present in the mixture. Peak annotations: 1 – lactate and β-hydroxybutyrate, 2 – acetate and acetoacetate, 3 – pyruvate, 4 – glucose-1-phosphate, 5 – succinate and malate, 6 – 2-oxoglutarate and glucose-6-phosphate, 7 – fructose-6-phosphate, 8 – fumarate, 9 – AMP and oxaloacetate, 10 – 3-phosphogylcerate, 11 – citrate, 12 – phosphoenolpyruvate, 13 – ADP, 14 – fructose-1,6-bisphosphate, 15 – ATP (small peak due to small ATP concentration in the mixture).

HPIC conditions employed, presumably due to chemical decomposition by OH^- of the phosphodiester bond. This problem has been reported earlier in the literature [1]. The peak appearing on the standards chromatogram after the addition of these compounds is the same as that of inorganic phosphate. Also, MS spectrum is in agreement with the aforementioned supposition. We have also analyzed how these compounds behave after they are directly injected into electrospray source and found that both phosphotrioses produce pseudomolecular ions that can be easily detected. Therefore, we concluded that both compounds are hydrolyzed during separation on the column in KOH gradient. The main products of the reaction – glyceraldehyde and dihydroxyacetone – have no or very little retention on anion-exchange column. Therefore, the only product that can be detected is inorganic phosphate. The quantitation of phosphotrioses in the sample can, however, be achieved by employing a different chromatographic method (for details, see Section 6.4).

6.2.2 MS Settings

Waters ZQ mass spectrometer has been used throughout the analysis. In order to optimize the conditions of detection, solutions of various metabolites in MeOH:Acn:H$_2$0 (2:2:1, v/v/v) (the solution used for metabolites extraction – for details, see Section 6.2.4) were injected directly into MS electrospray source. Various

TABLE 6.1   Optimization of MS Parameters

Compound	Optimal Capillary (kV)	Optimal Cone (V)
ATP	3	30
AMP	2	30
ADP	4	30
Citrate	3	5
Succinate	3	15
Lactate	2.5	10
Phosphoenolpyruvate	3	10
Glucose-6-phosphate	3	15
Acetate	3	15
Fumarate	3	10
Oxaloacetate	2	10
Phosphodihydroxyacetone	3	10
Phosphoglyceraldehyde	3	8
Malate	3	15
3-Phospoglycerate	4	15
Fructose-1,6-bisphosphate	3	15
Fructose-6-phosphate	4	10
2-Oxoglutarate	2	6

cone and capillary voltage settings have been applied to obtain optimal conditions for further analyses, taking into consideration, for example, metabolites' stability as well as signal-to-noise ratio. The optimized values of parameters have been summarized in Table 6.1. Other parameters were the same for all analyses: cone gas flow – $50\,dm^3\,h^{-1}$; desolvation gas flow – $400\,dm^3\,h^{-1}$; source temperature – $90\,°C$; desolvation temperature – $110\,°C$; negative electrospray ionization (ESI) mode. Concentration of metabolites was 1 mM, while injection rate was $10\,\mu l\,min^{-1}$.

The pilot analyses showed that the optimal voltage ranges are 5–30 V for cone and 2–4 kV for capillary. Therefore, for pivotal analyses of various metabolites mixtures, the following parameters were applied:

- Cone: 5, 10, 15, 30 V (four different channels have been collected simultaneously)
- Capillary: 2.5 kV
- Source temperature: $90\,°C$
- Desolvation temperature: $110\,°C$
- Cone gas flow: $50\,dm^3\,h^{-1}$
- Desolvation gas flow: $400\,dm^3\,h^{-1}$.

Desolvation gas flow has been changed in HPIC-MS analyses due to the increase in the mobile-phase flow rate (0.25–$0.38\,ml\,min^{-1}$ – for details, see Section 6.2.3) in comparison to the direct injection experiments.

As there is a wide variety of electrospray MS instruments on the market, the values shown in Table 6.1 should be treated as only beginning values for users' own tuning of the method. With the exception of adenosine phosphates, which are better quantitated by means of integrating UV chromatograms (for details, see Figure 6.4), most of the metabolites are prone to cone-voltage-induced fragmentation, and analysis requires low cone voltage settings in the 10–15 V range. The tuning is best performed by directing a flow from IC instrument to the electrospray source and adding the metabolite just before the source by connecting the infusion pump supplying the compound solution by T-coupling into the eluent.

6.2.3   HPIC-MS Settings

The pivotal HPIC-MS analyses have been performed applying HPIC system (1 column, 65-min gradient – for details see Figure 6.1b) coupled with ESI-MS. The parameters of mass spectrometer were similar to those described in Section 6.2.2 except for substantially higher flow of desolvation nitrogen (i.e., 700 dm^3 h^{-1}). As the outlet from the suppressor has been introduced directly into the MS instrument, the external suppression mode has been used throughout the analyses (the ASRS suppressor has been continuously regenerated with deionized water supplied by an external pump).

Direct quantitative analysis applying mass spectrometer is problematic due to the run-to-run differences in ionization efficiency. Those differences can be caused by numerous factors such as fouling of the instrument, coeluting compounds present in the matrix, other conditions pertaining to the instrument operation such as vacuum. Therefore, preparation of appropriate standard curves does not ensure satisfactory quantification of the analytes.

To deal with the problem, internal standards are widely used in MS analyses. The compounds used as internal standards should not be normally present in the studied material. The most accurate method is the application of standards of the same chemical structure as the compounds of interest, but containing stable isotopes such as ^{13}C or ^{2}D. The main advantage of such an approach is that due to the minimal difference in composition, the rate and extent of ionization of both compounds (compound of interest and modified standard) are very similar. Therefore, the signal ratios of labeled and unlabeled compounds are the same regardless of the ionization conditions or efficiency. Methods applying quantification in reference to the compounds containing stable isotopes have been developed and described in the literature. Canelas et al. [2] performed ion-exchange LC-MS-MS analyses of metabolites from *Saccharomyces cerevisiae* in reference to ^{13}C-labeled internal standards. Without doubt, such an approach provides the most robust and accurate method of quantitation when dealing with an MS instrument, but it is also clear that it is the most cumbersome and expensive one, as it relies on availability, purchase, and the addition of many labeled internal standards to the sample. Also, the evaluation of such an analysis is more time-consuming as the number of mass traces to integrate doubles.

On the other hand, some researches claim that it is possible to perform MS analyses without internal standards. Ni et al. [3] demonstrated a microscale method for metabolomic analysis of samples obtained from Langerhans islets using capillary

LC-ESI-MS in negative mode. The reproducibility of the method was sufficient to detect large, relative changes in metabolite concentrations associated with fuel changes without internal standards. Similarly, Wang *et al.* [4] presented a new method for quantifying proteomic and metabolomic profile data by LC-MS with ESI in human plasma. According to the authors, the approach is applicable to any fluid or tissue. The approach relies on linearity of signal versus molecular concentration and reproducibility of sample processing. It should also be noted that, in some special cases, analyst may only be interested in finding changes in ratios between different sets of metabolites, without their accurate quantitation. In such cases, it is possible to do so without internal standards.

While analyzing various metabolites in complex samples, such as plant or animal tissue extracts, it is often too challenging to provide labeled standards for all of them. Therefore, other approaches have been applied, such as the use of only a few of different internal standards. Schaub *et al.* [5] accounted for the differences in ionization efficiency by using naphthyl phosphate and carboxyphenyl phosphate as internal standards while analyzing extracts obtained from *Escherichia coli* with HPIC-MS. Similar applications have been prepared by Waters for quantitative analysis of proteins [6]. In this method, a known amount of trypsin-digested protein (e.g., yeast alcohol dehydrogenase) is added to the mixture of trypsin-digested proteins. The signals from the proteins of interest are automatically compared to those from the digested internal standard. It enables quantitative analysis of various proteins using only one internal standard.

In our HPIC-MS analyses, we routinely use two different internal standards – ^{13}C-pyruvate and tartrate. Standard curves for each of the analyzed metabolites are prepared so that the metabolite amount in picomole is plotted versus the ratio of metabolite peak area to the peak area of 14 nmol standard. A set of calibration curves is prepared for each standard employed. The amount of internal standard injected onto the column is the same throughout the standard curve range and in analyzed samples and amounts to 14 nmol (for details, see Figure 6.6). The single mass chromatograms for all metabolites and standards are extracted from the TIC chromatograms obtained during the sample separation on ion-exchange column and integrated. The peak areas pertaining to the metabolites are divided by those of internal standards, and metabolite amounts are inferred from the calibration curves. The results obtained from both sets of calibration curves are averaged. For best results, separation of each sample should be run thrice and results are averaged.

The software automation of the process is possible but complicated by the fact that concentrations of different metabolites differ substantially, shape of early peaks can be compromised by early eluting inorganic salts and peaks of some metabolites, which have the same mass such as hexose-phosphates: glucose-6-phosphate and fructose-6-phosphate are not fully resolved, making it necessary to aid the integration process by handpicking the peak boundaries. The analysis of obtained chromatograms thus requires some effort and time, but the payoff is the large number of metabolites measured in one run.

In metabolites that have more than one hydrogen that can easily dissociate from the molecules, multiply charged pseudomolecular ions can be observed.

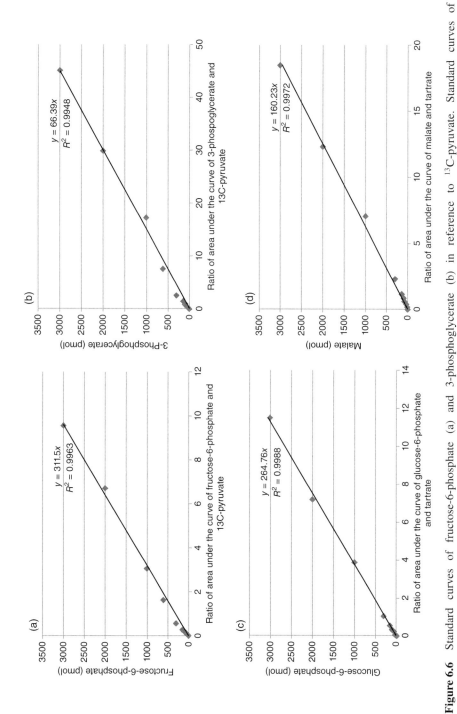

Figure 6.6 Standard curves of fructose-6-phosphate (a) and 3-phosphoglycerate (b) in reference to ^{13}C-pyruvate. Standard curves of glucose-6-phosphate (c) and malate (d) in reference to tartrate. All compounds were detected by ESI-MS. The amount of tartrate and ^{13}C-pyruvate was the same in all samples, while the amount of the analytes of interest increased up to 3000 pmol/injection.

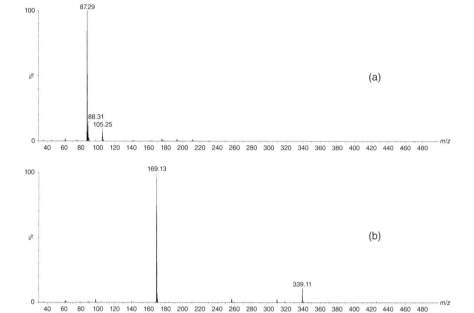

Figure 6.7   Mass spectra of peaks on TIC chromatogram corresponding to (a) pyruvate, signal at $m/z = 87$ from $[M-H]^-$ ion; (b) fructose-1,6-bisphosphate, $m/z = 339$ from $[M-H]^-$ and $m/z = 169$ from $[M-2H]^{2-}$ ion.

3-Phosphoglycerate, citrate, fructose-1,6-bisphosphate, ADP, and ATP are all likely to give rise to both $[M-H]^-$ and $[M-2H]^{2-}$ ions in mass spectrometer's source – compare Figure 6.7. Analysts should be aware of this and examine the relevant spectra, if the ratio of signal strength of differently charged species is constant. In case of doubt, it may be better to prepare calibration curves where peak areas of detected pseudomolecular ions arising from one compound are summed. In our practice, the use of calibration curves related to the ion present in greater abundance usually produced satisfactory results.

The amounts of adenyl nucleotides (AMP, ADP, and ATP) have been determined relying on spectrophotometric detection at 260 nm (UV/Vis variable wavelength detector) (see Figure 6.4). It was due to the fact that nucleotides are present in the cells in substantial amount. Therefore, UV/Vis detector provides adequate sensitivity to detect these compounds. Moreover, such a detector has several advantages over MS:

1. High level of run-to-run reproducibility
2. Linearity within a wide range of nucleotide concentrations
3. No need of the use of suppressor, which may compromise the validity of, for example, AMP determination
4. Better robustness of the method.

6.2.4   Extraction of Metabolites from Cells and Tissues

Appropriate methods of metabolites extraction enable reproducible and efficient analysis of intracellular metabolites.

An optimal method of metabolites extraction has to fulfill several criteria such as the following:

- Analysis-to-analysis reproducibility
- Fast and effective quenching of biochemical processes in the studied material and precipitation of proteins
- Limited number of steps to minimize degradation/loss of metabolites
- Minimization of the amount of accompanying anions (e.g., phosphate and chloride from cell culture media and/or cell/tissue washing buffers as their excess can deteriorate chromatographic separation and detector sensitivity)
- Reasonable efficiency.

In general, it is better to develop a method of limited extraction efficiency and good reproducibility. However, the limiting factor might be the sensitivity of the detectors while analyzing trace metabolites.

The issue of proper extraction of metabolites from cells and tissues has been discussed in the literature. Various authors have compared numerous methods of metabolites extraction from cells, utilizing different extraction media:

- Perchloric acid (PCA)
- Boiling methanol or ethanol
- Formic acid
- Cold acetonitrile, ethanol, or methanol
- Methanol/$CHCl_3$
- Water/acetonitrile
- Water/acetonitrile/methanol
- Cell lysis reagent.

It is advisable to acquaint oneself with the most widely used methods and try some of them out on material in question, before deciding on extraction protocol. At the end of this chapter, we present the method of our choice.

Dietmair *et al.* [7] compared a total of 12 extraction protocols (i.e., acetonitrile, freeze methanol, cold 50% methanol, methanol/chloroform, hot 80% methanol, cold methanol, hot ethanol, hot ethanol in HEPES buffer, cold ethanol, hot water, KOH, PCA). Of the 12 different extraction methods tested, cold extraction in 50% aqueous acetonitrile proved to be superior. The recovery of a mixture of standards was excellent, and the concentration of extracted intracellular metabolites was higher than for the other methods tested.

Ritter *et al.* [8] developed and optimized a multistep method employing methanol, $CHCl_3$, and tricine buffer. Extraction of metabolites was preceded by cell wash with phosphate-buffered saline. Slightly modified set of methods has been discussed by Canelas *et al.* [2], who investigated metabolites in *S. cerevisiae*. One of the methods applied was extraction with ice cold ($-20\,°C$) mixture of methanol:acetonitrile:water (v/v/v). The method has been also been utilized by Bennett *et al.* [9].

In another study, methanol/water/chloroform mixture was used to extract metabolites from Langerhans islets and *E. coli* [10]. Hot buffered ethanol was applied for extraction of metabolites from *S. cerevisiae* [11]. Schaub *et al.* [5] proposed an integrated model of sample transfer, quenching, and extraction of the metabolites from *E. coli* cells. An integrated model (simultaneous quenching and quantitative extraction of intracellular metabolites were realized by short-time exposure of cells to temperatures $\leq 95\,°C$, where intracellular metabolites are released quantitatively) enabled reliable and reproducible measurements. Sellick *et al.* [12] tested various extraction protocols (methanol, methanol/water, hot ethanol, KOH, PCA, methanol/chloroform) for metabolome analysis of Chinese hamster ovary cells. The extraction of metabolites using two extractions with 100% methanol followed by a final water extraction recovered the largest range of metabolites. For the majority of metabolites, extracts generated in this manner exhibited the greatest recovery with high reproducibility.

Various methods have also been applied for extraction of metabolites from plants. Li *et al.* [13] extracted metabolites from tobacco using methanol:water mixture (80:20, v/v) followed by sonication. The method enabled identification of 113 metabolites with high repeatability by applying HPLC-MS/MS.

Extraction of metabolites from adherent cells poses some additional problems. Different approaches of cell detachment have been applied prior to extraction. Some authors added the solvent onto the cell layer, thus performing cell lysis and extraction directly on the plate [9, 14, 15]. However, direct quenching and extraction raise the problem of cell count determination, a parameter required for the normalization of data generated in cell culture systems. Therefore, Dettmer *et al.* [16] compared trypsin/ethylenediaminetetraacetic acid (EDTA) treatment of cell layer followed by scraping cells in buffer prior to metabolite extraction versus direct scraping with a solvent. Seven different methods of extraction were tested. Based on overall performance, methanol/water (80%/20%) was chosen as a suitable extraction solvent. The authors proved that gentle scraping of the cells in a buffer solution and subsequent extraction with methanol/water resulted in on average a sevenfold lower recovery of quantified metabolites compared with direct scraping using methanol/water, making the latter one the method of choice to harvest and extract metabolites from adherently growing mammalian SW480 cells.

In our laboratory, we started out by extracting metabolites from biological material with the PCA. Briefly, the cells were harvested with 3.5% PCA, precipitated proteins were centrifuged, while supernatant containing metabolites of interest was neutralized with 3 M K_2CO_3. The neutralization step resulted in the formation of insoluble potassium perchlorate, which was separated in additional centrifugation step, and the supernatant was used for further analyses. This method has been widely used for extraction of nucleotides before analysis with HPLC (e.g., [17, 18]). Moreover, it has been often applied prior to enzymatic quantification of intracellular metabolites [19]. The major advantages of using PCA for metabolites extraction are the following:

rapid quenching of biochemical processes, easy removal of extracting reagent after neutralization, and limited number of steps.

However, the use of acidic solutions for metabolite extraction requires accurate neutralization and, thereby, leads to high salt concentrations, hampering subsequent chromatographic analysis [5]. Moreover, some metabolites may not be stable in a strongly acidic medium. We tried to employ this method while preparing the samples for IC-MS analyses from different biological samples. The major problem we have encountered was the excess of bicarbonate produced during sample neutralization. As bicarbonate anions bind to AS11HC column resin, their excess deteriorates the quality of the chromatograms. The problem can be partially solved by postcolumn application of in-line self-regenerating bicarbonate trap to reduce the amount of bicarbonate ions that reach the conductometric detector. However, as the quality of results was still unsatisfactory, we tested other methods of metabolites extraction.

The aforementioned methanol/water extraction method, slightly modified by the replacement of the standard washing buffer (i.e., Phosphate-buffered saline) with 300 mM mannitol to decrease the concentration of inorganic anions added to the sample, provided promising results as far as extraction and subsequent analysis were concerned, but as a method applying direct cell scraping raised the issue of proper normalization. Because use of such a method practically eliminates the feasibility of cell counting, it is necessary to utilize different means of standardization, that is, to relate metabolite amounts to the intracellular protein content. However, the Bradford method of protein assay [20] is compatible with only low concentrations of organic solvents (i.e., up to 10%). Moreover, 80% methanol caused incomplete protein precipitation; therefore, it was not feasible to determine the protein content in either the solution or the pellet. Thus, we modified the method by adding four volumes of acetonitrile (the final composition of extraction solution was ice-cold methanol:acetonitrile:water 4:4:2, v/v/v). Precipitated proteins were resuspended in 100 mM NaOH + 0.125% Triton X-100 (the concentrations compatible with the Bradford method) and quantified.

6.3 APPLICATION OF THE METHOD IN ANALYSIS OF METABOLITES IN PLANT AND ANIMAL MATERIAL

We applied the optimized method for analyses of various metabolites extracted from plant and animal material. The scope of the analyzed metabolites was as follows:

- Glycolysis and gluconeogenesis
- Krebs cycle
- Energy state of the cells (AMP, ADP, and ATP)
- Redox state of the cells (e.g., 3-hydroxybutyrate/acetoacetate and lactate/ pyruvate ratios).

In this part of the chapter, we present example chromatograms obtained from samples prepared from various material of animal cells, solid tissue, and plant leaf – together with a short description of the relevant extraction procedures.

6.3.1   Analysis of Metabolites from Cell Cultures (Primary Cultures as well as Established Cell Lines)

In order to measure intracellular levels of metabolites in adherent cells with the help of ion chromatography–mass spectrometry, the cells were washed twice with $0.3 \, \mathrm{mol \, l^{-1}}$ mannitol and extracted with ice-cold mixture containing methanol:acetonitrile:water (2:2:1 v/v/v). The obtained extracts were centrifuged (10,000g for 10 min). The supernatants containing metabolites were stored at $-80\,^{\circ}$C until further analysis. To measure the cell protein for normalization, the cell residues were resuspended in $0.1 \, \mathrm{mol \, l^{-1}}$ NaOH containing 0.125% Triton X-100 [21]. The results of the example analysis of so obtained material are summarized in Figures 6.8–6.10.

6.3.2   Analysis of Metabolites from Solid Tissues

The critical step of metabolite extraction from solid tissues is a proper homogenization of the biological material. Briefly, the piece of tissue (approximately 50–250 mg) was frozen with liquid nitrogen to stop all metabolic processes and mechanically disrupted in mortar cooled with liquid nitrogen. Then, the sample was homogenized in Potter's homogenizer applying extraction solution (methanol/acetonitrile/water 2/2/1 v/v/v) at $4\,^{\circ}$C (the amount of solution was approximately $5 \, \mathrm{\mu l \, mg^{-1}}$ of tissue). After that, the homogenate was sonicated for 10 min. Following centrifugation, the supernatants were stored at $-80\,^{\circ}$C until analysis.

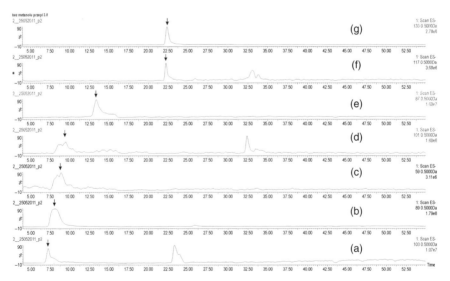

Figure 6.8   Analysis of metabolites extracted from primary cultured rabbit kidney cortex tubules. Selected mass chromatograms for metabolites of interest (metabolite content in picomole per milligram of protein $\pm$ SD in parenthesis) (a) β-hydroxybutyrate (703 $\pm$ 173), (b) lactate (9447 $\pm$ 1721), (c) acetate (not assayed quantitatively at the time of analysis), (d) acetoacetate (217 $\pm$ 56), (e) pyruvate (287 $\pm$ 3), (f) succinate (102 $\pm$ 31), (g) malate (4278 $\pm$ 322).

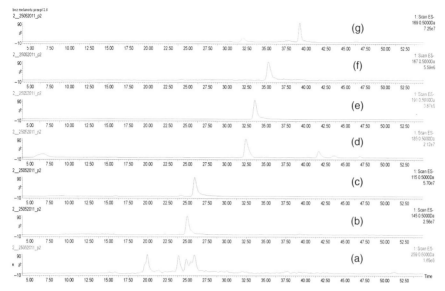

Figure 6.9 Analysis of metabolites extracted from primary cultured rabbit kidney cortex tubules. Selected mass chromatograms for metabolites of interest (metabolite content in picomole per milligram of protein ± SD in parenthesis): (a) hexose phosphates [fructose-6-phosphate (20 ± 3); glucose-6-phosphate (40 ± 31)], (b) 2-oxoglutarate (247 ± 23), (c) fumarate (684 ± 78), (d) 3-phosphoglycerate (233 ± 45), (e) citrate (6245 ± 978), (f) phosphoenolpyruvate (42 ± 7), (g) fructose-1,6-bisphosphate (289 ± 32).

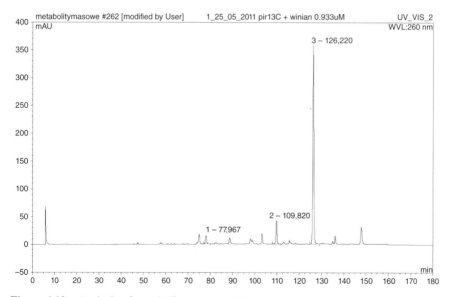

Figure 6.10 Analysis of metabolites extracted from primary cultured rabbit kidney cortex tubules. UV chromatogram at 260 nm. Selected peaks for metabolites of interest (metabolite content in picomole per milligram of protein ± SD in parenthesis): 1 – AMP (890 ± 15), 2 – ADP (2790 ± 310), and 3 – ATP (23,630 ± 1700).

In Figure 6.11 we present the MS chromatogram obtained during analysis of metabolites from rat liver sample, while in Figure 6.12, analysis of rat brain tissue.

6.3.3 Extraction of Metabolites from Plants

The extraction procedure used for leaf extraction was modified to eliminate chlorophyll from the analyzed sample.

Frozen in liquid nitrogen, pea leaves (50 mg) were powderized in mortar prior to their cold extraction in 1.5 ml mixture of chloroform:methanol:water (2:1:0.8 v/v/v). The solid residue was discarded after centrifugation (5 min, 100,000g). To 1 ml of the extract, 0.5 ml water was added and everything was mixed and centrifuged again to separate phases. The top water–methanol phase was used for polar metabolites analysis. The results are presented in Figures 6.13 and 6.14.

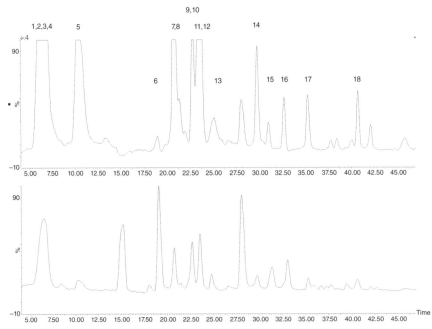

Figure 6.11 Analysis of rat liver tissue metabolites extract. Bottom chromatogram – total ion current. On top – summed chromatogram of masses of chosen metabolites. Peak annotations (metabolite content in nanomole per milligram frozen tissue $\pm$ SD in parenthesis): 1 – lactate (1.148 $\pm$ 0.191), 2 – β-hydroxybutyrate, 3 – acetate (0.169 $\pm$ 0.049), 4 – acetoacetate, 5 – pyruvate (0.219 $\pm$ 0.038), 6 – glucose-1-phosphate, 7 – succinate (0.576 $\pm$ 0.186), 8 – malate (1.910 $\pm$ 0.333), 9 – 2-oxoglutarate (0.003 $\pm$ 0.001), 10 – glucose-6-phosphate (0.110 $\pm$ 0.022), 11 – fructose-6-phosphate (0.105 $\pm$ 0.017), 12 – fumarate (0.507 $\pm$ 0.116), 13 – AMP, 14 – 3-phosphogylcerate (0.044 $\pm$ 0.006), 15 – citrate (0.010 $\pm$ 0.004), 16 – phosphoenolpyruvate (0.012 $\pm$ 0.003), 17 – ADP, 18 – ATP. The amounts of some annotated metabolites, for example, acetoacetate, were quantifiable but not calculated at the time.

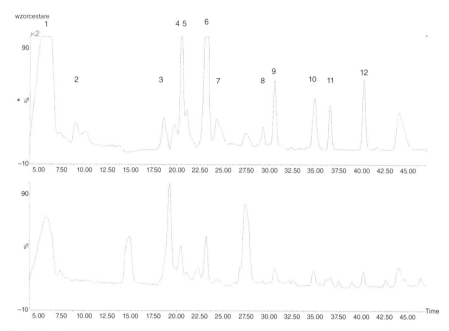

Figure 6.12 Analysis of rat brain tissue metabolites extract. Bottom chromatogram – total ion current. On top – summed chromatogram of masses of chosen metabolites. 1 – lactate, 2 – pyruvate, 3 – glucose-1-phosphate, 4 – succinate, 5 – malate, 6 – fumarate, 7 – AMP, 8 – 3-phosphoglycerate, 9 – citrate, 10 – ADP, 11 – fructose-1,6-bisphosphate, 12 – ATP. On summed masses chromatogram glucose-6-phosphate, fructose-6-phosphate, and 2-oxoglutarate are not indicated as their elution peaks did not rise above the baseline. They were, however, identifiable and quantifiable in single-mass chromatograms (not shown).

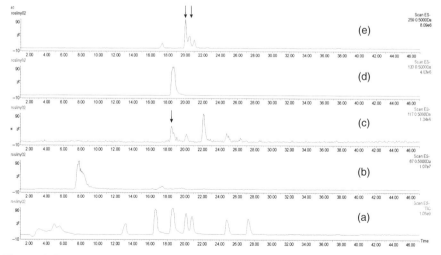

Figure 6.13 IC-MS analysis of metabolite extract from pea leaves. (a) Total ion current, (b) selected-mass chromatogram for *pyruvate*, (c) selected-mass chromatogram for *succinate*, arrow points to relevant peak, (d) selected-mass chromatogram for *malate*, (e) selected-mass chromatogram for *hexose phosphates* (arrows point to *glucose* and *fructose-6-phosphates*).

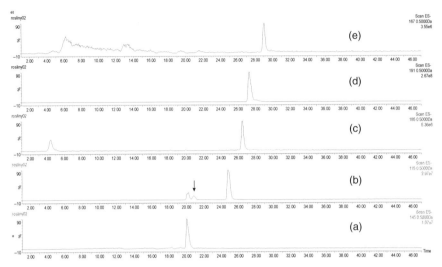

Figure 6.14 IC-MS analysis of metabolite extract from pea leaves. Selected-mass chromatograms for (a) 2-oxoglutarate, (b) fumarate, arrow points to relevant peak, (c) 3-phosphoglycerate, (d) citrate, (e) phosphoenolpyruvate.

6.4 CONCLUSIONS

The proposed IC-MS method enables reliable identification and quantification of most of glycolysis and citrate cycle intermediates. The method has been developed by applying protocols described in the literature, which have been appropriately modified and adjusted due to some issues and problems that occurred while analyzing the metabolites of interest. While applying the method in different laboratories, the specification of the ion chromatograph as well as mass spectrometer has to be taken into consideration. Therefore, instruments' settings optimized in our laboratory should be critically evaluated and adjusted by an analyst. Moreover, the method is not universal for all intracellular anionic metabolites. While applying the method, various conditions have to be considered, for example, the stability of the compound during the extraction process or analysis. In case of some metabolites, for example, phosphotrioses, which turned out unstable in basic conditions, complementary methods should be taken into account (e.g., applying normal-phase column and acetonitrile gradient instead of IC). Moreover, the method is not applicable to glucose, that is, the final product of gluconeogenesis. In this case, electrochemical detection may be taken into consideration.

To sum up, due to the large number of metabolites involved in pathways, it is often necessary to use various complementary analytical methods. The instability of certain intracellular metabolites makes it desirable to configure the equipment to perform the analyses simultaneously, applying different analytical protocols at the same time, if feasible. Complex analysis of the whole metabolic routes may help to resolve many questions concerning mechanisms of the regulation of metabolism and influence of both extrinsic and intrinsic factors thereon in cells, tissues, and organisms.

6.5   REFERENCES

[1] Bhattacharya, M., Fuhrman, L., Ingram, A. *et al.* (1995) Single-run separation and detection of multiple metabolic intermediates by anion-exchange high-performance liquid chromatography and application to cell pool extracts prepared from *Escherichia coli. Analytical Biochemistry*, **20**, 98.

[2] Canelas, A.B., ten Pierick, A., Ras, C. *et al.* (2009) Quantitative evaluation of intracellular metabolite extraction techniques for yeast metabolomics. *Analytical Chemistry*, **81**, 7379.

[3] Ni, Q., Reid, K.R., Burant, C.F., and Kennedy, R.T. (2008) Capillary LC–MS for high sensitivity metabolomic analysis of single islets of Langerhans. *Analytical Chemistry*, **80**, 3539.

[4] Wang, W., Zhou, H., Lin, H. *et al.* (2003) Quantification of proteins and metabolites by mass spectrometry without isotopic labeling or spiked standards. *Analytical Chemistry*, **75**, 4818.

[5] Schaub, J., Schiesling, C., Reuss, M., and Dauner, M. (2006) Integrated sampling procedure for metabolome analysis. *Biotechnology Progress*, **22**, 1434.

[6] Waters Manual. (2010) Application Solutions for High Definition Proteomics.

[7] Dietmair, S., Timmins, N.E., Gray, P.P. *et al.* (2010) Towards quantitative metabolomics of mammalian cells: development of a metabolite extraction protocol. *Analytical Biochemistry*, **15**, 155.

[8] Ritter, J.B., Genzel, Y., and Reichl, U. (2008) Simultaneous extraction of several metabolites of energy metabolism and related substances in mammalian cells: optimization using experimental design. *Analytical Biochemistry*, **373**, 349.

[9] Bennett, B.D., Yuan, J., Kimball, E.H., and Rabinowitz, J.D. (2008) Absolute quantitation of intracellular metabolite concentrations by an isotope ratio-based approach. *Nature Protocols*, **3**, 1299.

[10] Edwards, J.L., Edwards, R.L., Reid, K.R., and Kennedy, R.T. (2007) Effect of decreasing column inner diameter and use of off-line two-dimensional chromatography on metabolite detection in complex mixtures. *Journal of Chromatography A*, **1172**, 127.

[11] Groussac, E., Ortiz, M., and François, J. (2000) Improved protocols for quantitative determination of metabolites from biological samples using high performance ionic-exchange chromatography with conductimetric and pulsed amperometric detection. *Enzyme and Microbial Technology*, **26**, 715.

[12] Sellick, C.A., Knight, D., Croxford, A.S. *et al.* (2010) Evaluation of extraction processes for intracellular metabolite profiling of mammalian cells: matching extraction approaches to cell type and metabolite targets. *Metabolomics*, **6**, 427.

[13] Li, L., Zhao, C., Chang, Y. *et al.* (2014) Metabolomics study of cured tobacco using liquid chromatography with mass spectrometry: method development and its application in investigating the chemical differences of tobacco from three growing regions. *Journal of Separation Science*, **37**, 1067.

[14] Hofmann, U., Maier, K., Niebel, A. *et al.* (2008) Identification of metabolic fluxes in hepatic cells from transient 13C-labeling experiments: Part I. Experimental observations. *Biotechnology and Bioengineering*, **100**, 344.

[15] Teng, Q., Huang, W.L., Collette, T.W. *et al.* (2009) A direct cell quenching method for cell-culture based metabolomics. *Metabolomics*, **5**, 199.

[16] Dettmer, K., Nürnberger, N., Kaspar, H. *et al.* (2011) Metabolite extraction from adherently growing mammalian cells for metabolomics studies: optimization of harvesting and extraction protocols. *Analytical and Bioanalytical Chemistry*, **339**, 1127.

[17] Riss, T.L., Zorich, N.L., Williams, M.D., and Richardson, A. (1980) Comparison of the efficiency of nucleotide extraction by several procedures and the analysis of nucleotides from extracts of liver and isolated hepatocytes by HPLC. *Journal of Liquid Chromatography*, **3**, 133.

[18] Wynants, J. and van Belle, H. (1985) Single-run high-performance liquid chromatography of nucleotides, nucleosides, and major purine bases and its application to different tissue extracts. *Analytical Biochemistry*, **144**, 258.

[19] Bergmeyer, H.U. (1983) *Methods in Enzymatic Analysis*, Verlag Chemie GmbH, Weinheim, Basel.

[20] Bradford, M.M. (1976) A rapid and sensitive method for the quantitation of microgram quantities of protein utilizing the principle of protein-dye binding. *Analytical Biochemistry*, **72**, 248.

[21] Usarek, M., Jagielski, A.K., Krempa, P. *et al.* (2014) Proinsulin C-peptide potentiates the inhibitory action of insulin on glucose synthesis in primary cultured rabbit kidney-cortex tubules: metabolic studies. *Biochemical Cell Biology*, **92**, 1.

7

ANALYSIS OF PERCHLORATE ION IN VARIOUS MATRICES USING ION CHROMATOGRAPHY HYPHENATED WITH MASS SPECTROMETRY

JAY GANDHI

Metrohm USA, 4738 Ten Sleep Lane, Friendswood TX 77546, USA

7.1   INTRODUCTION

In 2003, Chemical and Engineering news (C&EN) [1] reported a cover story that Colorado River and Lake Mead are contaminated with perchlorate and extensive cleanup is required. This became highlight at United States Environment Protection Agency (USEPA). Perchlorate is an oxidant used primarily in solid fuel propellants for rockets, missiles, and pyrotechnics. Perchlorate water has been found across the southwestern United States. Some sources have been traced to the defense industry or to manufacturers that supply the defense industry. Perchlorate is a known thyroid hormone inhibitor for iodine uptake. Ion chromatography (IC) with conductivity detection can be used to measure perchlorate levels in drinking and wastewaters (as per United States Environmental Protection Agency (EPA) Method 314.0 enhanced [2]). The method is reliable to approximately $1–5\,\mu g\,l^{-1}$ in drinking water, but sensitivity decreases dramatically as the complexity of the matrix is increased (such as in surface and wastewaters). Both false-positive and false-negative results may occur due to matrix effects and coeluting substances detected by nonspecific conductivity

Application of IC-MS and IC-ICP-MS in Environmental Research, First Edition.
Edited by Rajmund Michalski.
© 2016 John Wiley & Sons, Inc. Published 2016 by John Wiley & Sons, Inc.

detection. Lower detection limits (DLs) for perchlorate are needed, since the EPA and state environmental agencies are seeking to target levels in the $1–2\,\mu g\,l^{-1}$ range. Reliability of the measurement in heavy matrix samples is also important. The use of a mass spectrometer as a detector for perchlorate at much lower DLs ($50–100\,ng\,l^{-1}$) has shown promise, but reliability issues and problems related to suppression of the electrospray ionization (ESI – the production of ions by evaporation of charged droplets obtained through spraying and electrical field) signals in typical matrices are well known. The key to reducing suppression is to ensure that analyte and high concentrations of matrix are well separated and do not enter the ion source and interface at the same time.

In addition to ion suppression in the source, the m/z attributed to perchlorate anion (99 and 101) have isobaric interferences that can be attributed to minor sulfate isotopes and organic materials that can be present and bleed from the suppressor solutions used for IC and the associated cation suppressor. The proper selection of separation column and suppressor is critical to reduce sample bleed and to provide efficient separation of high levels of interfering ions, particularly sulfate.

7.2   PRECAUTIONS UNIQUE TO ION CHROMATOGRAPHY–MASS SPECTROMETRY

At USEPA, early attempts were made to use ion chromatography instruments using volatile buffers (such as amine solution in the range of pH 12) along with anion-exchanger analytical column and no suppressor (nonsuppressed IC) hyphenated with single-quad mass spectrometer [3]. These attempts were not suitable as amine solution is well known for making instrumentation hardware "dirty" and it takes long time to clean and maintain cleanliness. Hence, traditional ion chromatography with nonvolatile buffer systems with suppressor device is usually used when performing ESI or any atmospheric pressure ionization (API) technique. Some IC mobile-phase reagents (such as strong inorganic acids) are not suited for direct introduction into API sources. The operator must be certain to avoid mobile phases that are not compatible with the stainless steel parts of the mass spectrometer. To avoid inorganic salt buildup, it is essential that a suppressor, unique to the IC technique, be employed. The suppressor removes cations from the eluent stream, after the separation column, and replaces them with a proton. In the API source, accumulation of salts from the mobile phase and any dissolved solids in the sample are eliminated. During system equilibration, prior to adding the suppressor to the flow path, it is important that the eluent from the IC be diverted by the integral valve of the mass-selective detector and not directed to the ESI source. This eliminates the possibility of any sodium hydroxide, sodium carbonate, or other mobile-phase constituent from entering the source while the suppressor system is equilibrating or otherwise off-line. In the event that a contaminating solution is introduced to the source, the mass spectrometer (MS) system should be vented, and surfaces up to and including the glass capillary should be cleaned. This will recover the performance lost due to nonvolatile eluent introduction in the MS system.

7.2.1 Instrumental and Operating Parameters

Major principle of using mass spectrometry is absolute confirmation of perchlorate ion.

Perchlorate molecule is made up of one chlorine (^{35}Cl) and four oxygen (^{16}O), which gives molecular mass to charge (m/z) of 99. Also, naturally occurring chlorine atom has presence of 25% isotopic abundance of mass 37 (^{37}Cl), which gives isotopic abundance of m/z 101 at 25%. When mass spectrometer is used for analysis, these isotopic abundance ratios can be easily monitored at 3:1 ratio for each perchlorate ion with no interference. However, as per USEPA method [4], matrix containing 1000 parts per million (ppm) of sulfate must be tested; hence, 4% sulfur isotope (^{34}S) mass of 34 yields m/z of 99, which heavily interferes with perchlorate ion if not chromatographically resolved and ultimately skews the monitoring of perchlorate isotopes. In order to minimize ionic suppression and maintain good quality of ion formation, it is strongly recommended to use oxygen-18 (^{18}O)-enriched perchlorate as an internal standard, which will yield m/z of 107 ion. Commercially available internal standard is added in 5 ppb concentration.

The analytical system consists of a Metrohm AG ion chromatograph and 6100 MSD Single-Quad (Agilent Technologies) Figure 7.1. A standard electrospray interface was used. Figure 7.2 demonstrates general operational principle of mass spectrometer detector (MSD). The two systems were synchronized by use of contact closure between the chromatographic IC system and the mass spectrometer. A complete list of instrumental parameters is listed in Tables 7.1 and 7.2. Figures 7.1 and 7.2 are created using IC option 3 and MSD option 1.

Uniqueness about MSD specific for perchlorate ion is that when fragmentor voltage is increased to 210–220 V (Table 7.2, option 2), one more oxygen is lost from

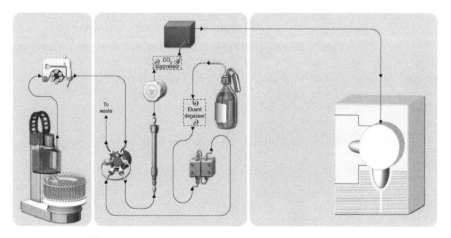

Figure 7.1 Typical ICMS setup using single-quad MSD.

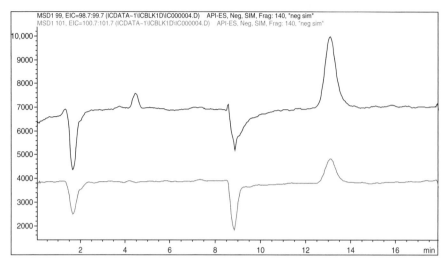

Figure 7.2 Trace of 500 ng l^{-1} perchlorate in reagent water.

TABLE 7.1   Ion Chromatography Parameters

	Option 1	Option 2	Option 3
Ion Chromatograph Model	Metrohm Professional 850 IC with MSMHC/MCS	Metrohm Professional 850 IC with MSMHC/MCS	Metrohm Professional 850 IC with MSMHC/MCS
Column	Metrosep ASUPP5-100 (100 mm × 4.6 mm ID)	Metrosep ASUPP5-250 (250 mm × 4.6 mm ID)	Metrosep ASUPP7-250 (250 mm × 4.6 mm ID)
Eluent composition	30 mM NaOH + 30% methanol	30 mM LiOH + 25% acetonitrile	10.8 mM Na$_2$CO$_3$ + 25% acetonitrile
Sample volume	100 µl	100 µl	100 µl
Column flow rate	0.7 ml min^{-1}	0.7 ml min^{-1}	0.7 ml min^{-1}
Tandem conductivity detection	Yes	Yes	Yes
Suppressor regeneration solution	200 mM nitric acid	200 mM nitric acid	200 mM nitric acid

TABLE 7.2   Mass Spectrometer Parameters

	Option 1	Option 2	Option 3
Mass Spectrometer Model	6100 Series Single Quad	6100 Series Single Quad	6400 Series Triple Quad
Tune	"Negative Mode Autotune"	"Negative Mode Autotune"	"Negative Mode Autotune" + "Optimizer for MRM"
Capillary voltage (V)	(−ve) 1400	(−ve) 1400	(−ve) 1400
Nebulizer pressure (psig)	30	30	30
Fragmentor voltage (V)	140	210	140
Drying gas	Nitrogen	Nitrogen	Nitrogen
Drying gas flow ($l\,min^{-1}$)	9	9	10
Collision energy	N/A	N/A	20
Single ions monitored	99, 101 (107 for ISTD)	83, 85 (89 for ISTD)	$99 \rightarrow 83$
			$101 \rightarrow 85$
			$107 \rightarrow 89$

perchlorate ion; hence, even using single-quadruple mass spectrometer, perchlorate ion can be easily monitored in the presence of very high sulfate interfering ion.

Instrumental parameters for the analysis of perchlorate by IC-MS were initially chosen to reduce or eliminate suppression due to coelution of matrix and analyte ions and to lessen the effective concentration of matrix in the electrospray interface. Choosing operating conditions in this way increased reliability and stability of the system, but at the cost of potential sensitivity. To decrease matrix suppression, a 4 mm ID column was selected over a 2 mm ID column. The larger diameter column reduced the effective concentration of matrix in the system by dilution effects. The larger ID column also allowed a 100-µl injection, as used for this work, and a larger injection volume can easily be accommodated. Capacity of the column is also far greater than that of smaller ID columns, resulting in improved peak shape of any matrix or high-concentration compounds and reduced tailing into the analyte peak. All of these factors ensure that the majority of the matrix is well separated from the analyte for the reduction or elimination of suppression effects. In addition to the ion-exchange column used for the separation, a suppression column was utilized to eliminate sodium and calcium in the sample matrix. While not a direct problem with the detection of analyte, these nonvolatile cations accumulate in the interface from mobile phase and interfere with the long-term stability of the system when high total dissolved solid (TDS) samples are analyzed. Complete removal of the metal cations also decreases the risk of suppression by ensuring that only protonated anions enter the mass spectrometer interface. The 6100 MSD ESI interface is designed for relatively high flows while maintaining high sensitivity, reaching DLs of less than $100\,ng\,l^{-1}$. Many ESI interfaces are extremely flow-sensitive and do not perform well at flows typical for 4 mm ID columns. The 6100 MSD ESI interface performs best at low flows but does not exhibit the same drastic decrease in sensitivity at higher flows as observed with other interfaces.

7.3   RESULTS AND DISCUSSION

The IC-MS trace of a $500\,ng\,l^{-1}$ (Environmental Resource Associates, USA, certified. Perchlorate proficiency testing standards were used as reference standards) perchlorate standard is shown in Figure 7.3. It demonstrates very good signal-to-noise for perchlorate (m/z 99) eluting at a retention time (RT) of about 13 ± 2 min. Also, make a note that the retention time of perchlorate on anion-exchange column is dependent on the level of organic solvent present because perchlorate behaves as hydrophobic ion versus typical inorganic ions behaves as hydrophilic ion. The Metrohm Professional IC uses a combination of three suppressors that can be changed during a run to ensure that any one suppressor does not become saturated with cations. While one suppressor is in operation, another is recharging with proton (H^+), while still another is rinsing with ultrapure water. The abrupt signal changes observed at 1.5 and 9 min are due to ultrapure water entering the flow path from the rinsed suppressor during automated operations of the suppression column system.

The sequential changing and reconditioning of the suppressors during analysis are extremely important in the analysis of high matrix samples, Figure 7.3. Demonstrates m/z 99 and 101 traces for $1\,\mu g\,l^{-1}$ perchlorate in a $1000\,mg\,l^{-1}$ matrix of sulfate, chloride, and bicarbonate.

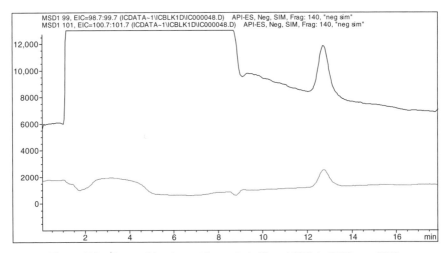

Figure 7.3   Trace of 1 ppb perchlorate (m/z 99 and 101) in 3000 ppm TDS.

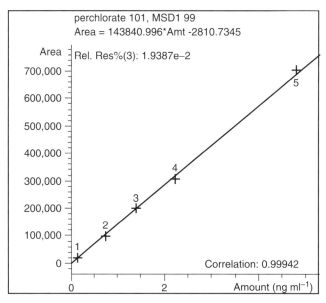

Figure 7.4 Calibration *m/z* 99.

Early in the chromatogram, the effect of the matrix can be clearly seen as both interference at the monitored mass and suppression of the signal in general. The suppressor change at 9 min ensures that a clean suppressor is in place for the perchlorate ion. This results in a very clean signal for perchlorate at about 13 min.

Figures 7.4 and 7.5 show the calibration data for both *m/z* 99 and 101 for perchlorate from 0.1 to $50 \mu g \, l^{-1}$. Calibration at both masses is linear over the measured range.

The *m/z* 99 single-ion chromatograms for a set of fortified matrix (from 150 ppm TDS to 3000 ppm TDS) each containing a $1 \mu g \, l^{-1}$ perchlorate are shown in Figure 7.6, a small RT change is noted for perchlorate between a perfectly clean standard and the matrix additions. This minor shift is nominal for IC due to initial overloading of the separation column. It does not interfere with the identification or determination of perchlorate. To avoid this ionic suppression issue and accommodate various MSD manufacturers in the USEPA method 332.0 [4], USEPA used synthetically enriched oxygen-18 perchlorate ($Cl^{18}O_4$) as an internal standard.

The results of a much more difficult test of the system, the analysis of $1 \mu g \, l^{-1}$ spikes in three different levels of matrix prepared according to EPA Method 314, are listed in Table 7.3. The samples were run sequentially with a blank and calibration verification (CCV) run after each set of nine samples. Recoveries of analyte at the $1 \mu g \, l^{-1}$ level are very good for all matrices, with an average recovery of better than

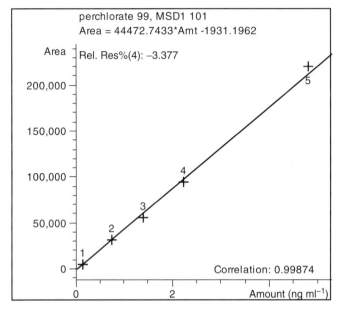

Figure 7.5   Calibration m/z 101.

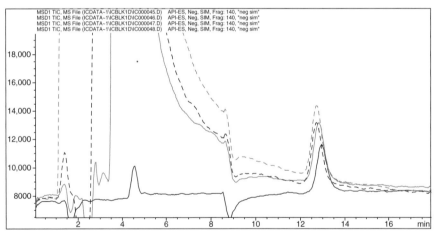

Figure 7.6   Perchlorate fortified at $1 \, \mu g \, l^{-1}$ in various matrix concentrations (150–3000 ppm
TDS).

TABLE 7.3 Average Recovery of 1 µg l^{-1} Perchlorate Fortified in Various Concentrations of Matrix

Number	Sample ID	QA Calculations True Value (µg l^{-1})	m/z 99 (%)	m/z 101 (%)
1	Reagent blank			
2	0.15 ppb std	0.159	101.73	105.95
3	0.25 ppb std	0.245	98.12	98.25
4	0.5 ppb std	0.585	93.75	94.34
5	1.0 ppb std	1.029	100.35	93.73
6	2.5 ppb std	2.575	103.50	105.72
7	5 ppb std	5.99	95.61	99.29
8	10 ppb std	10.87	103.81	106.66
9	Reagent blank			
10	CCC (0.5 ppb)	0.599	101.54	99.93
11	750 ppm TDS + 0.799 ppb – 1 (n = 5)	0.799	94.54	101.38
16	CCC (5 ppb)	5.99	99.09	95.07
17	1000 ppm TDS + 0.759 ppb – 1 (n = 5)	0.759	90.71	83.81
22	CCC (0.5 ppb)	0.599	100.94	81.58
23	1600 ppm TDS + 1 ppb – 1 (n = 5)	1.023	94.23	103.53
28	CCC (5 ppb)	5.99	98.02	94.42
29	1800 ppm TDS + 1 ppb – 1 (n = 5)	1.115	98.96	99.93
34	CCC (0.5 ppb)	0.599	101.87	99.51
35	2000 ppm TDS + 1.3 ppb – 1 (n = 5)	1.311	105.04	111.23
40	CCC (5 ppb)	5.99	105.34	102.94
41	500 ppm TDS + 0.711 ppb – 1 (n = 5)	0.711	98.25	96.53
46	CCC (10 ppb)	10.87	105.37	106.67
	Average (150 injections)		100.34	98.67

95% for both monitored ions in all matrices. The recovery data in matrix demonstrate that the system is not affected by the presence of potential interferences in the system at very high concentrations. These matrices demonstrate favorable recovery where perchlorate is present. More importantly, perchlorate was found in samples known not to have perchlorate or in samples containing high levels of interference acknowledged to complicate perchlorate determination when using conductivity detection.

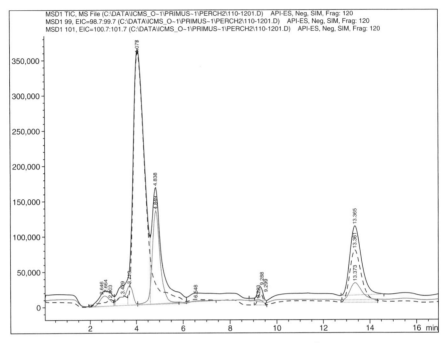

Figure 7.7   Extracts of lettuce fortified with $1 \mu g \, l^{-1}$ perchlorate.

One of the major interferences known in the electroplating industry is the compound named *para*-chloro benzene sulfonic acid (PCBSA). Typically, PCBSA surfactant [5] coelutes with perchlorate on few known anion-exchange columns with aromatic backbone and interferes as false high-positive perchlorate detection by suppressed conductivity detection. This issue was also resolved using Metrosep ASUPP7-250 (aliphatic backbone anion exchanger) analytical column.

The chromatograms of the lettuce analysis and the lack of interference around perchlorate are shown in Figure 7.7. The dip in the blank and spike lettuce samples is due to large amounts of another eluting material entering the electrospray source, suppressing the ionization of the background signal. Conventional conductivity detection is not useful for this sample since the area around perchlorate elution is overwhelmed by large amounts of coeluting material. Hence, this illustrates the applicability of the method for the analysis of vegetables. Perchlorate at various trace levels in vegetables, especially water containing were found to be significantly higher than irrigation source waters due to bioaccumulation of perchlorate. Between 2003 and 2007, the author and his colleagues performed perchlorate analysis in various fruits and vegetables such as spinach, strawberries, green onion, lemon, orange, cantaloupe, water melons [6]. This developed method was later adopted by USEPA-Office of Solid Waste to meet the requirements for perchlorate analysis in soils, seawater, and solids [7, 8].

Currently, many laboratories are using Metrohm Ion Chromatography coupled with Triple-Quad (QQQ) mass spectrometer for perchlorate analysis. This method using same analytical column and appropriate eluent chemistry is also known to be adopted for other analytes such as bromide/bromate, nitrate/nitrite, and oxyhalides in drinking water [9–12].

ACKNOWLEDGMENT

The author is very grateful to Dr Doug Lipka, Rick McMillin, Dr Melvin Ritter, Mr Johnson Mathew, Dr Joe Hedrick, and Dr Carl Zhang for supporting the project.

7.4 REFERENCES

[1] Chemical and Engineering News, 2003, Volume 81, Issue 33, "*Cover Story - PERCHLO-RATE Tracking pollutant's path down the Colorado River*" 1.

[2] USEPA Method 314.0 enhanced, ODW/OGW, *Determination of Trace Level Perchlorate in various water matrices by ion chromatography and suppressed conductivity detection.*

[3] USEPA Method 331.0, *Determination of Trace Level Perchlorate in various water matrices by LCMS or LCMSMS.*

[4] USEPA Method 332.0, *Determination of Trace Level Perchlorate in various water matrices by ICMS or ICMSMS.*

[5] Mathew, J., Gandhi, J., and Hedrick, J. (2005) Trace level perchlorate analysis by ion chromatography-mass spectrometry. *Journal of Chromatography A*, **1085** (1), 54.

[6] LCGC magazine Application Notebook, June 2004, page 33, *Determination of Perchlorate in various vegetable extracts by ICMS.*

[7] USEPA-OSW-SW846 Method 6850, *Determination of Trace Level Perchlorate in high ionic water, soil and solids by LCMS or LCMSMS.*

[8] USEPA-OSW-SW846 Method 6850, *Determination of Trace Level Perchlorate in high ionic water, soil and solids by ICMS or ICMSMS.*

[9] Metrohm USA Application works, AW US6 0066-2004, *Trace Level Perchlorate analysis in various water matrices using ICMS.*

[10] Agilent Inc. Application Note, 5989-0816EN, *Trace Level Perchlorate analysis in various water matrices using ICMS.*

[11] LCGC magazine Application Notebook, February 2004, page 36, *Determination of Perchlorate in waters by ICMS.*

[12] Agilent Inc. Application Note, 5989-7907EN, *Trace Level Perchlorate analysis in various water matrices using ICMSMS.*

8

SAMPLE PREPARATION TECHNIQUES FOR ION CHROMATOGRAPHY

WOLFGANG FRENZEL

Technische Universität Berlin, Department of Environmental Technology, Strasse des 17. Juni 135, Berlin 10623, Germany

RAJMUND MICHALSKI

Institute of Environmental Engineering, Polish Academy of Sciences, M. Skłodowskiej-Curie Street 34, Zabrze 41-819, Poland

8.1 INTRODUCTION

Since the inception of ion chromatography (IC) in the middle of the 1970s, this method has underwent a tremendous development, and today, IC is the leading analytical method for the determination of ionic constituents of all kinds (anions, cations, inorganic and organic ions, small and large (bio)molecular ions) in almost any type of sample matrix. Several books [1–3] and uncountable scientific papers have been published on ion chromatography. In addition, manufacturers of ion chromatographic instrumentation offer a large number of well-proved application examples.

Due to the multitude of stationary phases, with the availability of different detection principles (more recently, for instance, elemental and molecular mass spectrometry, the main subject of this book, has been very successfully coupled to IC) even apparently difficult and challenging separation problems in ion analysis can generally be solved satisfactorily. The outstanding characteristics of IC are the possibility to determine simultaneously many (equally charged) ions in relatively short time,

Application of IC-MS and IC-ICP-MS in Environmental Research, First Edition.
Edited by Rajmund Michalski.

its high selectivity, the wide dynamic working range, microanalytical features, sufficient sensitivity for the majority of applications, and low consumption of reagents. Manufacturers of IC offer a broad spectrum of apparatus covering low-price systems for simple routines and educational purposes, dedicated analyzers for particular kinds of samples (e.g., soft drink analyzer) as well as complex modular-configured instruments permitting continuous process analysis, the application of gradient techniques, hyphenation to different detectors, and, more recently, the integration of sample preparation steps in a fully automated manner.

Nevertheless, new analytical problems in various scientific disciplines, emerging candidate ions considered (or already proved) to be environmental pollutants (e.g., perchlorate, bromate, trace metal species of various oxidation states or chemical forms present in different sample matrices), which need to be monitored in a variety of samples, increasing requirements concerning the quality of analytical results, and an increasing pressure on laboratories to reduce costs of analysis are challenges similarly to scientists and manufacturers dealing with IC in its widest sense. Current scientific developments in the IC field chiefly concern the following topics:

- Improving the stationary phases with regard to separation power without sacrificing the total analysis time.
- Increasing the sensitivity and selectivity by using other detection principles apart from the mainly used conductivity measurement. The role of mass spectrometry (MS) as a detector for IC is to be especially highlighted in this respect.
- Solving specific separation and detection problems for the determination of individual ions (e.g., trace oxyhalides, diverse sulfur and phosphorous species), for ion analysis in special matrices (e.g., electroplating baths, food, and beverages), or in particular areas of application (e.g., biomedical research, forensic science, fermentation monitoring).

Manufacturers of IC instrumentation are taking great efforts to offer users a wide range of instrumentation, which, with respect to their flexibility, robustness, and cost, generally satisfy the considerably different requirements in the various fields of application.

Real-world applications of IC vary greatly, but there is virtually no scientific field in which this method cannot be used for the routine determination of ions or for solving special problems. The most important fields of application are related to environmental analysis (with needs for ion determination in all kinds of water, sludge, soil, sediment, and air), the food and beverages industry, and the clinical and pharmaceutical sectors. Some further fields in which IC has established itself as a versatile method are the monitoring of cooling water in power plants, semiconductor and electroplating industries, mining and metal processing, paper manufacturing, and biotechnology.

Ion determination in the broad range of samples occurring in the different areas of application requires an adequate selection of separation conditions as well as appropriate detection methods. Aqueous samples with a moderately high total ion content that are not loaded with an organic matrix cause few problems in IC and can usually be directly injected. In the analysis of more complex matrices, where different types of possible interferences may occur, attempts should always be made to first exhaust the separation power of the chromatographic system and to make use of the selectivity of different detectors applicable before a sample pretreatment step is considered. Problems related to overlapping signals caused by inadequate separations (i.e., insufficient stationary-phase selectivity) can often be solved by selecting different stationary phase, optimization of the mobile phase, or – often very effective – to use a selective detector or by coupling various detectors with different selectivity. Inductively coupled plasma with optical emission spectrometry (ICP-OES) or inductively coupled plasma with mass spectrometric (ICP-MS) detection, for instance, provides very high selectivity for trace element determination even without IC separation so that no interferences of the typically ionic and also of nonionic constituents eluting from the separation column are to be expected. Here, it is more the other way round; IC separation can be regarded as a kind of sample preparation for the respective atomic spectrometric methods. The reasons for hyphenating IC with OES or element-MS typically are metal speciation studies where chromatography is used for species separation according to oxidation state, valences, or status of complexation; the respective atomic spectrometric methods serving as sensitive multielement detectors [4–7].

Molecular MS offers high sensitivity and is also very selective so that chromatographically unresolved compounds can often be resolved unequivocally by their mass-to-charge ratio. A very impressive example is the simultaneous determination of 14 amines in air using cation chromatography with tandem MS detection [8]. Using conductivity detection, many of the amines are not resolved but due to their characteristic mass-to-charge ratios, MS permits interference-free identification and quantification. By evaluation of isotope ratios of target elements of the ion species or molecule fractions, a high degree of certainty about the identity of a particular ionic species can be achieved. Moreover, molecular MS provides identification potential for unknown compounds by typical fragmentation patterns.

The presence of uncharged low-molecular-weight compounds in the samples to be analyzed is generally not problematic in IC determinations of anions and cations since they are eluted within the chromatographic hold-up time and do not interfere. However, evaluation of very early eluting ions is sometimes adversely affected by the injection peak caused by the sample solvent (e.g., the water dip in suppressed conductivity detection using carbonate eluents) or system peaks caused by disturbances of stationary phase ion equilibria.

Looking into the practice of IC in many applications, users more often than not meet with difficulties that are not related to the ion chromatographic separation and detection, rather than the incompatibility of the sample with the IC system. Just to mention a few typical problems:

- Particulates and high-molecular-weight compounds (such as proteins, humic matter, fat, and surfactants) need to be removed from liquid samples before injection.
- Samples with extreme high and low pH must be neutralized.
- Solid samples need to be extracted or decomposed.
- Gaseous analytes need to be transferred into the liquid phase prior to injection.

Considering solid sample analysis, sample pretreatment involves quite often several steps before a suitable liquid is eventually obtained ready for injection into the IC system. Other problems encountered and calling for sample preparation are too low analyte concentrations or the presence of high excess of some ions when low concentrations of other ions are to be determined (e.g., seawater or brine analysis).

It is therefore surprising that sample preparation for IC analysis is still a subject of limited interest, at least when the tremendous number of scientific publications concerning IC and the minor role sample preparation plays in these papers are considered. Sample preparation can be regarded the Achilles heel of IC – as also in many other instrumental analysis methods – since the working steps involved (together with the sampling process) represent the largest source of errors in the whole analytical chain. The entire advanced analytical process can be invalidated if any unsuitable sample preparation method has been employed before the sample is introduced into the chromatographic system [9]. Also, sample preparation usually cannot be easily automated. It is therefore often the limiting factor concerning sample throughput and hence an economic issue for analytical routine laboratories.

In this chapter, a general overview of sample preparation methods is given and typical problems adversely affecting IC analysis as well as possible solutions to these problems are discussed. Currently applied procedures in routine work are reviewed and new developments in sample preparation for IC analysis are presented. Emphasis is also given to postcolumn treatment of the eluent required when molecular MS is used as IC detector.

8.2 WHEN AND WHY IS SAMPLE PREPARATION REQUIRED IN ION CHROMATOGRAPHY?

The enormous performance power of current IC with respect to chromatographic resolution, simultaneity, sensitivity, and speed can be demonstrated best by the analysis of standard solutions or samples containing a limited number of ions at similar concentration levels in a matrix-free aqueous solution. However, in practice, the user normally encounters real samples whose composition may vary considerably and which in many cases can only be analyzed by IC after a more or less complex and complicated sample preparation procedure.

Some typical aims of sample preparation are as follows:

- Removal of potential interferences and/or isolating the analyte(s) from the sample, thus increasing the selectivity of the method
- Increasing the concentration of the analyte(s) and thus the overall sensitivity of the determination
- Converting the original sample matrix into a suitable form for analysis
- Converting the analyte(s) into a more suitable form for determination (e.g., by derivatization).

The most important sample preparation methods for chromatographic analysis (including ion chromatography), taking into consideration the sample form (liquid, solid, or gaseous) are as follows [10, 11]:

1. Liquid samples (filtration, dilution, pH adjustment, standard addition, derivatization, liquid–liquid extraction, solid-phase extraction (SPE), distillation, microdiffusion, and membrane-based separations)
2. Solid samples (drying, homogenization, dissolution, extraction/leaching, digestion, ashing, and combustion)
3. Gaseous samples (absorption in liquids, adsorption on solid phases, membrane-based sampling, and chemical conversion).

Irrespective of the pretreatment steps actually applied, provision of a robust and reproducible sample preparation method is required that takes into account the variability of the sample matrices and the possible influence of matrix constituents on the results obtained. With respect to quality assurance, sample preparation is a very crucial issue and is often underestimated in analytical practice.

In practice, sample preparation means, for example, that samples with too high analyte concentrations must be appropriately diluted prior to injection in order to fall into the dynamic working range of the detector. On the other hand, if analyte concentrations are low and the sensitivity of the detection system is insufficient, a quantitative determination is often only possible after preconcentration of the analyte ions. Problems are also caused by samples in which the ratio of the concentrations of the ions to be determined (simultaneously if possible) differ by orders of magnitude. This is, for example, the case in the determination of trace anions in seawater, brine, the simultaneous determination of chloride (high), nitrite (low), phosphate (high), and other ions (often low) in strongly polluted wastewater [12, 13]. The same holds true for the analysis of extracts of solid samples when salts are added to release ionic compounds (e.g., the use of $CaCl_2$ as extractant in soil analysis).

Greater difficulties may even occur in the analysis of samples containing substances that influence the chromatographic separation due to overlapping signals, lead to alterations of retention times, or create baseline fluctuations as a result of changes in ion-exchange equilibria between eluent and stationary phases. This applies not only to samples with high acidity or alkalinity but also to samples containing non-ionic (but highly polar) organic substances that interact with the stationary phase and may behave similar to the ions of interest. Even more serious are alterations of the stationary phase caused by matrix components through irreversible adsorption, which may cause variations of the retention times and resolution and eventually render the separation column unusable. This is likely to occur when humic material, lignins, organic dyes, fats, surfactants, and proteins are present. Many (even supposedly "clean") samples contain undissolved fractions in the form of colloidal matter and (invisible by eye) microparticulates that need to be removed before injection.

In principle, IC cannot be used directly for the analysis of solid samples. Either the samples must be completely dissolved or at least the ions to be determined must be released. Very often, the dissolution of solid samples or the release of the analyte cannot be carried out by water alone, so that the solution to be analyzed by IC also contains additional substances (acids, bases, salts), which are then present as new matrix components. A direct result of this is that, in the preparation of solid samples, initial consideration must be given to the compatibility of the extraction solution with the IC separation. Also, after dissolution, extraction, or digestion, further steps for matrix elimination or pH adjustment are often required.

Gases can only be analyzed when the components to be determined have been previously transferred to an aqueous solution. This is commonly done, for example, by passing the sample gas through an absorption solution and analyzing this solution by IC or by adsorbing the gaseous analyte onto a solid phase and then eluting it with a suitable liquid and carrying out the IC determination on the eluate. Alternative methods for gas sampling are based on diffusive sampling (passive sampling, denuder techniques). Here, the trapped analytes also need to be dissolved in a suitable solution prior to subsequent IC determinations.

In order to cope with the various problems and inherent difficulties that might be connected with ion determination in real samples of different origins and compositions, a variety of sample preparation procedures are available to the user. In the following sections, these methods and procedures are presented, instrumental requirements outlined, and examples of application given.

8.3 AUTOMATION OF SAMPLE PREPARATION (*IN-LINE* TECHNIQUES)

Since the mid-1980s, ion chromatography has gradually become more and more automated. The introduction of sample changers was a first step permitting unattended operation of IC and hence improving the productivity. A major step forward was the implementation of PC control of hardware functions, but similarly or even more for the recording and evaluation of the detector signals, including peak integration, retention time measurements, and data storage. In this latter field, the current IC software provides the user many attractive features, which make IC operation convenient and increasingly reliable (e.g., through "intelligent" performance control features).

With respect to sample preparation, sample changers are nowadays available, which (controlled by software) permit automated sample dilution of out-of-range samples and reinjection, the addition of reagents and – if desirable – to perform standard additions. Also, special single-use vials for sample changers are available with integrated microfilters (filter caps) to remove particles prior to sample injection. More advanced IC systems also provide the self-preparation of standard solutions for calibration from a single stock solution.

The integration of in-line filters and solid sorbent microcolumns between injection valve and separation column (also a kind of sample preparation prior to the actual separation) is a highly recommended (if not mandatory) measure to improve the lifetime of columns and to avoid or at least reduce the occurrence of extraneous signals in the chromatogram caused by unspecific response to nonionic matrix constituents.

The implementation of other more advanced and complex kinds of sample preparation techniques into IC systems in an automated manner is generally based on the application of a secondary flow through system. Loan is made here from the developments and long-lasting experience of implementation of in-line sample preparation into continuous air-segmented flow analysis and flow injection analysis.

By the use of syringe, piston or peristaltic pumps in combination with switching and injection valves connected with appropriate tubing, any kind of liquid (samples, standards, cleansing solutions, reagents, etc.) can be flexibly moved through a system of capillary tubing in forward/backward mode at variable flow rates (including temporarily stopped flow). Separation and sample cleanup devices such as SPE microcolumns or membrane modules can be inserted into these flow systems, and in this way, a high degree of automation of the sample preparation procedures is obtained.

There are at least two different principle modes of operation possible to use flow systems for sample preparation in hyphenation with IC [11]. Both modes can generally be accomplished with the same basic hardware components; the choice depends on the sample preparation technique to be implemented and other factors such as sample volume available, composition and complexity of the sample matrix, and the analyte(s) to be determined (i.e., anions, cations, single analyte ion, or simultaneous determinations).

The first configuration is based on the provision of a relatively large amount of sample solution, which serves to flush the entire auxiliary flow system used for matrix removal, neutralization, solvent exchange, or analyte preconcentration. In this variant, there must be a sufficiently large volume of sample available so that it is warranted that the previous sample has been rinsed out of the system before the loop content of the injection valve is transferred into the IC system. In the case of SPE and the use of ion-exchange microcolumns for sample preparation, the retention capacity can get quickly exhausted with associated problems of matrix breakthrough, reduced lifetime of the sorbents, or low and variable recovery rates in analyte preconcentration systems.

In an alternative configuration, only a small volume of sample is inserted into the auxiliary flow system and this sample plug is transported by ultrapure water through the sample preparation part to a matrix retention column placed on the way to the injection loop of the IC system or a trapping/preconcentration microcolumn placed in the loop of the injection valve. This variant has the advantage that only the sample volume necessary for the chromatographic determination passes through the sample preparation part. Since the sample plug is sandwiched in between the transfer solution (usually ultrapure water), the system is less prone to cross-contamination. In addition, the capacity of solid-phase microcolumns for matrix retention or ion exchangers is not exhausted so quickly, and hence, breakthrough problems with adverse effects on the performance are reduced.

In the following sections dealing with the various sample preparation techniques, in-line configurations will also be presented together with the resulting features and relevant practical applications. Reference is repeatedly made to the two basic configurations mentioned above. A comprehensive presentation of in-line techniques and several examples of practical applications is given by the two leading manufacturers of IC [14, 15] as well as in a review on this subject [11].

8.4   SAMPLE PREPARATION METHODS

The sample preparation methods existent for a wide range of problems to be solved are manifold, and the procedures and instrumental requirements for sample preparation differ considerably in detail. The working steps involved range from relatively simple measures such as sample dilution, filtration, or pH adjustment up to complicated and sometimes even multistep procedures such as analyte extraction from solid materials, analyte preconcentration, and sample cleanup steps for matrix elimination.

The choice of the sample preparation method to be used depends, on the one hand, on the physical state and the composition of the sample; on the other hand, it depends to a considerable extent on the analyte ions to be determined. It is worth mentioning already here that in many instances, a particular sample preparation procedure cannot be applied for multiion determinations for the one or other reason. A procedure optimized for the determination of a single ion or few ions can be inappropriate for the determination of other ions in the same sample. In consequence, the attractive feature of IC to simultaneously determine many ions may get lost. Examples are the trace determination of oxyhalides in tap water after ion-exchange preconcentration, which only works well after previous removal of excess chloride and sulfate (e.g., using SPE-cartridge precipitation with Ag and Ba ion loaded ion exchangers) or the determination of perchlorate and iodide in different matrices after a multistep procedure with analyte trapping, matrix removal, and several flushing steps involved [16, 17].

The criteria for the selection of an appropriate and promising procedure for IC are connected to the ion separation principle (i.e., anion or cation exchange, ion exclusion), the separation conditions actually used for a particular separation problem, and, for many users, the availability of particular IC configurations (e.g., conductivity suppression, gradient techniques, types of detectors).

Sample pretreatment for the determination of anions frequently requires (or allows to use) different measures compared to those used for the determination of cations. It is also obviously much more difficult to carry out sample preparation when a large number of ions are to be determined simultaneously rather when only a few or even a single ion is of interest. At this stage, it is worth mentioning that IC – irrespective of its potential for simultaneous measurements of many inorganic and organic ions – is increasingly applied to the determination of only a few ions or sometimes only to a single ion. One reason often is that alternative methods (e.g., spectrophotometry, electroanalytical techniques) are not sufficiently sensitive and/or selective. This is certainly true for the determination of organic acids but also for oxyhalides, sulfur, and phosphorous species other than sulfate, sulfite, and sulfide, and o-phosphate as well as a number of other ions at trace levels such as iodide or bromide.

With respect to the choice of a particular sample preparation technique, the actual concentrations of the ions to be determined may, under certain circumstances, be a crucial and decisive factor. The quality of the separation column used with regard to its resolution, capacity, pH stability, and compatibility to the presence of organic solvents as modifiers can also be decisive for the needs of sample preparation and/or for the sample pretreatment procedure(s) eventually selected. Finally, aspects such as sample volume available, labor, time, and cost of application of the respective procedures are to be considered.

Apart from the undesirable increase in overall analysis time and costs connected with sample preparation, the risk of analytical errors also increases when sample pretreatment is applied. This is particularly valid when a series of different sample preparation steps is to be performed prior to sample injection into the IC system. Contamination and loss or conversion of analytes may falsify the required qualitative and quantitative information or even cause its complete loss. This must be taken into account in the choice of the preparation steps and, if possible, avoided or controlled by appropriate measures. A comprehensive discussion of issues related to quality assurance in sample preparation (not only) for IC has been presented by Seubert *et al.* [11].

Considering the large number of possible IC applications, there are only a few types of samples that can be directly analyzed without any sample preparation. The optimistic and ideal situation of "*dilute and shoot*" is only rarely given in the real-world application of IC. If samples are already present in aqueous form, then sample preparation may have various aims. Adjustment to a particular pH value is only necessary in rare occasions as many liquid samples – in particular those relevant in environmental analysis – are in a medium pH range and the common eluents for IC possess a certain buffering effect; this means that interferences in the chromatographic separation do rarely occur. In contrast, strong acids and bases cannot usually be analyzed directly so that the sample solution must at least be partially neutralized prior to injection. In anion chromatography, this causes problems as, after the addition of an acid (with alkaline samples), the acid anion cannot obviously be determined in the sample solution but it may also hinder the chromatographic separation of other anions. In the opposite case, the addition of a base (in the analysis of strong acids) prevents the determination of the added cation and leads to increased salt loading, which may also cause interferences in chromatographic separations. An often-used means to overcome the problems described is the use of ion exchangers (in microcolumn or membrane-based configurations), which in their acidic or the basic forms can neutralize alkaline or basic solutions, respectively, without any reagent addition and without dilution.

Apart from the relatively simple addition of substances for neutralization, matrix adaptation, and/or modification of the chemical form of the species to be determined by derivatization, in many other cases, it is necessary to carry out further sample preparation and separation steps with or without phase transition before interference-free IC analysis is possible. This holds for ion determination in organic solvents, samples with high load of interfering matrix components, any kind of solid sample, and gas analysis applications of IC. The basics, possible applications, and implications as well as instrumental needs and configurations of common sample preparation methods and procedures are described in the following sections.

8.4.1 Filtration and Ultrafiltration

The aim of filtration as a sample preparation step is to remove particulate matter from the liquid sample, which otherwise can block the capillary flow system (including the injection valve) or accumulate on the column head, leading to increased back pressure and reduced operation time. The prefiltration step of samples does not substitute the highly recommended implementation of in-line filters and/or precolumns in IC systems since they are an additional measure of safety preventing the deterioration of the separation column. Common pore sizes of filtration membranes appropriate for sample preparation are in the range of 0.2–0.45 µm. For biologically active samples, the use of sterile filters with a pore size of 0.2 µm is advisable as they also retain microorganisms and therefore prevent the conversion of certain analytes in the filtrate by microbial oxidation or reduction. With respect to the proper filter media (a large variety is commercially available), the selection is guided by compatibility with the sample solvent (typically, it is water and almost any filter media can be applied), freedom of blanks, the possible adsorption risk of analytes on the filter surface, practicability, and costs. Analyte loss through adsorption on the filter material or the filter cake formed during the filtration process should always be considered since this can lead to significant underestimations in particular in trace ion analysis. It is generally advisable to withdraw the first few milliliters of the filtrate and to test contamination and adsorption losses by running blank samples and perform recovery tests of spiked samples, respectively. Considering the minute volumes required for IC analysis, miniaturized formats (in the form of disposable syringe filters) are to be preferred.

Manual operation with the sample liquid passing through the filter by positive pressure is possible but even for samples with only moderately high particle load, it can be tiring and may take a few minutes per sample to receive sufficient filtrate volume. Even after dilution, milk samples (with no particle load but fat droplets in the form of a fine dispersion) are hard to press through a 0.45 µm filter.

Automation of the filtration step is possible via purpose-made disposable autosampler vials with implemented membrane filters. This is a common option for many liquid chromatography systems. For samples heavily loaded with particles, it is advisable to carry out a coarse filtration before membrane filtration or to separate off the bulk of the solid substances by centrifugation.

Ultrafiltration (UF) is a separation technique that utilizes membranes that retain not only (micro)particles but also colloids and high-molecular-weight compounds. UF membranes have much smaller pore sizes than membrane filters for microfiltration (the range is approximately 0.001–0.1 µm), yet they are generally categorized according to their specific molecular-weight cut-off. This is a value that refers to the molecular weight (which is related to the molecular size) of a compound being able (or just unable) to pass through the membrane. Membrane cut-off values of commercially available UF membranes are in the range of 1–500 kD. Membrane materials most frequently employed in ultrafiltration are cellulose acetate, regenerated cellulose, cellulose ester, polycarbonate, polysulfone, and polyethersulfone.

As far as both the apparatus and time required are concerned, ultrafiltration is clearly more complicated and time-consuming than common membrane filtration; the costs involved are also higher. The small pore size of UF-membranes requires a considerable pressure gradient, which is commonly achieved by vacuum suction or ultra-centrifugation. For the latter, dedicated miniaturized disposable UF units are available consisting of two containers that are screwed together and are separated by the membrane, thus considerably simplifying the entire procedure [18, 19]. In recent papers, an essentially reagent-free sample preparation using UF has been described for perchlorate determination in milk samples, omitting any further treatment as required with other sample preparation procedures for milk analysis [20].

Ultrafiltration can also be applied in flow-through systems where the sample (feed) stream in a sandwich-like device is flowing along one side of the membrane and a suction pump creates an appropriate underpressure so that only part of the liquid is passing through the membrane and either collected in a vial for further processing or directly provided to the injection valve of an IC system [21]. The tangential flow of the sample solution along the upper side of the UF membrane (only about 5–10% of the sample solution passes actually through the membrane) prevents the quick buildup of a filter cake so that possible clogging of the membrane due to deposition of particles and/or adsorption of high-molecular-weight compounds is considerably reduced. Furthermore, contamination is less likely to occur in the in-line mode of ultrafiltration and possible changes of performance can at any time be checked by running a blank or standard solution.

An alternative to sandwich units with flat membranes are tube-in-tube configurations with hollow fiber UF membranes. They have a much larger exchange area, which causes a higher permeate flow and thus faster separation. The useful life of continuously operated UF strongly depends on the complexity of the sample matrix but can in many instances exceed several days of operation without membrane replacement. Typical applications of ultrafiltration prior to IC analysis are the determination of anions and cations in different types of polluted water, beverages of all kind, biological samples (blood, serum, urine), extracts of plants, dairy products, soil and waste material, and also fermentation broth and process liquids. Also, ion determination in oil–water emulsions and dispersions has been described using in-line ultrafiltration in combination with ion chromatography [21].

In certain cases, ultrafiltration is carried out with a number of different membranes connected in series with decreasing molecular-weight cut-off in order to achieve a fractionation according to molecular weight. This is, for instance, of interest in trace metal determination studies of environmental and biological samples where the different fractions can be associated with solubility, potential (bio)availability, and mobility of metals in the respective solid matrices [22, 23].

8.4.2   Solid-Phase Extraction (SPE)

SPE is currently one of the most powerful, versatile, and probably also the most widely applied techniques used for liquid sample preparation in connection with different analytical method including all forms of liquid chromatography [24–26]. SPE is a sample preparation process by which compounds that are dissolved or suspended in a liquid are separated from other compounds in the mixture according to their physical and chemical properties. SPE (sometimes termed disposable chromatography) relies on basically the same principles as that of liquid chromatography. However, the aim of SPE in contrast to liquid chromatography is not to separate all analytes present in a sample from each other rather than to isolate specific analytes or to remove matrix constituents that interfere in the subsequent determination. The versatility of SPE allows this technique to be used for purification, preconcentration, ion exchange, solvent exchange (e.g., transfer of the analyte from an organic sample to an aqueous medium), desalting, and derivatization (the analyte is retained on the sorbent, derivatized, and then eluted). A large variety of sorbent materials for SPE exists. Depending on the chemical properties of the components to be preconcentrated or purified and of the matrix they are contained in, there is a large number of sorbents available, such as the following:

- Nonpolar, reversed phase (e.g., octadecyl, octyl, ethyl, phenyl, graphitic carbon, styrene/divinylbenzene copolymers)
- Polar, normal phase (e.g., cyano, amino, diol, silica gel, florisil, alumina)
- Cation and anion exchangers (e.g., quaternary amine, carboxylic acid, sulfonic acid)
- Specialty phases (mixed-mode SPE, restricted-access materials, molecular imprinted polymers to mention just a few).

Sorbent materials can be categorized according to the base solid or with respect to the active/functional surface group being responsible for the retention of particular compounds. The majority of sorbents used in SPE are based on silica, but organic polymers, carbonaceous materials as well as metal oxides are also applied.

The selectivity of SPE is governed by the affinity of the respective solutes dissolved or suspended in a liquid sample for the solid material. The choice of the appropriate sorbent is hence the key to successful application of SPE. SPE procedures have many advantages (particularly when compared to liquid–liquid extraction), which explain the widespread use and large variety of applications. Some of the attractive features are as follows:

- High recovery rates
- Preconcentration capabilities
- Efficient purification
- Easy automation, including in-line configurations
- Compatibility with all determination methods for liquid samples
- Reduced consumption of sample and organic solvents.

8.4.2.1 *Basic Procedures and Retention Mechanisms* SPE systems are available commercially both in column or cartridge form and in the form of extraction discs (membrane-embedded sorbent materials) [24, 26]. SPE procedures normally involve passing the sample solution through a small column (e.g., solid-phase cartridge) that is filled with a sorbent material suitable for the separation, which retains either the analyte ions or the interfering matrix components. In the case of analyte retention and preconcentration, the analyte ions are eluted from the column in a subsequent step and the eluate is then analyzed by IC. Additionally, phase conditioning prior to sample application and removal of matrix constituents by intermediate washing steps is often included in the overall SPE procedure. If matrix components are retained at the solid phase, then the determination is carried out on the solution passing through the sorbent bed.

The sample solution can be passed through the sorbent bed manually, semiautomatically, or in a fully automated manner. A particularly elegant and efficient way of carrying out automated sample preparation by SPE is provided by the so-called column-switching techniques. In this case, analyte preconcentration or matrix elimination occurs in an auxiliary flow system operating in the low-pressure range that is connected in-line with the IC instrument. Here, the injection valve can be regarded to be an interface between the two flow systems.

The true versatility of SPE can only be realized and achieved by first understanding and then properly controlling the delicate interplay of physicochemical interactions that exist between the sample/matrix components and the solid phase. As in LC, the separation by SPE can be optimized by judicious choice of the stationary sorbent phase and the proper modification of the mobile phase during sorption (i.e. the sample solution), wash, and elution steps, respectively.

In principle, when SPE methods are used, care should be taken that the sorbents used have a sufficiently high degree of purity as there is a serious contamination risk, particularly in trace determinations. Hydrophobic packing materials require conditioning with an organic solvent that is miscible with water in order to ensure wettability of the sorbent. When ion exchangers are used, the ion-exchange capacity of the sorbent used must be matched to the separation problem using appropriate cartridge dimension. Finally, care should also be taken that the contact time between the components to be separated and the sorbent material is sufficiently long. Regardless of the type of sorbent used and the practical execution, optimization of the SPE taking into account the composition of the sample matrix is essential. This includes the determination of the reproducibility and the recovery rate by spiking experiments as well as checking contamination and possible analyte loss.

The use of SPE for matrix modification and the separation of interfering substances are extremely versatile [11, 27, 28]. The neutralization of acidic and alkaline samples can be carried out by means of ion-exchange resins in basic or acidic form, respectively. Ion exchangers can also be used for the selective separation of interfering ions. Examples of SPE sample preparations are the separation of a high chloride or sulphate content with cationic exchangers in the Ag or Ba forms, respectively [29], or the exchange of multivalent cations (which may cause interference in anion chromatography) for alkali metal ions. Interfering concentrations of carbonate/bicarbonate can be eliminated by the use of cation exchangers in their fully protonated form.

Another important application of SPE is the removal of matrix ions that would interfere with analyte determination by overloading the column or by producing overlapping peaks. Examples of this are the removal of cations when analyzing mineral water for anions and the removal of anions such as chloride and carbonate from water samples to permit interference-free bromate determination [30]. Finally, ion-exchange SPE can also be advantageously used to retain ions from samples containing organic solvents (or even from pure water) or other dissolved nonpolar constituents, which are incompatible with IC analysis.

Nonspecific sorption on inorganic and organic polymer phases has achieved great importance in the analysis of samples containing organic substances. Depending on their polarity and molecular size, these can adversely affect IC analysis in various ways. In the worst case, it can cause irreversible damage to the stationary phase. Hydrophobic packing materials such as polystyrene-divinylbenzene, poly-methacrylate resins, or chemically modified silica gels, which are used in HPLC as reverse-phase stationary phases, are particularly suitable for the separation of hydrocarbons, surfactants, aromatic compounds, and long-chain organic (fatty) acids. Polyvinylpyrrolidone has been shown well suitable for use as the sorbent phase for the separation of humic materials, tannins, lignins as well as organic dye compounds, phenolic materials, aldehydes, and aromatic acids. Other more rarely used sorbents are carbonaceous materials (e.g., charcoal, graphitic carbon), inorganic oxides and silicates (e.g., Al_2O_3, magnesium silicate, zeolites), as well as a wide variety of other organic polymers that have proven advantageous for specific separation problems [31].

Nonpolar SPE can also be used for the preconcentration of heavy metals after derivatization forming metal complex compounds with subsequent determination by cation-exchange chromatography. Alternatively, the so-called chelate-forming ion exchangers can be used for this purpose. These are extraction reagents covalently bound to an organic polymer carrier that selectively binds individual metal ions or groups of metal ions. In a second step, the metal cations are eluted with a small volume of mineral acid and can then be directly injected into the IC instrument for trace metal determinations.

8.4.2.2 *In-Line SPE-IC*

Implementation of SPE cartridges or microcolumns into IC is common practice to avoid the introduction of unwanted matrix constituents to the separation column. To this end, it is generally recommended to use a precolumn filled with appropriate sorbent material (sometimes the stationary-phase material used for separation, sometimes reversed-phase material) placed between the injector and the separation column.

For sample preparation, SPE columns are either hosted in the injection loop of the IC system or located in an auxiliary flow system, which is connected in-line to the IC instrument. A general requirement for in-line SPE is the reversibility of the entire process since a sorbent column once installed should be desirably usable or many repetitive retention/elution cycles. In the following, in-line SPE configurations and applications are described based on ion exchange and unpolar (reversed-phase) mechanism. Distinction is also made between applications with analyte retention (generally connected with preconcentration) and matrix removal by SPE microcolumns.

The implementation of ion-exchange microcolumns into flow systems hyphenated to IC serves, as in batch mode, to neutralize samples, to remove possibly interfering sample constituents, or to obtain ion preconcentration. For the last purpose, the most popular arrangement is the insertion into the injection loop of the IC system. After a defined volume has been passed through the trap column, the injection valve is switched and the preconcentrated analytes are eluted by the mobile phase onto the analytical column where separation takes place. The capacity of the retention column must be selected so that the mobile phase of the IC method is capable of permitting spontaneous elution in order to prevent peak broadening.

For sample neutralization, the preferential way of in-line SPE is to inject a certain volume (typically 10–100 µl) of sample solution into an aqueous stream of the auxiliary flow system, which is guided through the ion-exchange microcolumn (for matrix removal) and further through an ion trap column placed in the injection loop of the IC system. For the determination of anions in strong alkaline solutions, for instance, a fully protonated cation exchange microcolumn is located between sample introduction and IC valve, and the anions present in the outcoming solution are retained on the anion trap column from which they are eluted onto the separation column by the mobile phase. In a similar manner (with microcolumns of opposite charge), cations can be determined in strong mineral acids. Regeneration of the neutralization columns can be done after each sample analysis or when the capacity is assumed to be exhausted.

The in-line removal of chloride or sulfate with silver- or barium-loaded cation exchangers, respectively, can be done with basically the same instrumental configuration. It might be mentioned that serial arrangement of different microcolumns can be accomplished in the auxiliary flow system providing the simultaneous removal of several potential interfering compounds.

8.4.2.3 *Matrix Elimination Using In-Line SPE*

The term matrix elimination is used here for procedures where nonionic sample constituents are separated from the ionic analytes of interest and for the determination of ions in nonaqueous solvents that are incompatible with IC and cannot hence be directly injected [11].

A simple way to perform in-line matrix elimination is to place an ion-exchange concentrator column for the respective ions in the loop of the IC valve and to pass through a defined sample volume. Thereafter, the injection valve is switched and the preconcentrated ions are transferred to the separation column by the mobile phase. The maximum sample volume acceptable depends on the total ion content of the sample and the capacity of the ion exchange trap column. For relatively low analyte concentrations, the sample solution can be aspirated directly from the sample changer and passing through the trap column. When elevated levels of ions are present and/or the matrix is expected to adversely affect (or even damage) the trap column, the injection of small volumes (typically 10–500 µl are used) is advisable into an auxiliary flow system with subsequent transfer of the sample plug to the concentrator column with ultrapure water as the carrier solution. The transfer solution is also used to rinse the concentrator column after passage of the sample solution in order to prevent any matrix constituents to reach the separation column.

Typical examples of applications of this approach are the determinations of anions or cations in hydrogen peroxide solutions or in polar and nonpolar organic solvents [11, 32]. Hydrogen peroxide solutions of up to 50% can be analyzed directly in this way. For the determination of anions in organic solvents, the use of the transfer techniques is also generally recommended. In this case, the ionic contaminants can be determined in polar solvents (acetone, acetonitrile, methanol, ethanol, etc.) as well as very nonpolar solvents (e.g., halogenated hydrocarbons such as dichloromethane).

As described earlier for the implementation of ion-exchange microcolumns for neutralization, SPE columns with a variety of different sorbent materials can also be placed in an auxiliary flow system for matrix removal. Using nonpolar (reversed phase) sorbents, hydrophobic sample constituents are retained and the ionic compounds passing through can be subsequently trapped and injected into the IC system as explained before. By proper selection of the sorbent phase, many organic compounds can be retained and interference-free ion determination becomes possible even for complex samples. However, the sorbent materials themselves sometime release contaminants, which may interfere in ion determination. Because the sorptive capacity of solid-phase microcolumns is limited, timely exchange for a fresh one is required. In-line regeneration – as is possible with ion exchangers – is generally not crowned with success.

Solid-phase removal of interfering (organic) compounds can be successfully performed for large series of samples in the determination of anions or cations in natural waters containing elevated levels of humic acids, wastewaters containing a high organic fraction, dye solutions, and aqueous extracts of plant, soil, and other solid samples after removal of particles by filtration or centrifugation [33]. The application of any matrix elimination technique should be accompanied by recovery tests with spiked samples, since the retention of ions on the trap columns is sometimes different compared to that of ions present in pure aqueous solutions. Determination of anions and cations at the low ppb level is possible even in complex sample without sacrificing the performance of IC separations [34].

8.4.3   Liquid–Liquid Extraction

Liquid–liquid extraction is of little importance in the context of IC analysis as the transfer of the analyte into an organic solvent phase means that either stripping or evaporation of the solvent is required before separation by ion chromatography can be carried out. The working steps associated with liquid–liquid extraction require a great deal of expenditure on both (mostly manual) workload and time and are connected with a high risk of contamination and loss of analytes. The few examples to be found in the literature mostly refer to the preconcentration of heavy metal ions after conversion with organic complexing agents and subsequent separation by cation-exchange chromatography [35]. The analysis of ions in oil samples or solvents immiscible with water can be carried out after extraction with water or the chromatographic eluent. However, this involves a posttreatment step in which the remaining solvent contained in the aqueous phase must be removed (e.g., by SPE) before the sample can be analyzed by IC. This also means that only the extractable ion fraction is determined. The total content of a particular element is only accessible after complete mineralization (e.g., with a combustion apparatus).

8.4.4 Gas-Phase Separations

Separation via the gas phase is a method with a very high degree of selectivity as only few analytes amenable to IC determination can be converted into a volatile form. This means on the contrary that the attractive option of simultaneity offered by IC is lost to some extent. Nevertheless, there are quite a number of interesting applications involving this type of sample preparation in combination with IC.

Gas-phase separations can be carried out in different manner with regard to the necessary instrumentation. Although classical distillation is easy to perform, it usually requires large sample volumes, takes long time, and cannot be automated. A very elegant technique that has almost been completely forgotten is the so-called microdiffusion technique [36]. It involves placing the sample solution and an absorption solution in two separate flat vessels (Petri discs, for example) beside each other under a common air tight cover. Under isothermal conditions, transfer of the volatile analyte (or a previously generated volatile form of it) occurs from the liquid sample solution to the absorption solution via the gas phase. The separation process is relatively slow, but a large number of samples can be prepared in parallel in this way with low cost.

A completely different possibility of utilizing gas separation is the use of hydrophobic microporous membranes. The methods known as gas diffusion and pervaporation are based on the fact that aqueous solutions cannot wet hydrophobic membranes. If the sample and the absorption solutions are separated by such a membrane, volatile compounds diffuse through the air-filled pores of the membrane and are enriched in the absorption solution. This type of gas-phase separation has been used in miniaturized batch procedures (sometimes termed membrane distillation). The required hardware is commercially available in a single-use format. Alternatively, gas diffusion and pervaporation can be implemented into continuous-flow or flow injection systems. The latter permits ready automation with in-line coupling possibility to IC (see Section 8.4.6.3).

Interesting examples of the use of gas-phase separation by distillation, microdiffusion techniques, or gas diffusion via hydrophobic membranes are the determination of cyanide, sulfide, sulfite, and nitrite, which can be volatilized as gases (HCN, H_2S, SO_2, NO_x) from the matrix-loaded sample and then converted back to their respective anionic forms in a suitable absorption solution for IC determination [37].

In a similar way, ammonium (and also Kjeldahl nitrogen) and some short-chain amines can be isolated after alkalinization of the sample solution via the gas phase and determined in an acidic collector solution by cation-exchange chromatography. It is worth mentioning that gas-phase separations can be applied also directly to samples heavily loaded with particulate matter without any additional sample preparation. However, the release of analytes from such samples may be affected by the adsorption of the analyte(s) onto the solid material.

8.4.5 Precipitation

Removal of matrix components by precipitation is another means to prepare liquid samples for IC analysis. Few examples can be found in the literature, which is certainly connected with the tedious manual procedures involved but probably even more with the risk of analyte loss caused by coprecipitation of the analyte(s). The formation of low solubility salts can be applied to remove, for example, chloride or sulfate by precipitating the two anions with silver or barium ions, respectively. It must be borne in mind, however, that the respective metal cations are introduced into the sample solution, which may cause problems in the subsequent anion determination using IC. The use of barium- or silver-loaded SPE cartridges to facilitate on-cartridge precipitation of sulfate and chloride, respectively, is an often applied means omitting the necessity of precipitate removal. Precipitation of proteins from biological and food samples by acidification or addition of an organic solvent (acetonitrile and methanol are often employed) is common practice in many HPLC procedures for analysis of dairy products and biological samples of different kind. For the determination of ionic species using IC, this procedure is obviously not very popular since it is proposed in only a few publications [38].

For trace determination of perchlorate in plants protein precipitation has been accomplished by heating the sample. After centrifugation, the supernatant solution is exposed to alumina and eventually passed through an SPE cartridge filled with polydivinylbenzene sorbent. The cumbersome procedure cannot be applied to other ions than perchlorate since substantial losses have been observed [39]. In this particular case, this is regarded to be an advantage as ionic interferences in perchlorate measurement using conductivity detection are significantly reduced.

Another example of a precipitation reaction as a cleanup step prior to IC is the determination of anions in meat products [40]. The meat sample is cut into coarse pieces, borax solution is added, and then the sample is blended with a high-frequency homogenizer. The resulting dispersion is heated and treated with zinc sulfate/potassium hexacyanoferrate solutions (Carrez reagent) to precipitate the proteins. After centrifugation, the supernatant solution is filtered through a $0.45\,\mu m$ filter and injected into the IC system. In order to avoid interference by excess sulfate (rendering impossible phosphate determination), Zn acetate can be used as an alternative to sulfate. Carrez precipitation has also been applied as sample preparation procedure for nitrate determination in milk [41].

8.4.6   Membrane-Based Separations and Sample Treatment

Membrane separation plays an important role in several technical processes such as water treatment, recovery of precious metals, purification of pharmaceuticals, and other industrial products, and also in medical treatment, such as hemodialysis. Despite some efforts of the IUPAC [42], the terminology of membrane-based separation techniques is not yet harmonized, and sometimes, different terms are used in the literature for the same basic separation process (and the same term is used for different techniques). In order to avoid confusion among the readers, the classification as used here is given in Table 8.1 together with the fundamental principles and kind of membranes involved.

In the context of analytical chemistry in general and ion chromatography in particulate, the use of membrane separation (despite filtration) is still very limited. However, there are a few prominent examples of using membranes in IC away

TABLE 8.1 Compilation and Nomenclature of Membrane-Based Separation Techniques

Designation	Driving Force	What Gets Separated	Aim of Application	Membrane Characteristics
Micro/ ultrafiltration	Pressure gradient (positive or negative)	Dissolves analytes	Removal of particulates, colloidal matter and high-molecular-weight compounds	Hydrophilic (or wetted) micro- and nanoporous polymers
Passive dialysis	Concentration gradient; diffusion controlled	Dissolved low-molecular-weight compounds	Removal of particulates, colloidal matter, and high-molecular-weight compounds	Hydrophilic nanoporous, or homogeneous polymers
Donnan dialysis	Ion exchange; concentration gradient; diffusion controlled	Ionic species of opposite charge to membrane	Removal of ions of opposite charge to membrane, neutralization of (strong) acids and bases	Ion exchange membranes with positively or negatively charged groups
Electrodialysis	Ion exchange; electrical field; migration	Ionic species of opposite charge to membrane	Removal of ions of opposite charge to membrane, neutralization of (strong) acids and bases	Ion exchange membranes with positively or negatively charged groups
Gas diffusion (pervaporation)	Vapor pressure and solubility differences; diffusion controlled	Dissolved gases and volatile compounds	Removal of particulates, colloidal matter, nonvolatile compounds	Hydrophobic microporous or homogenous polymers

from sample preparation, that is, suppressor modules for conductivity detection and automated eluent generation systems, both utilizing ion-exchange membranes either in passive mode or in the form of electrodialysis [43].

The variety of available membranes of different compositions and structures permits the realization of different separation principles and thus to generate a certain degree of selectivity of separation for analytical applications [44]. It should be considered, however, that membrane separation – with the exception of certain kinds of supported liquid membranes [45, 46] – can generally provide only group separations of compounds of similar physical or chemical properties from others with distinctly different properties.

Separation of particles and colloidal matter from liquids, removal of high-molecular-weight compounds, separation of volatile compounds, and separation (or more precisely exchange) of ions of opposite charge are representative examples of what can be done conveniently via membrane-based sample preparation. With respect to the kind of membranes, distinction is made according to pore size, microstructure (homogeneous, microporous, fibrous), physicochemical behavior (hydrophilic, hydrophobic), reactivity (e.g., nonpolar, polar, and ion-exchange membranes), and shape (flat and hollow fiber membranes).

Hydrophilic fibrous and microporous membranes with variable pore size are typically applied for coarse and microfiltration, respectively, whereas hydrophilic homogenous polymer membranes are used in ultrafiltration as well as passive dialysis. Hydrophobic microporous and thin-walled nonporous polymer membranes (e.g., made from PTFE, PVDF, silicone) can be used for separation of volatiles and dissolved gases as is, for instance, done in eluent degassing systems and the more recently introduced CO_2-suppressor systems for anion-exchange chromatography [47].

In the following sections, the principles of different dialysis techniques as well as gas-diffusion separation are briefly outlined and the features of the respective techniques for sample preparation in IC analysis are presented. A very interesting aspect of membrane separation is the relatively easy way to automate the procedures by their implementation into continuously operating flow systems. The hyphenation of dialysis and gas diffusion with IC and relevant applications are also presented.

8.4.6.1 *Technical and Instrumental Aspects*

8.4.6.1 Technical and Instrumental Aspects Irrespective of the kind of membrane technique considered, a principal distinction can be made between their application in batch- and continuous-flow configurations. In both cases, the sample solution is separated from a receiver solution by a semipermeable membrane. Transfer of analytes or unwanted matrix components across the membrane occurs if and as long as a driving force exists. The driving force can be a concentration gradient, electrostatic attractions, electromigration (created by an applied electrical potential), or solubility differences of the analyte(s) in the contacting phases. The transfer rate (mass transfer per time unit) depends not only on the magnitude of the gradient but also on the membrane characteristics and the convection/diffusion conditions on either side of the membrane. In optimizing membrane separation procedures, it should be clearly defined whether a high mass transfer (limiting case is the complete depletion from the sample and total recovery in the receiver solution), a high transfer rate, or a high concentration in the receiver solution (connected with mass transfer and the liquid volume of the receiver solution) is desired [48]. Considering matrix removal, the aim is to achieve a high depletion from the sample, whereas for analyte separation, a high recovery and possibly even preconcentration in the receiver solution are generally preferable.

In batch systems, the sample and receiver liquids (separated by the membrane) are held in place for a predefined time, typically until (distribution) equilibrium is achieved. Depending on the kind of separation process involved, the result of a batch separation can be equal concentrations of the analyte on both sides of the membrane (as is, for instance, the case in passive dialysis after equilibrium has been attained) but also a reverse concentration gradient for the analyte compared to the initial situation (e.g., uphill transport in Donnan dialysis). By chemical conversion of the analyte in the receiver solution, the concentration gradient can be maintained and eventually complete transfer achieved. It must be warranted, however, that no back diffusion of the trapping reagent or derivate formed in the receiver solution occurs. This is given, for example, in gas-diffusion separation with microporous membranes when volatile analytes are converted in the receiver solution into nonvolatile ionic species.

In order to enhance the speed of the batch separation process, the membrane surface to liquid volume ratio should be high. Additionally, agitation of the liquid phases (by stirring or shaking) and heating of the solutions are advantageous to enhance mass transfer in the liquid phase and reduce the diffusion layer thickness in the vicinity of the membrane. The format of batch units available for membrane separation is quite different. It comprises relatively large-sized bag-type configurations, miniaturized systems requiring only minute sample volumes, and probe-type configurations, which can be used similarly for membrane separations in the laboratory or for at-site, *in situ* sampling. Batch systems are generally associated with tedious manual operation steps and therefore do not appear attractive for routine applications. However, some of the miniaturized membrane units commercially available can be combined with pipetting stations permitting the parallel processing of many samples in an automated manner. Since the separation units are generally used only once, carryover effects are absent. Costs of commercially available disposable units are, however, relatively high.

The implementation of membranes into flow-through systems is a subject of major concern in industrial applications and many configurations have been developed in the past aiming at optimum efficiency of the separation process. The design of the separation unit and the flow conditions of the sample and receiver streams are important parameters decisive for the transport rate of analytes across the membrane, the completeness of separation (depletion and recovery), the resulting concentration in the receiver solution (which is eventually brought to detection), and the consumption of sample and receiver liquid. Sandwich and tubular configurations with flat or hollow fiber membranes, respectively, are the two types of separation units commonly applied for analytical purposes (Figure 8.1).

The geometric dimensions of membrane modules reported in published work vary significantly. In order to attain high recovery, thin-flow channels and a large available

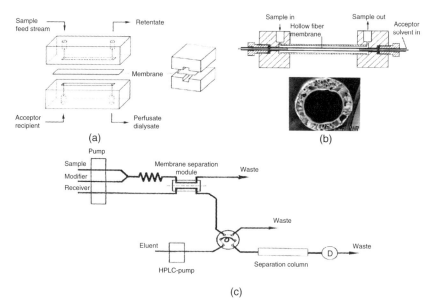

Figure 8.1 Schematic representation of (a) planar sandwich and (b) tubular hollow-fiber membrane separation units. In (c) is shown the hyphenation of the flow-through units with an IC system. The injection valve can be regarded as a kind of interface between the two flow systems.

membrane area are advantageous. The absolute flow rates and flow rate ratio of sample and receiver solutions have an effect on sample and reagent consumption, the residence time, and thus also on the degree of separation and the resulting concentration of the analyte [49, 50]. Flow-through systems incorporating membrane separation can be operated separately from the chromatography system (and the receiver solution or matrix-depleted sample solution collected in vials) or directly hyphenated to the injection valve of the IC system. The latter is easily accomplished by connecting the respective outlet tubing of the separation unit to the inlet of the injection valve (see Figure 8.1c).

Irrespective of the layout of the separation unit, a conceptual distinction can be made between exhaustive, equilibrium, and kinetically controlled separations. In the exhaustive mode, the total amount of analyte present in the sample is recovered and eventually contained in the receiver liquid. This is an interesting option whenever only a small amount of sample is available; yet in practice, it requires long residence time of the sample in the separation unit. In order to achieve distribution equilibrium of membrane separation in flow-through systems, the contact time between sample and receiver solution must generally be long. In practice, separation units with relatively large membrane surface and thin flow channels are used and the flow rates are kept low (or solutions are even temporarily stopped). In addition, membranes with high permeability are selected. For a certain material and type of membrane, the permeability is related to the thickness of the membrane. One advantage of both the exhaustive and equilibrium separation modes is that the separation process itself does not need to be calibrated and that small variations in the experimental parameters do not cause significant alterations of the resulting analyte concentration. The kinetically controlled situation of membrane separation is the simplest to achieve since only constant flow rates of sample and receiver solutions must be warranted. However, for a fixed experimental setup (design of flow-through unit and membrane), the degree of separation, that is, the recovery and resulting analyte concentration in the receiver, depends sensibly on flow conditions and needs to be regularly checked. Also, the analyte recovery is typically low, resulting in a reasonably high degree of dilution.

8.4.6.2 Dialysis Techniques Dialysis techniques include passive dialysis, electrodialysis, and Donnan dialysis, which differ with respect to the driving force causing separation, the kind of membranes used, and – as a result of this – the selectivity and area of application. The fundamental procedures, however, are common to all three techniques in that analytes or interfering species present in the sample cross the semipermeable membrane and are trapped in a receiver solution. In some applications, the matrix-depleted sample solution and in others, the analyte-enriched receiver solution are brought to detection. With respect to the mode of operation, batch and flow-through systems are to be discerned, both requiring specific hardware and offering particular features and limitations.

Passive Dialysis In passive dialysis, separation occurs due to a molecular sieving effect; hence, selectivity is defined solely by the different size of the analyte or matrix compound(s) to be removed. The typical applications, similarly to UF, are the separation of small ions or neutral molecules from colloidal matter and high-molecular-weight compounds. Membranes used are similar to UF membranes and are also characterized by their molecular-weight cut-off. However, in contrast to UF, in passive dialysis (ideally), no flow of solvent through the membrane occurs so that particles and high-molecular-weight compounds are not deposited on the membrane surface and considerably longer lifetime is obtained. The driving force in passive dialysis is the concentration gradient across the membrane. In batch procedures, the resulting concentration of analytes after attainment of equilibrium is equal on both sides. A common configuration for batch dialysis widely applied in practice (mainly for desalting of enzyme solutions and biological fluids prior to HPLC) is the so-called dialysis tubing, which can be cut to size by the user, closed by a clamp, and then exposed to the sample solution. The typical time to achieve equilibrium in batch mode is about 4–8 h; quite often, batch dialysis of many samples is performed in parallel overnight to save working time. Using commercially available miniaturized disposable units with large membrane surface to liquid volume ratio (e.g., Slide-a-Lyzer® and Float-A-Lyzer of Thermo Instruments and Spectrum-Lab, respectively), the procedure is simplified compared to the use of classical dialysis tubing and slightly faster. The prices of these units are, however, considerable.

Flow-through dialysis has a long tradition in air-segmented flow analysis and flow injection analysis for efficient in-line sample cleanup [51–53]. The hyphenation of dialysis with LC was first published in the early 1980s, and in the following years,

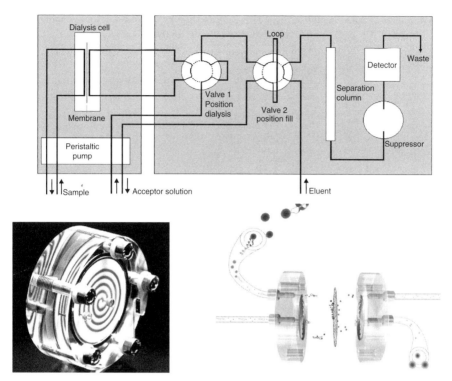

Figure 8.2 Schematic representation of the patented flow-dialysis system used for in-line sample preparation prior to ion chromatography. The two insets show the large surface-to-volume ratio spiral dialysis cell and the principle of dialysis excluding microparticles and high-molecular-weight compounds from membrane transfer (*Metrohm AG, Switzerland*).

a slowly increasing interest was observed [54–56]. In-line dialysis–IC coupling was probably first explored in the author's laboratory in the early 1990s and later on examined in detail by many coworkers in the frame of diploma theses and internal reports. Only few papers have been published in the scientific literature since then [57–60]. A dialysis system for in-line sample preparation in combination with IC is shown in Figure 8.2.

In the flow-through mode utilizing peristaltic or syringe pumps for liquid propelling the sample and receiver solutions can be easily manipulated and replaced or replenished at will. This permits, for instance, to provide the sample solution continuously on one side of the membrane while the acceptor stream is halted. After sufficiently long waiting time, the concentration in the receiver solution approaches that present in the original sample. Only then the receiver solution is transferred into the injection loop of the IC system. The patented Metrohm dialysis system [61] operates in this manner with the advantages that no sample dilution occurs and (after initial optimization of all experimental parameters) the dialysis process needs no calibration.

TABLE 8.2   Typical Applications of Dialysis Sample Preparation and Needs for Pretreatment Prior to Sample Admission to In-Line Dialysis IC

Kind of Sample	Interfering Matrix Constituents	Pretreatment Prior to In-Line Dialysis
Wastewater (with sludge)	(Micro)particles and organic load (humic matter)	None; centrifugation
Soil extracts	Particles and organic load (humic matter)	Coarse filtration or centrifugation
Industrial effluents	Particles, organic load (e.g., surfactants, lignins)	None; coarse filtration or centrifugation
Beverages and juices (with pulp)	Pulp and colorants	None; centrifugation
Extracts of dairy products	Particles and proteins	Coarse filtration or centrifugation
Emulsions and dispersions (cutting and drilling oil)	Oil and other organic compounds	None; dilution
Body fluids	Proteins	None; prevention of coagulation

The time required to attain equilibrium is about the same as for the ion separation by IC so that the sample preparation of one sample is done while the previous one is analyzed, that is, no extra time is required.

The marketing of in-line dialysis by Metrohm company has led to extensive experience with this system in routine laboratories. The company has published numerous application notes impressively demonstrating the many areas of application as well as the simplification of the analytical procedure compared to alternative means of sample preparation [14]. In many instances, the combination of microfiltration (for removal of particulates) and SPE (for removal of high-molecular-weight compounds) can be substituted by automated in-line dialysis. In Table 8.2, a compilation of sample types, possible interferences, and requirements of sample preparation left (if any) prior to admission of the sample solution to the dialysis system is presented.

Dialysis probes are another way to perform dialysis in a conceptually different manner to the batch and flow-through operations described above [60, 62]. Instead of two compartments or flow channels, one for the sample and the other for the receiver solution, typical dialysis probes consist of a cylindrical body with a thin channel at the bottom covered by the membrane. Connections are drilled to the two ends of the channel to permit the provision of a perfusion liquid flowing behind the membrane (Figure 8.3). The contact to the sample solution is made by dipping the probe into the solution contained in a beaker. Analytes present in the sample are thus transferred into the perfusion stream (due to the driving force of the concentration gradient) and are carried away continuously. The perfusion liquid can be either sampled in a vial or directly introduced into the injection loop of the IC system (in-line configuration).

Microdialysis is a special format of passive dialysis, which has found enormous application in neurophysiological science, pharmacokinetic investigations, and drug

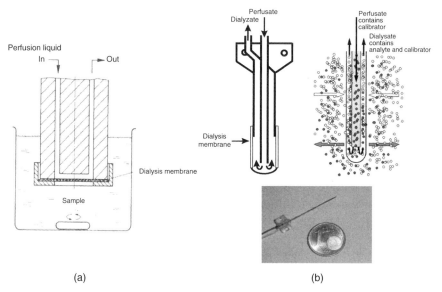

(a) (b)

Figure 8.3 Configuration of dialysis probes suitable for selective sampling by direct immersion into the sample solution. (a) Flat-membrane probe with channel length in the range of a few millimeter; (b) Cannula-type microdialysis probe (scheme, photograph and principle of operation).

discovery studies. More recently, increasing interest in environmental analysis and process applications became evident [63–65]. Capillary dialysis tubing of 50–500 μm inner diameters is employed in different probe configurations. The most common type is the concentric cannula probe that has a typical membrane length of only 1–3 mm and is operated with typical perfusion flow rates of 1–10 μl min^{-1}. Alternative designs are loop-type, side-by-side, shunt-type, or linear hollow-fiber configurations. Due to the large surface-to-volume (lumen of the hollow fiber) ratio of the latter, the experimental conditions can be set so that partitioning equilibrium can be achieved between the outer environment and the perfusion liquid without halting the flow. This simplifies calibration and analyte dilution does not occur. The minute devices can be employed for probing microenvironments. On-line connection of the probe to the chromatographic systems is not without difficulties, but can be solved with some technical skill [66]. Few papers have been published demonstrating the features of hyphenating microdialysis with ion chromatography [67–70].

Donnan Dialysis and Electrodialysis In Donnan dialysis, the sample is separated from a receiver solution of relatively high ionic strength by an ion exchange membrane, which – depending on the charge of the membrane – is permselective for anions or cations only [71]. The ionic strength gradient across the membrane establishes a chemical potential that allows ions possessing a charge opposite to the charge of the membrane groups to pass through the membrane phase. Since the permeation of counter ions is hindered due to electrostatic repulsion, the transfer of ions present in the receiver solution into the sample solution must be balanced for electroneutrality reasons by the transfer of sample ions in the opposite direction. Typical applications of Donnan dialysis for analytical purposes are matrix normalization and analyte pre-concentration [72, 73].

Matrix normalization in turn can be achieved either through selective addition of a counterion or purposive removal of ions of defined charge [74, 75]. Acidic samples can, for instance, be neutralized by using an anion-exchange membrane module and an alkaline solution on the other side of the membrane. A relevant problem of this application is the partial transfer of ions of opposite charge to that of the membrane (nonideal permselectivity), which can lead to contamination of the sample with ions of the receiver solution. Analyte preconcentration via Donnan dialysis is possible (the so-called uphill transport [71, 73, 76]), yet not very attractive in combination with IC. The high load of ionic species introduced adversely effects IC separation.

In electrodialysis, at least a three-chamber system is required with the central chamber containing the solution to be treated and the two adjacent chambers equipped with electrodes that serve as anode and cathode of an electrical circuit. The chambers are separated from each other by ion-exchange membranes, the polarity of which being connected with the purpose of application. The driving force of ion transport is the overlap of a concentration gradient (as in common Donnan dialysis) and the electromigration of ions caused by the electrical field. Electrodialysis can be used for water purification, for desalting, and for neutralization of acidic or alkaline solutions [77]. In the latter application, hydroxonium or hydroxyl ions present in the anode or cathode compartment, respectively, are transferred to the central chamber containing the sample solution. In return, cations or anions originally present in the sample are moving through the ion-exchange membrane toward the outer cell compartments. The hydroxonium and hydroxyl ions can be delivered by a suitable acid or base or, alternatively, electrolytically generated from pure water. The latter approach has the advantage that counterions of the respective acid or base, which might be transferred to the sample compartment as a result of nonideal permselectivity, do not contaminate the sample.

Apart from sample preparation, electrodialysis became a viable alternative to common Donnan dialysis for conductivity suppression and has found application also for in-line generation of eluents for anion and cation chromatography from pure water (e.g., Reagent-Free IC, trademark of Dionex Corp.) [78]. Sample preparation via electrodialysis has also been reported. The most relevant application for subsequent IC analysis is neutralization of strong acids or bases. In early work, batch systems were used but later on advantage has been taken from the flow-through configuration, which can easily be hyphenated to IC [79, 80].

Determination of halogenides, phosphorus, nitrogen, and sulfur after dry ashing (e.g., Schöniger oxygen flask combustion followed by adsorption in alkaline solution) or wet digestion (acid treatment or alkaline fusion) using IC is generally difficult to accomplish because of the high load of salts, acids, or bases introduced. Also, ionic species being introduced in the decomposition process (acid anions or base cations) cannot understandably be determined. A few interesting papers have been published where electrodialytic sample treatment has been successfully applied for such kind of samples [81].

8.4.6.3 Gas Diffusion and Pervaporation

The permeability of many organic polymer materials for gases has been utilized for the separation of gases and volatile compounds in many technical and analytical applications [82, 83]. The terminology of membrane-based gas separations, however, is not consistent and terms such as gas diffusion, gas dialysis, gas permeation, membrane distillation, and pervaporation are sometimes applied for similar basic processes. In this section, we only refer to gas separation processes in which two liquid phases (donor and receiver solution) are separated by a semipermeable membrane; this is named gas diffusion here. The use of gas-permeable membranes for gas sampling and analysis in combination with IC is discussed in Section 8.8.

Membranes used for gas diffusion are of either hydrophobic microporous structure (mostly made of fluoropolymers or polypropylene) or nonporous (typically made of polydimethylsiloxane). When hydrophobic microporous membranes are used in contact with (nonwetting) aqueous solutions, the pores of the membrane resemble an air gap and compounds to be separated are released at one boundary layer and are redissolved at the other. A limitation of microporous membranes is that they cannot be used with nonpolar (organic) solvents and may lose their performance when samples contain surfactants, oil, or fat. Homogeneous silicon membranes do not pose such problems; however, they constitute a third phase (similarly to an organic solvent) and analyte transfer is by dissolution, permeation, and redissolution. Selectivity is thus governed by partitioning, and the separation process resembles liquid–liquid extraction with coupled back extraction. True membrane extraction techniques are out of the scope of this contribution, but it might be mentioned that a large number of papers have been published on this subject, particularly for nonpolar organic compounds and their determination using HPLC or GC [84, 85].

Prominent examples of gas-diffusion separation not related to sample preparation are continuous in-line eluent degassing in liquid chromatography and CO_2-removal devices following the common (chemical or electrodialytic) suppressor systems applied in anion IC to reduce the background conductivity further [47].

Since the separation process in gas diffusion is based on volatility differences between the analytes and potentially interfering matrix constituents, a relatively high selectivity can be achieved. The transfer of dissolved ionic constituent, the majority of organic solutes and colloidal matter and (micro)particles is completely excluded. Analytes amenable to gas-diffusion separation are dissolved gases, compounds of sufficient volatility, and any component that can be converted intentionally into a gas or volatile compound.

Many of the applications of gas-diffusion separation as a sample preparation technique fall into the third category and refer to the determination of anions or cations that are in acid–base equilibrium with a gaseous species; for example, sulfite, sulfide, cyanide, nitrite, carbonate, ammonium, short-chain amines, and carbonic acids.

The long-lasting experience with gas diffusion in flow systems (air-segmented, flow injection, and sequential injection [48, 52, 86]) has led to optimization of the

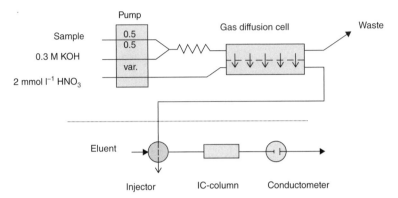

Figure 8.4 Coupling of gas-diffusion separation with ion chromatography for the determination of ammonium and short-chain ammines. KOH is used to convert the analytes to ammonia and volatile ammines, which diffuse across the membrane and are trapped in a slightly acidic nitric acid solution. Separation of the N-containing cationic species is eventually done by cation-exchange chromatography.

analytical procedures with respect to the instrumental configuration (design of the separation unit, flow rates) as well as the optimum chemical conditions (matrix modification of the sample/donor stream, composition of the receiver solution). The translation to the coupling with IC is relatively simple and has been realized in several papers [87–90].

The reconversion of the gaseous species into the ionic form is accomplished using slightly acid or alkaline receiver solutions for the trapping of basic (e.g., ammonia) or acidic (e.g., sulfur dioxide, hydrogen sulfide, hydrogen cyanide) gases, respectively. These solutions are well compatible with the IC eluents commonly applied in cation and anion chromatography. A typical manifold configuration for gas-diffusion IC is shown in Figure 8.4 exemplifying the simultaneous determination of ammonia and volatile amines. The method has been adapted to a variety of samples including the analysis of municipal wastewater, soil extracts, and industrial effluents [91]. One of the beauties of gas-diffusion separation is the easy way to achieve analyte preconcentration (vide infra). However, considering analyte determination by IC, a limitation can be seen in the high selectivity of gas-diffusion separation. The IC feature of simultaneity is almost lost since typically only a single analyte or few analytes are accessible by gas diffusion at the same time.

8.5 TRACE ANALYSIS AND PRECONCENTRATION FOR ION CHROMATOGRAPHIC ANALYSIS

Trace analysis problems exist in many scientific disciplines, and the demand for correspondingly sensitive determination methods including ion analysis has led to various attempts to improve IC detection limits. With the currently available instrument configurations in IC (usually equipped with conductivity detection with or without suppressor technique), numerous ions can be routinely determined without great difficulties even in the low microgram per liter concentration range. Using larger injection volumes can increase the sensitivity by up to about one order of magnitude; this is, however, only possible for samples with low ionic strength since otherwise band broadening occurs.

A different way of using IC for trace analysis is an instrumental improvement using detectors with high response factors and low noise level leading to improved signal-to-noise ratios. In this way, and also by the use of different detectors that are basically more sensitive than conductivity detection (e.g., amperometry, fluorescence, and in particular mass spectrometry), the detection limits for the determination of many ions can be improved considerably. However, this is often accompanied by a reduced spectrum of components that can be determined simultaneously and/or an increase in the complexity of the necessary instrumentation (and hence an increase in the costs). Postcolumn derivatization and spectrometric detection are also applied for selectivity and sensitivity enhancement in IC [92].

By the use of preconcentration of the target ions in a pre-chromatographic step, it is possible to achieve considerable improvements in sensitivity and often, at the same time, matrix elimination. The latter is of particular importance for preconcentration of trace components in difficult matrices. There are different means available for ion preconcentration; the most attractive are based on SPE and the use of membranes, both of which are described in the following two sections.

8.5.1 Preconcentration Using SPE

SPE with appropriate sorbents is very versatile and is by far the most frequently applied method for preconcentration [93–95]. The selectivity of the enrichment step can be influenced by the type of sorbent so that particular ions (alone or together with other ones) can be preconcentrated and separated from other matrix components. In principle, very high preconcentration factors can be achieved by first loading the sorbent phase with relatively large volumes of the sample solution and then elute the analytes with a small volume of extraction solvent. For example, numerous anions in relatively pure water samples can be preconcentrated simultaneously without severe difficulties on anion exchangers and after elution determined by anion chromatography. However, there is always a risk that, even if the ion retention capacity of the sorbent is adequate, not all the ions to be determined will be retained quantitatively. Samples with a higher total ionic strength function themselves as an eluent during preconcentration; this means that only ions with a high affinity to the ion exchanger (e.g., multivalent ions or easily polarizable ions) will be retained quantitatively. A further general problem results from the fact that a sufficiently strong elution agent must be used for eluting the preconcentrated ions to avoid large elution volumes and band broadening. It must also be borne in mind that the eluting solvent represents a new matrix that must be compatible with the IC determination. A further aspect to be considered in all kinds of SPE is the retention kinetic of the respective compounds to be separated. Too fast flow rates during the loading/preconcentration step may cause kinetic breakthrough even though the ion retention capacity of the ion-exchange microcolumn is not exhausted.

With respect to practical aspects, trace preconcentration can be carried out in a conventional off-line procedure with ready-to-use SPE-syringe or cartridge systems (see Section 8.4.2). This has the advantage of being very flexible without need for further instrumentation. More important from analytical point of view is the fact that the SPE unit is generally used for a single sample only and then disposed. Therefore, samples with elevated levels of matrix constituents can be processed, and poisoning of the sorbents by irreversible retention of sample constituents does not pose a serious problem.

When a large number of samples have to be processed, the manual procedure is, however, tiring and the numerous necessary sample contacts with the various vessels and the laboratory atmosphere are a serious reason for possible contamination. Integration of the preconcentration step into the analytical procedure of IC determination (in-line preconcentration) is relatively easy to accomplish by implementation of a small preconcentration/trap column in the injection loop of the IC system. A discrete sample volume is then passed through the column at a constant flow rate, and on switching the valve, the retained components are eluted by the mobile phase and separated on the column. In order to perform well, the sorbent material must be stable at high pressures and the capacity of the sorbent as well as the retention strength for the analytes must be sufficiently high. On the other hand, the interaction between the analyte and the preconcentration column must not be too strong as the mobile phase must be able to elute the analytes in a plug-like manner to prevent band broadening. In a concrete application, the selection of an appropriate sorbent and column dimensions is important and optimization of the various parameters must be carefully made. As an alternative to the positioning of the trap column in the injection valve, it can also be implemented in a secondary low-pressure flow system and the eluate transferred to the injection loop by direct coupling to the IC instrument. Although inline preconcentration techniques are very attractive with respect to automation capability, low risk of contamination, and reproducibility of the entire procedure, the repeated use of the same sorbent for a larger series of samples requires careful consideration of possible sorbent fouling. In some cases, intermediate flushing with a suitable solvent is necessary to remedy the sorbent, but in other cases, irreversible matrix retention leads to failure and requires exchange of the SPE microcolumn.

The preconcentration step for ionic compounds is highly selective with respect to the charge of ions; yet, ions of the same charge are generally retained simultaneously. This is an advantage for multiion determination in combination with IC separations but is limited to samples with moderately high total ionic strength and situations where the concentrations of the different ions are in a similar range. This applies, for example, to the quality control of ultrapure water used in the semiconductor industry, to the monitoring of boiler water in power plants and, with certain constraints, to the analysis of clean environmental waters.

Whereas in numerous cases multicomponent preconcentration is desirable, in some applications, selective preconcentration of a single ion or few ionic species is intended. Both can be accomplished by proper selection of the experimental conditions (selection of sorbent, modification of sample composition, interme-diate rinsing of the trap column for elution of unwanted constituents, etc.). This concept has, for instance, been used in the preconcentration of perchlorate in high salinity waters where after the passage of a discrete sample volume through the anion-exchange trap column interfering matrix ions (mainly chloride, sulfate and carbonate) are washed out by a diluted NaOH solution before the more strongly retained perchlorate is eluted onto the separation column and quantified by suppressed conductivity detection [96].

The domain for trace metal determinations clearly is atomic spectrometry in its many variants. However, for relatively clean aqueous samples, cation chromatog-raphy is a viable alternative offering simultaneous determinations in relatively short time. After preconcentration using chelating ion exchangers, very low detection limits can be achieved with conductivity detection. The hyphenation of IC with ICP-MS after preconcentration provides even more selective and sensi-tive metal determinations, and in addition, it opens the pathway for trace metal speciation [97].

8.5.2   Membrane-Based In-Line Preconcentration

In the majority of membrane-based separation techniques, the ultimate aim is to remove matrix constituents, which may cause interferences in the subsequent deter-mination step. However, some of the techniques described have also the potential to preconcentrate analytes and thus to improve the detection limits of the analytical method applied. Trace enrichment using ion-exchange microcolumns suffers from the disadvantage of being matrix dependent. This is less critical in membrane-based techniques for analyte preconcentration such as Donnan and electrodialysis and gas-diffusion separation. One attractive common feature is that the membrane effectively prevents the transfer of the majority of matrix constituents and poisoning is generally not a problem.

In passive dialysis, by nature, preconcentration cannot be achieved; at best, the original analyte concentration is maintained. However, a very interesting and promis-ing configuration is the coupling of passive dialysis with subsequent SPE. A suitable experimental setup is shown in Figure 8.5.

In this system, the analytes separated by dialysis are continuously transported to a microcolumn filled with a suitable sorbent placed in the injection loop of the IC system. After a preset time, the injection valve is switched and the enriched ana-lytes displaced by the eluent. The advantage of this hyphenation is, on the one hand, that ions present in the dialysate are preconcentrated (and thus overall sensitivity improved) and, on the other hand, that matrix constituents such as proteins, humic compounds, microparticles, colloidal matter, which may interfere in the SPE step, are excluded [98]. The lifetime of the SPE column can thus be significantly pro-longed. This concept has been published years ago and commercialized by Gilson (the so-called ASTED system) for analyte preconcentration of dialysates prior to HPLC determination [99].

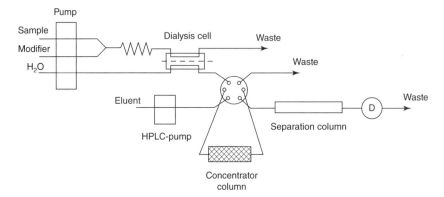

Figure 8.5 Manifold configuration for membrane separation coupled to trace enrichment using sorbent extraction.

In Donnan dialysis, preconcentration of analytes is the result of uphill transport of the target ion against a concentration gradient. This is possible only by a simultaneous downhill transport of other ions of the same charge. Ions originally present in the sample are thus exchanged for ions delivered from the receiver channel. If the receiver volume is made much smaller than that of the sample (or, in flow-through systems, the flow rate of the sample is much higher than that of the receiver), preconcentration can be principally achieved. Since the receiver contains a high excess of ions, the analytes after preconcentration are present in a high ionic strength matrix. This generally leads to severe complications in the IC determination so that the attractiveness of this approach is very limited.

Several examples of Donnan dialysis of preconcentration of cations and anions prior to IC determinations can be found in the literature [72, 73, 76]. A closer inspection of the published work reveals only moderate preconcentration factors of about 10 at maximum. In the classical format of electrodialysis, preconcentration capabilities are principally the same as for Donnan dialysis. Transfer rates are slightly faster, and in principle, pure water can be used as the receiver solution, thus omitting the problems involved in Donnan dialysis with the high ionic strength matrix of the receiver solution. In a modification of electrodialysis, a four-chamber system was configured with two channels serving as donor and receiver stream separated with a cellulose acetate membrane as in common passive dialysis. Two outer compartments (with fixed solution inside) were equipped with electrodes and separated from the flowing streams by ion-exchange membranes [100]. With proper selection of the current applied the electromigration of ions causes faster dialysis, and by selecting an appropriate flow rate ratio between the sample(donor) and receiver stream, preconcentration factors of about 5–20 have been achieved.

The very high selectivity of distillation and gas-diffusion separation can be regarded as advantage or drawback, depending on the intended purpose of application and the kind of sample matrix. For volatile analytes, gas diffusion is a very attractive means of matrix separation and analyte preconcentration in a single step. The preconcentration factors achievable under typical experimental conditions are in the range of 5–20. When gas diffusion is applied in a flow-through system, this can be achieved within 5–10 min. In direct hyphenation with IC, one sample can thus be pretreated by gas diffusion while the previous one is processed by IC; hence, no extra time is required [49].

8.6 IN-LINE PRESEPARATIONS USING TWO-DIMENSIONAL ION CHROMATOGRAPHY (2D-IC)

Preseparations by two-dimensional ion chromatography are based on linking two individual IC systems in a way that a section of the effluent of the first IC system after passing the detector is transferred into the second IC system. The overall system consists of two high-pressure pumps, two separation columns, and two detectors so that the complexity and costs of 2D-IC are high. Commercial availability is given with appropriate software. Generally, the first dimension column is of high capacity where preseparation takes place. By proper timing, it is possible to cut off any section of the first dimension chromatogram and hence to reinject a plug containing only a certain fraction of the ions initially present in a sample. The sectional cut can be made by the use of a common injection valve of appropriate volume or, alternatively and more often applied, using a trap column retaining the target ion or a number of ions of the selected fraction. The trapped analytes are eluted by the mobile phase of the second dimension IC system. The trapping configuration has the advantage that due to the analyte compression effect, band broadening in the second separation is reduced. It is worth mentioning that 2D-IC works well only when the trap column offers reasonable selectivity for retention of the analyte in the presence of the mobile-phase ions or when a suppressor systems is used in the first dimension so that the ions of interest are present in a low ionic strength medium. In suppressed anion chromatography with common carbonate or hydroxide eluents, the resulting weakly dissociated carbonic acid or water, respectively, shows low or no concurrence with the retention of the target anionic analytes.

A typical application of this sophisticated, yet rather demanding and expensive technique is the determination of low levels of ionic compounds (sometimes only a single one) in the presence of a high excess of other ions (or a single ion) of the same charge that elute in close vicinity to the target ion. In such cases, common one-dimensional IC suffers often from insufficient resolution and difficulties to evaluate the small signals from low-level analyte concentrations on the tail of the previous peak or the partial overlap with the following. Prominent examples are the trace determination of bromate and perchlorate in water using conductivity detection rather than postcolumn derivatization techniques or IC-MS [101]. For both ions, 2D-IC methods have been developed and recently been accepted as official methods of analysis by USEPA [102, 103]. Another interesting application is the trace determination of total phosphorus after peroxodisulfate digestion [104]. In this case, the high excess of sulfate formed by hydrolysis of the digestion agent hinders evaluation of the phosphate signal using one-dimensional IC. By cutting out the phosphate elution region of the first separation and concentrating this section on an anion trap column phosphate can be determined interference-free in the second dimension. By clever column switching and diverting of the matrix and analyte fractions in this particular application, 2D-IC was performed with a single separation column.

2D-IC can be also applied to the determination of ions in strong acids or bases as an alternative to in-line ion-exchange or electrodialysis neutralization. Anion determination in hydrochloric and hydrofluoric acids has been demonstrated to become feasible by injection of the prediluted acids into the first dimension IC system and, after diverting the matrix anions (i.e., chloride and fluoride, respectively) to waste, to trap the later eluting anions (e.g., nitrate, bromide, sulfate, phosphate). The separation of these anions in the second dimension is no longer adversely affected by the excess acid (no baseline fluctuations and system peaks) or an overlap with the remaining small amount of matrix anions present in the cut section.

Application of 2D-IC for cation separations is less popular. One interesting approach has been proposed by Rey *et al.* [105], who have used a clever column-switching technique with two cation-exchange columns of different functionality in series.

The ability to couple separation columns with different selectivity as well as different separation mechanisms opens up unique possibilities for improved resolution and simultaneous determination of a larger number of analytes as is possible in one-dimensional IC. The coupling of ion exclusion (IE) in the first dimension with ion-exchange separation in the second dimension, for instance, has been utilized for trace anion determination in concentrated hydrofluoric acid [106]. Another example is the determination of weak organic acids by ion exclusion and determination of inorganic anions by reinjection of the void volume of IE separation (after preconcentration on an anion trap column placed in the injection loop of the second dimension IC system) onto an anion-exchange column [107]. A combination of nonsuppressed and suppressed conductivity detection has been used for detection of the organic and inorganic species, respectively.

Considering the enormous progress in comprehensive two-dimensional and even multidimensional liquid chromatography, the high potential to transfer some of the concepts to IC is obvious, and it is likely that this will find attraction whenever there are needs for improved separation of ionic compounds that cannot easily be solved in another way.

8.7 SAMPLE PREPARATION OF SOLID SAMPLES

The determination of ionic compounds or substances that can be converted into an ionic form in solid samples is an important issue, and application of IC has gained increasing attraction as a member player in the orchestra of available analytical methods. Solid sample analysis includes soil, sludge, sediment, dust, waste, geological material, various industrial products and materials, as well as biological samples, and all types of solid and semisolid foodstuff. Even oils, fats, emulsions, and similar samples can be included under the term semisolids, as similar considerations are to be applied to the necessary sample preparation steps prior to IC determination.

Solid samples are obviously not directly amenable to IC analysis. Depending on the kind of solid material and the intention of analysis, sample preparation includes a series of operations. After sampling, sample homogenization by blending and shredding, drying, pulverization, and other measures are required to obtain a solid subsample, which is eventually used for analysis. The analysis of solids by IC requires either the transfer of the whole sample (complete dissolution) or at least the release of the ionic species of interest into an aqueous phase. This can be carried out in very different manner, depending on the solubility of the substance to be analyzed and the ionic compounds to be determined.

Important criteria and considerations for the selection of the possible sample preparation techniques are as follows:

- The completeness of the ion transfer into the aqueous phase. In some cases, only the immediately soluble fraction (operationally defined by the extraction procedure) is obtained.
- The possible selective extraction of only the ionic compounds (salt fraction). This can be of value since matrix components remaining undissolved do not interfere in the subsequent IC determination.
- The compatibility of the solvent or extraction media with the subsequent IC analysis. This applies to both the pH of the extractant solution and the presence of substances that could affect the separation by overloading the column, the production of additional signals (baseline perturbations or system peaks), or by causing poisoning of the separation column.
- The apparatus and time required to carry out the procedure. Possible automation of the entire procedure is of particular interest in this respect.
- The danger of losses by adsorption, volatilization, or the uncontrolled conversion of the ions to be determined into a different chemical form.
- The risk of contamination by auxiliary reagents used and/or vessel contact.

Generally, conventional sample preparation methods for solid samples are as follows [108]:

1. Fusion methods:
   - Alkaline with $NaOH$, KOH, Na_2CO_3, K_2CO_3
   - Acidic with $KHSO_4$, $K_2S_2O_7$
   - Fluorination, chlorination, sulfurization.
2. Combustion methods:
   - Burning the sample in air/oxygen
   - Oxygen bomb, calorimeter bomb
   - Combustion in a stream of oxygen
   - Combustion in an oxyhydrogen flame.
3. Wet chemical digestions:
   - Open acid digestion with reflux

- Pressure digestion
- UV pyrolysis.

In principle, the sample preparation methods for solid substances can be classified according to treatment with a liquid/solvent and combustion procedures (dry ashing). The use of liquids can include complete dissolution of the entire sample, extraction or leaching of the soluble part, and wet ashing using (strong) acids or alkaline fusion. It can be assumed that a mild treatment of the samples (i.e., dissolution and extraction) will provide information about the content of ions present in the sample as salts. Conversion of particularly easily oxidizable substances (e.g., nitrite, sulfite, sulfide, cyanide) can, however, occur even in the course of a gentle extraction and must be taken into consideration when interpreting the results. The situation following drastic treatment with strong acids, a fusion process, or thermal decomposition of the solid sample is completely different. In these cases, quantitative determination of the ions by IC only provides information about the total content of all compounds that have been converted into the particular ionic species under the given digestion or combustion conditions. This means, for example, that halogenides, nitrate, phosphate, or sulfate concentrations in the digestion solution can provide no unambiguous information about the content of the respective contents in the solid sample, as other (particularly organic) halogen, nitrogen, phosphorus, and sulfur compounds may be present in ionic form after wet ashing or fusion in the resulting liquid or after combustion and absorption of the released gases in an appropriate liquid.

In the following, a variety of sample preparation techniques used for solid sample treatment are described and the respective features of the procedures in combination with IC as the final determination method outlined.

8.7.1 Dissolution and Aqueous or Acid Extraction

Complete or partial dissolution is the simplest method of sample preparation for the determination of ionic constituents in solid samples by IC. No special apparatus is required, and virtually all common plastic or glass containers can be used as extraction vessels. Sodium contamination can be a problem when glassware is used. Dissolution of the sample or extraction of the ions to be determined is normally carried out at room temperature, but can be accelerated by gentle warming or by heating the solvent to the boiling point. Care must be taken that thermally unstable ions are not destroyed or lost by evaporation. In practice, vigorous shaking, stirring, and ultrasonic treatment are used to provide intense contact between the solid and liquid phases.

Owing to its high polarity, water is a suitable extraction solvent for ionic compounds in solid samples and should be used whenever possible. In the extraction of soil samples and evaluation of hazardous waste sites, for example, according to the German Standard Method (DIN 38 414 part 4), the dry residue obtained is treated with pure water. After extraction for 24 h, the solution is filtered and injected into IC (if necessary via a reversed-phase cartridge) for anion determination.

Using water as extractant, no problems occur with respect to the compatibility with any IC separation method (in contrast to acids or bases that might interfere in

various manners), and with regard to its purity, water also meets the highest requirements encountered in trace analysis problems. As an alternative to pure water, it is
possible to use the mobile phase of the respective chromatographic separation system
as the extractant. However, with alkaline eluent solutions typically applied in anion
chromatography, the risk of precipitation of sample constituents is given. Diluted
nitric acid (a common eluent in cation-exchange chromatography) is well suitable
as a solvent for cationic species in many solid samples. Salt solutions should only be
used as extractants for special applications. The introduction of the respective cations
and anions of the extractant obviously precludes the determination of these ions, and
in addition, the elevated ionic strength of the solutions may overload the separation
columns and/or prevent the evaluation of other compounds eluting in neighborhood
of ions introduced through the extractant. Lithium and calcium chloride as well as
sodium formate and sodium bicarbonate solutions are sometimes applied as extractants for soil analysis (in particular for phosphate determination). In the determination
of plant-available phosphorus and potassium, the use of calcium acetate/lactate solution as the extract is also an established procedure.

The determination of less-soluble compounds such as sulfates, phosphates, and
carbonates of the alkaline earth metals or condensed phosphates may not always be
quantitative in purely aqueous extractants. In such cases, it may be necessary to use
dilute acids for extraction and to make compromises with regard to the separation
performance of the column and possible interference to the detection of the individual ions. The use of organic acids as extractant solutions (mainly formic, acetic,
and tartaric acids are employed) has the advantage that in anion chromatography,
the common inorganic anions to be determined are generally well resolved from the
respective organic acid anion peaks. The addition of polar organic solvents (methanol,
acetonitrile, acetone) sometimes improves the extractability of ions, particularly for
all kinds of samples of vegetable or animal origin. This may, however, occasionally
create interferences in the chromatographic separation in the form of system peaks.

Acid extraction is also suitable for cation analysis, in particular, for the determination of alkaline and alkaline earth elements. For example, in the determination of
potassium in propylene glycol, the sample is stirred in $1\,\mathrm{mmol\,l^{-1}}$ HNO_3 in a plastic
vessel for 1 h. After passing through a reversed-phase cartridge (for removal of the
organic matrix), the solution is then injected into the IC system.

8.7.2 Wet-Chemical Acid Digestions

If an aqueous extraction is not capable of releasing the ionic components to be determined or if the total amount of a particular element is of interest, wet-chemical digestion procedures are often applied. These are based on the treatment of the solid sample
with either concentrated acids or a mixture of acids or with alkaline compounds in a
fusion process. Although open-furnace heating is possible, acid digestion is nowadays
usually carried out in closed-vessel systems at elevated temperatures (the so-called
pressure digestion). Microwave digestion is the most suitable way to do so since it is
fast, prevents loss of analytes due to evaporation, and the polymeric vessel materials

(due to the extensive experience in trace metal analysis) are very robust and pure (little risk of contamination).

A serious problem with sample treatment by strong acids is the limited compatibility of the solutions obtained with subsequent IC analysis. The high acidity of the digestion solution prevents direct injection as this could damage the stationary phases. Very acidic samples also influence the ion-exchange equilibria on the separation column and can cause significant retention time displacements, additional so-called system peaks, or baseline fluctuations. In addition, the presence of the added acid anion(s) means that these ions obviously cannot be determined. Moreover, quantification of other anions is also often affected. Ion determination is possible in some cases after considerable dilution. Another route to prevent the adverse effects of high sample acidity after digestion in anion determination is neutralization of the sample. This can be done by passing the sample through a cation-exchange cartridge in the fully protonated form or – very elegantly – by the use of membrane-based cation-exchange units with combined electrodialysis.

Problems encountered with the high acidity are usually less critical in the determination of cations but the high hydroxonium ion content can also have serious effects on baseline stability adversely affecting in particular the early elution alkali ions and ammonium. In principle, further treatment of the acidic digestion solution can solve the problems as described earlier, but this requires additional working steps and analysis time.

8.7.3   UV Photolytic Digestion

Sample treatment with the aid of ultraviolet radiation in the presence of hydrogen peroxide is a very mild digestion method suitable only for the destruction of dissolved organic matter as well as condensed inorganic compounds (e.g., polyphosphates). The digestion of the dissolved organic matrix itself occurs by OH radicals generated by a high-pressure mercury lamp as a wideband UV irradiator. Since the penetration depth of the UV radiation in water is low, the sample solution must be provided in the form of a thin film. Special apparatus are commercially available permitting several samples to be digested at the same time. The digestion is done typically under atmospheric pressure so that evaporation losses occur of the aqueous samples requiring filling up the sample vessels after digestion to a defined volume.

Automation of UV–photolytic digestions is also possible in flow-through systems. This is widely employed in continuous-flow and flow injection analysis using quartz tubing for sample transmission coiled around an UV lamp. Typical applications are the determination of total amounts of nitrogen, phosphorus, sulfur, cyanide, and dissolved organic carbon.

Since only hydrogen peroxide is added to the sample, which is destroyed almost completely during the digestion process, no interferences from reagents occur (in contrast to wet acid digestion procedures). Also, contamination problems are minimal. However, UV digestions as all other digestion, fusion, and combustion methods not only remove interfering compounds but might also convert compounds into different chemical forms. This is for instance true for some ions that are unstable even under the mild conditions of UV photolysis. Nitrite, cyanide, and sulfite as well as oxoanions of chlorine (ClO^- and ClO_2^-), for instance, are oxidized and carboxylic acids are lost completely. The latter is of advantage, for example, in the determination of fluoride in the presence of formic and/or acetic acid, but must be kept in mind when interpreting results.

Despite several attractive features, UV digestion is surprisingly seldom used as sample preparation method for IC determinations. In-line digestions are feasible and can thus lead to fully automated analysis of samples with elevated concentration of dissolved organic matter.

8.7.4 Fusion Methods

An alternative to the wet digestion of solid samples with strong acids is fusion under alkaline conditions. The solid sample is mixed with a suitable fluxing agent (generally NaOH, KOH, Na_2CO_3, K_2CO_3) and heated in a crucible until it melts. After cooling down, the fused mass is treated with a suitable solvent (generally water) and the dissolved ionic species are determined in this solution. Problems also exist here with the limited compatibility of the fusion solution with IC analysis (particularly the high pH of the resulting solution). However, the fluxing agents frequently used in the fusion process such as sodium hydroxide and sodium carbonate are also components of the mobile phase typically used in anion chromatography. After appropriate dilution, it is often possible to carry out anion determination with little difficulties. Cation determination is strictly limited owing to the high alkali metal ion content (from the fluxing agent). Fusion is mainly used for the analysis of geological materials, soil, sediments, as well as glasses and ceramics [109]. The procedure is usually laborious, takes a long time, and involves a high risk of errors due to contamination and possible losses of analytes.

8.7.5 Dry Ashing and Combustion Methods

Other possible ways of preparing solid samples for IC analysis are dry ashing using air as the oxidizing agent or combustion in an atmosphere of pure oxygen. Combustion can take place in closed systems (e.g., Schöniger flask, Berthelot oxygen bomb, cold-plasma ashing) or in a stream of oxygen as is commonly used for determination of the sum of organic halogen compounds (adsorbable organic halogens – AOX, extractable organic halogens – EOX, etc.) [110–113].

Dry ashing and several of the combustion methods are preferentially used for biological samples, pharmaceutical preparations, polymeric substances, coal and other (also liquid) fuels, as well as foodstuffs and other samples with high organic matrix content. In the ashing procedure, the dry and homogenized solid sample is mineralized in a crucible or combustion boat by heating it in a muffle furnace for several hours at a temperature of typically 300–800 °C. The ashes are then dissolved in water or, if necessary, in a dilute mineral acid. After microfiltration, the solution is ready for IC analysis.

In dry ashing, the elements nitrogen, sulfur, and also any halogen compounds that are present in organic form are partially or completely volatilized as gases and cannot be determined in this way. Ashing aids can accelerate the mineralization process and other additives are sometimes necessary in order to minimize the volatilization of particular components. The determination of iodine in biological materials can, for instance, be carried out free of losses in the presence of sodium carbonate. Dry ashing is particularly useful for the determination of cations, in particular those of transition metals, as sample preparation is easy to carry out and due to the absence of acids and other additives in the sample treatment step no interferences occur in IC analysis caused by the sample treatment step. The Schöniger digestion method is widely used in practice. Here, the sample is burned in a closed glass or quartz flask in pure oxygen and the released combustion gases (e.g., SO_2, NO_x, HX) are taken up in an absorption solution contained in the flask. The advantages of this method are that it is simple, rapid, and favorably priced. A disadvantage is that only relatively small amounts of the sample can be burned (approx. 0.1 g), which strictly limits the sensitivity of the method. As an alternative to the Schöniger method, combustion can also be carried out at high pressure in an oxygen bomb. Oxygen pressures of up to 40 bar allow sample weights of 1 g and more [113].

In other combustion analysis methods (e.g., the Wickbold method), a continuous gas stream is passed over the sample contained in a boat or crucible that is heated to a temperature between 300 and 1000 °C to induce thermal decomposition [111]. The gas stream is then passed through an absorption solution to collect the decomposition products. A very important and frequently applied application of IC for the analysis of the absorbing solution is the determination of individual halogens (speciation) in the AOX and EOX methods. In contrast to the AOX/EOX methods where only the sum of organic halogens (excluding fluoride) is measured by amperometric titration, IC as the determination method provides halogen speciation and in addition information about the adsorbable or extractable organic fluorine [110, 112].

Determination of total sulfur and nitrogen through conversion to SO_2 or NO_x, respectively, and absorption of the released gases in appropriate solutions is also possible with IC for final determination of the ionic products (i.e., sulfate and nitrate).

In recent years, automation of the combustion step and ion chromatography determination of released gases in the absorber solution has been accomplished and is termed combustion ion chromatography (CIC) [114]. The fully automated sample processing capabilities from sample introduction into the furnace to data evaluation by ion chromatography software has found numerous real-world applications. Besides solid materials, liquids and gases can be analyzed. Sound information about this ingenious technique can be found in the literature and in web pages of manufacturers of respective apparatus.

With regard to the choice of absorption solution used in the methods mentioned earlier, both ready solubility of the released gases and compatibility with the IC analysis are important. Water, weak buffer solutions, and the mobile phase to be used in the subsequent chromatographic separation are basically suitable. As in some cases, elements are not converted into a uniform oxidation stage during combustion (e.g., simultaneous formation of NO and NO_2 from nitrogen compounds, SO_2 and SO_3 from sulfur compounds, Br_2 and $HBrO_3$ from bromine compounds), it is reasonable or necessary to use an absorber different than pure water so that conversion (by oxidation or reduction) to a single ionic form is achieved. This is why hydrogen peroxide is often added to the aqueous absorption solution serving as oxidizing agent for sulfite or hydrazine sulfate for reduction of bromic acid to bromide.

8.8   AIR ANALYSIS USING ION CHROMATOGRAPHY – APPLICATION TO GASES AND PARTICULATE MATTER

In general, the analysis of gases is not without problems, because air and exhaust gas samples are usually heterogeneous multiphase systems containing a large variety of different compounds, some of them at very low concentration levels. Apart from the gas phase, aerosols may also be present in the form of solid nano- and microparticles (airborne particulate matter) and liquid droplets (fog, rain, snow, etc.). A selective collection of the individual phases without artifact formation and determination of the ionic contents in the respective fraction is difficult to achieve and requires dedicated sampling strategies.

In many cases, constituents of particulate matter are determined after sedimentation or by collecting the particulates by filtration or deposition on an impactor plate, extracting the ions from the solid with water or weak acids, and then determining the ionic components by IC. The sample preparation procedures depend on the intended purpose of analysis and are basically the same as those discussed previously in the section about solid sample analysis. Analytes frequently determined in particulate matter using IC are chloride, nitrate, sulfate, carbonic acid anions, but also alkaline and alkaline earth metal ions and ammonium. For these analytes, aqueous extraction (e.g., in an ultrasonic bath) is a suitable means of sample preparation [115]. For trace metal analysis, acid extraction or microwave digestion is commonly employed prior to IC. A problem associated is the high acidity of the digests, which requires further treatment as discussed earlier.

Continuous sampling and analysis of airborne particulate matter are also possible but require rather complicated apparatus. One way is to collect the particles on a filter, which is sequentially extracted by a small volume of water while the filter remains in place of the dedicated sampler configuration. In another approach, particle-into-liquid conversion is achieved by introducing the aerosol into a chamber saturated with water steam and collecting the condensed water at the outlet. This solution is then transferred to the IC system [116]. Simultaneous determination of water-soluble ionic anions and cations has been accomplished with two IC systems running in parallel [117].

The liquid aerosol fraction in air pollution analysis is typically collected as wet deposit of rain, fog, snow, or hare. In these samples, ion determination can be carried out as common for other aqueous samples with low needs for sample preparation (only filtration is always recommended and analyte preconcentration required for trace analysis).

The determination of substances present in gaseous form by IC is almost always preceded by preconcentration through absorption in a suitable liquid or by adsorption of the components on a solid sorbent material or on a reagent-impregnated filter. Adsorption on a solid phase is followed, as in the analysis of dust collected on a filter, by elution with an extraction solvent, which is then analyzed by IC. In order to avoid the simultaneous collection of the dispersed aerosols, these are normally separated by filtration. From the point of view of the apparatus required, absorption in washing bottles or impingers as well as sorption on solid sorbents is relatively simply done using an air pump and an appropriate gas metering device. However, the volume flow and the dimensions of the absorption bottles or adsorber tubes must be adapted to suit the particular application, that is, the gaseous compound under investigation and the expected gas-phase concentration. During sampling in washing bottles, the gaseous substances are often converted into ionic components, which can subsequently be determined by IC. The choice of the absorption solution used is of great importance as its composition influences the selectivity of the conversion. For example, ammonia is always collected as ammonium in acid solution, nitrogen dioxide is almost always collected as nitrite, sulfur dioxide is collected in water as sulfite, but it can be converted into sulfate when hydrogen peroxide is added. The composition of the absorption solution also plays an important role in the completeness of the gas separation and conversion. Only in very few cases can the separation of several gaseous substances take place simultaneously, so that IC is often applied to the determination of a single ion only. Nonetheless, this can be of advantage compared to alternative detection methods such as spectrophotometry or potentiometry with ion-selective electrodes due to the higher selectivity and sensitivity achieved by IC. The same considerations apply in principle to adsorptive sampling on solid sorbents, except that in this case the additional elution step introduces further problems with respect to selectivity and risk of contamination.

Other ways of gas sampling that differ by principle and permit gas–particle separation without particle filtration are based on a diffusion-controlled separation. These techniques utilize the fact that gases have a diffusion coefficient much greater than that of particulate matter. In the so-called passive sampling, which has found wide-range applications in workplace monitoring as well as atmospheric measurements, collection of the gaseous analyte takes place on an adsorber layer in free contact with the atmosphere solely based on diffusion (no pump used) [118, 119]. The driving force behind analyte collection is the concentration gradient between the atmosphere and the adsorber surface, which is kept during sampling due to the removal of analyte in the immediate vicinity of the sorptive surface. Ideally, the surface should behave as sink with zero concentration in the immediate vicinity. Such devices are available in the form of tubes or badges and provide integrative measurements of the pollutant content in the atmosphere with only minimal apparatus. A vast amount of literature on passive sampling is available and commercial products are offered by several companies.

The final step in the analytical procedure is the determination of the collected analyte, which is often transformed into an ionic species. This holds true, for instance, in the determination of SO_2, NO_2, NH_3, volatile amines, organic acid vapors where the

Back-up filter

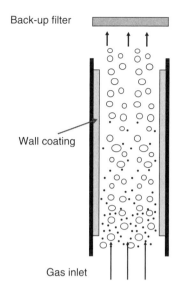

Wall coating

Gas inlet

Figure 8.6 Scheme of a static diffusion denuder for artifact-free separation of gases and particulate matter. The tiny black dots and the open circles symbolize gas molecules and aerosol particles, respectively.

final products are sulfite or sulfate, nitrite or nitrate, ammonium, or the corresponding salts of the amines and organic acids, respectively. Passive sampling of ozone can be done by reaction of ozone with nitrite immobilized on a suitable carrier surface and subsequent determination of the nitrate formed in a well-defined stoichiometric reaction. Ion chromatography is readily applicable to the determination of the respective product ions, and several applications can be found in the literature [120].

Denuders are an alternative (yet active sampling) approach for diffusion-controlled gas sampling, which have attracted considerable attention particularly in the collection and determination of reactive atmospheric gases [121, 122]. They can be configured in many different ways. In the simplest case, diffusion denuders consist of a tube whose inner walls are coated with a suitable adsorber material. A scheme of a static diffusion denuder for artifact-free separation of gases and particulate matter is shown in Figure 8.6.

As the gas being measured is propelled through the tube, the gaseous air pollutants migrate, diffusion-controlled, to the wall where they are collected. The driving force for transport to the absorber surface is, as in passive sampling, the concentration gradient between the flowing gas and the adsorber surface. Particles contained in the air pass unaltered through the tube and can be collected separately on a back-up filter placed at the outlet of the diffusion denuder. Alternative geometric configurations of diffusion denuders are proposed mainly aiming at higher sampling rates and thus better sensitivity or improved time resolution in the determination of the respective gases.

Application of denuder tubes with IC as the method of determination of the collected gaseous compounds is the determination of SO_2, NO_2, HNO_2, HCl, NH_3, among others. Denuder sampling has also been introduced into official methods of analysis [123]. All the variants for gas sampling described so far are batch methods involving many manual working steps and provide only a relatively poor temporal resolution. Sampling times are typically in the order of minutes to hours for washing bottles and absorber tubes and many days to several weeks for passive sampling and denuder tubes.

Continuous operation of diffusion controlled gas sampling is possible with the so-called wet effluent scrubbers and membrane-based devices [124, 125]. In the former, the gas sample is in contact to a liquid absorber solution, which is propelled through the wet scrubber at a low flow rate. Tubular and parallel-plate wet scrubbers have been designed. With the proper design of the wet scrubbers and adherence to laminar flow for the gas phase as in static denuders, only gaseous compounds can reach the absorber solution, whereas particulate matter is passing through and can be collected downstream.

Irrespective of the geometrical configuration (i.e., tubular or parallel-plate design), the liquid leaving the wet scrubber can be collected in vials, delivered to a flow-through detector or feeding the injection loop of a liquid chromatography system. The potential of wet scrubbers for continuous determination of reactive gases at trace and ultratrace levels has been impressively demonstrated by Dasgupta's group at the University of Texas, Arlington [125]. For a number of gaseous compounds, fully automated wet scrubber systems have been developed by his group but also reported by others using IC combined with suppressed conductivity detection for separation and quantification [126].

Two highly innovative commercially available gas analyzers are based on continuous wet effluent scrubber sampling combined with ion chromatography [127, 128]. One of them is MARGA *Monitor for Aerosols and Gases* offered by Metrohm company (Figure 8.7).

Both systems provide, in addition to gas collection, the sampling of particulate matter by a steam jet system (see above) and sequential measurement of the water extractable ion fraction. Using two IC systems in parallel anions and cations can be determined in both gas and particle phases. Specifications of the two instruments are similar and include determination of relevant species at the low microgram per cubic meter level, a time resolution of about 1 h, and unattended operation for several days. A different configuration of continuously operating diffusion denuders is based on membrane separation where the gas is flowing past a gas-permeable membrane that has an absorption solution on its other side. The hydrophobic microporous membranes mostly applied constitute a barrier for the liquid yet provide free diffusion for the gaseous analytes. The absorber liquid can be transferred, either continuously or after the gaseous analyte has been preconcentrated for a given period, to the injection valve of the IC instrument. A particular advantage of membrane-based diffusion scrubbers compared to the membrane-free variant can be seen in the additional selectivity toward particle collection and the ease to control liquid flow. The applications of membrane-based denuders are basically the same as for wet-effluent devices.

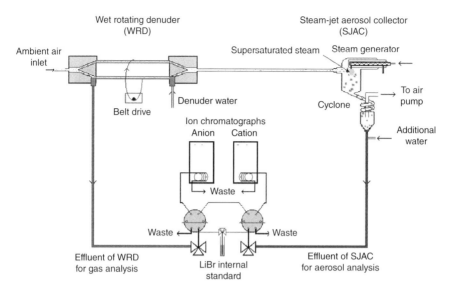

Figure 8.7 Fully automated system for simultaneous measurement of gaseous ionogenic compounds and particle-bound ions (MARGA). Analytical cycle permits separate gas and particle phase measurements with a time resolution of less than 1 h (redrawn with courtesy of Metrohm AG).

8.9   POSTCOLUMN ELUENT TREATMENT PRIOR TO MS DETECTION

In column liquid chromatography, the eluent leaving the separation column is commonly directed toward a suitable flow-through detector for registration and quantification of the analytes. In the case of IC, the most popular detection principle still is conductivity measurement. Distinction is made between direct (nonsuppressed) and suppressed conductivity detection. The introduction of the latter is part of the ingenious work by Stevens, Bauman and Small [129], which is regarded to be the invention of what is today meant by IC (i.e., HPLC for ionic species). Today, several other detectors are available and are applied as an alternative or in series or parallel with conductivity detection for IC aiming at improved selectivity or sensitivity. The composition of the mobile phase is connected with the choice of the stationary phase (e.g., completely different eluents are applied in anion and cation chromatography) but for anion and cation separations, the selection also depends on the respective separation column applied and the analytes to be separated. Finally, compatibility with the detector is an important consideration with respect to the choice of the mobile phase.

   The more recent introduction of IC-MS is certainly a milestone in IC detection [130, 131]. From the perspective of IC, the MS detector offers extremely high selectivity (coeluting analytes can be resolved by MS), the unique feature of unequivocal confirmation of analytes by their mass-to-charge ratios, and even possibilities of identification of unknown compounds. Also, sensitivity is generally excellent and outperforms that of many other detection methods. From MS point of view, IC can be seen as a means of sample preparation since the analyte mixture present in the original sample is separated so that under ideal conditions the individual ions are provided to the MS detector one after another. Possibly interfering sample matrices in MS measurements are already removed in the sample preparation procedures made prior to sample injection into the IC system.

The history of HPLC-MS hyphenation with its many conceptual and instrumental problems and innovations and the more or less successful application in practice explains to some extent why the introduction of IC-MS as a viable method has been delayed so much (early trials to combine ion exchange separation with MS detection date back to the 1960s). The important key to successful LC-MS and also IC-MS coupling was the almost simultaneous advent of two ingenious ionization techniques, that is, electrospray and chemical ionization under atmospheric pressure. Both interfaces allow for relatively high liquid loads when mobile phases with sufficiently high volatility are applied. Depending on the actually used components (IC system, MS system, and interface), sometimes the entire mobile phase can be admitted to the MS, or at higher IC flow rates, a flow split must be made in order to achieve total vaporization. Since the analytes in IC are ions or highly polar compounds, they are ideal candidates for ESI-MS detection; but chemical ionization under atmospheric pressure is also possible in some instances or even to be preferred.

Common IC eluents for anion and cation separations are not all well compatible with the MS detector (e.g., the high salt content may lead to clogging of the ion source and/or ion suppression; acidic eluents used in cation IC can destroy the ESI ion source). Despite these problems and limitations, moderately volatile eluents containing formic and acetic acid or ammonium salt have been successfully applied in several applications. By using IC with conductivity suppression, the mobile phase is transformed into either pure water (when hydroxide or mineral acid eluents are applied in anion- and cation-exchange chromatography, respectively) or into carbonic acid when the most common carbonate eluent for anion exchange chromatography is used. This is a comfortable situation for MS detection and is therefore to be preferred whenever applicable.

One additional problem in hyphenating IC with MS is the ion suppression effect of a large excess of one ion in the presence of traces of other nearby eluting ions. In such cases, for interference-free determination, either the excess ions have to be removed prior to sample injection into the IC system or the mobile phase leaving the IC system is diverted to waste while the fraction with excess ions is eluted. A larger number of applications of IC-MS are dedicated to trace analysis of ions that cannot or only with low sensitivity detected by conductivity measurements. This is, for instance, the case in the determination of disinfection by-products (chlorate, chlorite, bromate, haloacetic acids) but also for perchlorate and iodide [132, 133]. In several of these applications, either chloride and sulfate are removed prior to sample injection using SPE or the fractions containing these two ions after chromatographic separation are diverted to waste.

When using MS detection for IC applications, an organic solvent is sometimes required to assist desolvation and/or to improve ionization efficiency. To this end, the organic solvent is fed into the IC effluent by an auxiliary pump before entering the ion source [134]. Typical solvents used are methanol and acetonitrile. Improvement of detectability and also better stability can be achieved in the determination of small inorganic ions by the addition of ion association agents [135]. They can be provided in-line similarly to or even together with the organic modifiers.

Further progress and exploration of IC-MS hyphenation will certainly contribute to improve existing applications and to open up new possibilities particularly for ionic and highly polar organic compounds. The needs for sample preparation prior to IC separation, however, remain.

8.10   CONCLUDING REMARKS

Certainly, sample preparation does not belong to the favorite activities of analytical chemists; instead, it is often regarded as an unpleasant necessity. Sample preparation is usually the most laborious part of the analytical chain and constitutes (together with the sampling step) the most significant source of errors. But analytical chemists and in particular analysts in routine labs are well aware that sample preparation is often an indispensable and very demanding task when analyzing real samples. Any mistake made in sampling and sample preparation cannot be remedied by whatsoever sophisticated analytical methods are used for the final determination. Even routine processes (dilution, pH adjustment for sample preservation) as well as conditions during storage and handling (temperature, air contact) can cause alterations of the originally present analytes. The risk of contamination, analyte conversion, or losses during sample handling must always be kept in mind as this is the only way in which the desired quality of analytical results can be ensured.

With respect to ion determination using ion chromatography, sample preparation is necessary in many instances for different reasons. The ultimate aim is to receive eventually the analytes in dissolved form ready for injection into IC without distortion of the separation and without alterations of the stationary phase. Sample preparation methods can be distinguished according to the aggregate state of the sample, that is, liquid, solid, gaseous.

Automation (of both batch and in-line procedures) of sample preparation has gained increasing interest since it offers many advantages. The analysts in routine laboratories are freed from the many laborious and time-consuming activities connected with sample preparation. In addition, sample manipulation occurs under very repetitive conditions so that the overall reproducibility is generally increased and often also the accuracy is improved since systematic errors are less likely to occur. The awareness of automation advantages and increasing demands by users has more recently led to a number of developments, some of them being adopted by manufacturers of IC apparatus and companies providing dedicated solutions for sample preparation in general. Much of the sample preparation equipment available for other types of HPLC and even GC are similarly useful for IC analysis problems. In-line SPE, membrane-based sample preparation and CIC are convincing examples with respect to the many positive attributes they offer.

The variety of sample preparation methods available is huge, and it is not always easy to find a suitable method for the problem at hand. For one and the same type of sample, there are often several (also strategically) different possibilities to perform sample preparation so that the user has to make a decision based on different criteria. What often sounds so straightforward and simple requires in practical applications careful investigation and optimization of the many parameters affecting the performance of the respective sample preparation procedure.

An inherent problem in multiion analysis (one of the great advantages of IC over alternative ion analysis methods) is that a sample preparation procedure suitable for one or few analytes may be inappropriate for other analyte ions. Regardless of the choice of sample preparation method, each of the working steps involved must be optimized for each problem (with reference to the ions to be determined and the existing matrix).

With respect to trace ion analysis, analyte preconcentration is often the only way to achieve the required detection limits. SPE is the favorite technique presenting high enrichment factors. However, adverse effects on analyte retention due to matrix compounds must be considered and in many instances a preseparation of matrix constituents is required.

Considering the hyphenation of IC with MS, which at first glance appears an ideal combination of two methods with outstanding performance characteristics, there are sometimes needs to modify the analyte containing eluent prior to introduction into the ion source of the MS instrument. Ion suppression and alteration of the ion source may occur and this detracts from the use of some common IC eluents. Hence, in the initial sample preparation, this should be considered so that interferences can be avoided. Two-dimensional IC is an extremely sophisticated means to overcome some limitations in the separation power of common IC. When coupled with MS as the detector of the second dimension, high chromatographic resolution is combined with mass spectrometric resolution and the high sensitivity of the MS detector. Certainly, we can expect further developments in this respect as can already be seen in other 2D-LC-MS applications.

Future progress in IC might include the development of new stationary phases offering improved separation selectivity and detection modes different from those that are currently available. Moreover, an increasing use of IC with front-end hyphenation using advanced automated in-line sample preparation techniques will lead to the extension of the application to more complex samples.

8.11   REFERENCES

[1] Haddad, P.R. and Jackson, P.E. (1990) *Ion chromatography: principles and applications. Journal of Chromatography Library*, **46**, 1.

[2] Weiss, J. (2004) *Handbook of Ion Chromatography*, 3rd edn, Wiley-VCH.

[3] Fritz, J.S. and Gjerde, D.T. (2009) *Ion Chromatography*, 4th edn, Wiley-VCH.

[4] Seubert, A. (2001) On-line coupling of ion chromatography with ICP-AES and ICP-MS. *Trends in Analytical Chemistry*, **20**, 274–287.

[5] Coetzee, P.P., Fischer, J.L., and Hu, M. (2003) Simultaneous separation and determination of Tl(I) and Tl(III) by IC–ICP-OES and IC–ICP-MS. *Water SA*, **29**, 17.

[6] Ammann, A.A. (2002) Speciation of heavy metals in environmental water by ion chromatography coupled to ICP-MS. *Analytical and Bioanalytical Chemistry*, **372**, 448–452.

[7] Michalski, R., Jablonska, M., Szopa, S., and Łyko, A. (2011) Application of ion chromatography with ICP-MS or MS detection to the determination of selected halides and metal/metalloids species. *Critical Reviews in Analytical Chemistry*, **41**, 133.

[8] Verriele, M., Plaisance, H., Depelchin, L. *et al.* (2012) Determination of 14 amines in air samples using midget impingers sampling followed by analysis with ion chromatography in tandem with mass spectrometry. *Journal of Environmental Monitoring*, **14**, 402.

[9] Namiesnik, J. (2002) Trace analysis – challenges and problems. *Critical Reviews in Analytical Chemistry*, **32**, 271.

[10] Smith, R. (2003) Before the injection – modern methods of sample preparation for separation techniques. *Journal of Chromatography A*, **1000**, 3.

[11] Seubert, A., Frenzel, W., Schäfer, H. *et al.* (2004) *Sample Preparation Techniques for Ion Chromatography*, Metrohm AG, Herisau, Switzerland.

[12] Singh, R.P., Abbas, N.M., and Smesko, S.A. (1996) Suppressed ion chromatography analysis of anions in environmental waters containing high salt concentration. *Journal of Chromatography A*, **733**, 73–91.

[13] Gros, N. (2013) Ion chromatographic analysis of sea waters, brines and related samples. *Water*, **5**, 659–678.

[14] http://www.metrohm.com/en/products/ion-chromatography/ic-misp/

[15] http://www.dionex.com/en-us/products/sample-preparation/lp-71409.html

[16] Kirk, A.B., Kroll, M., Dyke, J.V. *et al.* (2012) Perchlorate, iodine supplements, iodized salt and breast milk iodine content. *Science of the Total Environment*, **420**, 73.

[17] Kirk, A.B., Dyke, J.V., Martin, C.F., and Dasgupta, P.K. (2007) Temporal patterns in perchlorate, thiocyanate and iodide excretion in human milk. *Environmental Health Perspectives*, **115**, 182.

[18] http://www.advantecmfs.com/filtration/ultrafiltration.shtml

[19] http://www.interchim.fr/cat/CentrifugationUltraFiltration.pdf

[20] Dyke, J.V., Kirk, A.B., Martinelango, P.K., and Dasgupta, P.K. (2006) Sample processing method for the determination of perchlorate in milk. *Analytica Chimica Acta*, **567**, 73–78.

[21] http://misp.metrohm.com/index.html

[22] Singhal, R.K., Preetha, J., Karpe, R. *et al.* (2006) The use of ultra filtration in trace metal speciation studies in sea water. *Environment International*, **32**, 224.

[23] Wennrich, R., Mattusch, J., Morgenstern, P. *et al.* (1997) Size and phase fractionation of water components by membrane filtration – distribution patterns for arsenic, iron and manganese in aqueous effluents of tin ore settling plant. *Fresenius Journal of Analytical Chemistry*, **359**, 161–166.

[24] Simpson, N.J.K. (2000) *Solid-Phase Extraction. Principles, Techniques and Applications*, Marcel Dekker, Inc., New York.

[25] Fritz, J.S. (1999) *Analytical Solid-Phase Extraction*, Wiley-VCH, New York.

[26] Thurman, E.M. and Millls, M.S. (1998) *Solid-Phase Extraction: Principles and Practice*, John Wiley & Sons, New York.

[27] Henderson, I.K., Saari-Nordhaus, R., and Anderson, J.M. (1991) Sample preparation for ion chromatography by solid-phase extraction. *Journal of Chromatography A*, **546**, 61–71.

[28] Slingsby, R. and Kiser, R. (2001) Sample treatment techniques and methodologies for ion chromatography. *Trends in Analytical Chemistry*, **20**, 288–295.

[29] Slingsby, R.W. and Pohl, C.A. (1996) Approaches to sample preparation for ion chromatography. Sulfate precipitation on barium-form ion exchangers. *Journal of Chromatography A*, **739**, 49.

[30] Zakaria, P., Bloomfield, C., Shellie, R.A. *et al.* (2011) Determination of bromate in sea water using multi-dimensional matrix-elimination ion chromatography. *Journal of Chromatography A*, **1218**, 9080.

[31] Bhagat, P.R., Pandey, A.K., Acharya, R. *et al.* (2008) Molecular iodine selective membrane for iodate determination in salt samples: chemical amplification and preconcentration. *Analytical and Bioanalytical Chemistry*, **391**, 1081.

[32] Kim, D.H., Lee, B.K., and Lee, D.S. (1999) Determination of trace anions in concentrated hydrogen peroxide by direct injection ion chromatography with conductivity detection after Pt-catalyzed on-line decomposition. *Bulletin of the Korean Chemical Society*, **20**, 696.

[33] Cataldi, T.R.I., Margiotta, G., Del Fiore, A., and Bufo, S.A. (2003) Ionic content in plant extracts determined by ion chromatography with conductivity detection. *Phytochemical Analysis*, **14**, 176.

[34] Jones, W.R. and Jandik, P. (1989) Elimination of matrix interferences in ion chromatographic analysis of difficult aqueous samples. *Journal of Chromatographic Science*, **27**, 449.

[35] Jackson, P.E. and Haddad, P.R. (1988) Studies on sample preconcentration in ion chromatography VII. Review of methodology and applications of anion preconcentration. *Journal of Chromatography A*, **439**, 37.

[36] Conway, E.J. (1939) *Micro-Diffusion Analysis and Volumetric Error*, Crosby Lockwood & Son, London.

[37] Cheng, J., Jandik, P., and Avdalovic, N. (2005) Pulsed amperometric detection of sulfide, cyanide, iodide, thiosulfate, bromide and thiocyanate with microfabricated disposable silver working electrodes in ion chromatography. *Analytica Chimica Acta*, **536**, 267.

[38] Bansleben, D., Schellenberg, I., and Wolff, A.C. (2008) Highly automated and fast determination of raffinose family oligosaccharides in Lupinus seeds using pressurized liquid extraction and high-performance anion-exchange chromatography with pulsed amperometric detection. *Journal of the Science of Food and Agriculture*, **88**, 1949.

[39] Ellington, J.J. and Evans, J.J. (2000) Determination of perchlorate at parts-per-billion levels in plants by ion chromatography. *Journal of Chromatography A*, **898**, 193.

[40] Siu, D.C. and Henshall, A. (1998) Ion chromatographic determination of nitrate and nitrite in meat products. *Journal of Chromatography A*, **804**, 157.

[41] Vlacil, F. and Vins, I. (1985) Determination of nitrate in milk by ion chromatography. *Die Nahrung*, **29**, 467–472.

[42] Koros, W.J., Ma, Y.H., and Shimidzu, T. (1996) Terminology for membranes and membrane processes (IUPAC Recommendations). *Pure and Applied Chemistry*, **68**, 1479.

[43] Strong, D.L. and Dasgupta, P.K. (1989) Electrodialytic membrane suppressors for ion chromatography. *Analytical Chemistry*, **61**, 939.

[44] Frenzel, W. (2011) Membrane-based sample preparation for ion chromatography – fundamentals, instrumental configurations and applications, in *Ion Chromatography 2011* (ed R. Michalski), SWSZ Publishing House, Katowice, Poland, pp. 7–23.

[45] Jönsson, J.A. (2009) *Membrane extraction in preconcentration, sampling and trace analysis*, Handbook of Membrane Separations, CRC Press, Boca Raton, p. 345.

[46] Mishra, D., Deepa, S., and Sharma, U. (1999) Carrier-mediated transport of some main group metal ions across various organic liquid membranes. *Separation Science and Technology*, **34**, 3113.

[47] Ullah, S.M.R., Adams, R.L., Srinivasan, K., and Dasgupta, P.K. (2004) Asymmetric membrane fiber-based carbon dioxide devices for ion chromatography. *Analytical Chemistry*, **76**, 7084.

[48] Frenzel, W. (1993) Separation and preconcentration of volatile compounds using gas-diffusion flow injection analysis. *Laboratory Robotics and Automation*, **5**, 245.

[49] Kiesow, T. (2007) In-line membrane-based sampling and sample preparation in hyphenation with ion chromatography. PhD thesis. Technical University of Berlin.

[50] Kolev, S. and van der Linden, W.E. (1992) Influence of the main parameters of a parallel-plate dialyser under laminar flow conditions. *Analytica Chimica Acta*, **257**, 317.

[51] van Staden, J.F. (1995) Membrane separation in flow injection systems: Part 1. Dialysis. *Fresenius Journal of Analytical Chemistry*, **352**, 271.

[52] Fang, Z. (1993) *Flow Injection Separation and Preconcentration*, VCH, Weinheim.

[53] Luque de Castro, M.D., Priego-Capote, F., and Sanchez-Avila, N. (2008) Is dialysis alive as a membrane-based separation technique? *Trends in Analytical Chemistry*, **27**, 315.

[54] van de Merbel, N.C. and Brinkman, U.A.T. (1993) Online dialysis as a sample preparation technique for column liquid chromatography. *Trends in Analytical Chemistry*, **12**, 249.

[55] Linget, C., Netter, C., Heems, D., and Verette, E. (1998) Online dialysis with HPLC for the automated preparation and analysis of amino acids, sugars and organic acids in grape juice and wines. *Analusis*, **26**, 35.

[56] Chiap, P., Hubert, P., and Crommen, J. (2002) Strategy of automated methods involving dialysis and trace enrichment on-line sample preparation for the determination of basic drugs in plasma by liquid chromatography. *Journal of Chromatography A*, **948**, 151.

[57] Kritsunankul, O., Pramote, B., and Jakmunee, J. (2009) Flow injection on-line dialysis coupled to HPLC for the determination of some organic acids in wine. *Talanta*, **79**, 1042.

[58] De Borba, B.M., Brewer, J.M., and Camarda, J. (2001) On-line dialysis as a sample preparation technique for ion chromatography. *Journal of Chromatography A*, **919**, 59.

[59] Steinbach, A., Wille, A., and Rick, A. (2008) Analysis of food samples with ion chromatography after in-line dialysis. *GIT Laboratory Journal*, **11**, 41.

[60] Frenzel, W. (1997) Online dialytic sampling and sample preparation in combination with flow analysis and chromatography. *GIT Labor-Fachzeitschrift*, **41**, 743.

[61] Frenzel, W. and Schäfer, H. (1998) Dialysis method for preparation and purification of analyte from sample solutions prior to chromatographic analysis. European Patent No. EP 820804.

[62] Mandenius, C.F., Danielsson, B., and Mattiasson, B. (1984) Evaluation of a dialysis probe for continuous sampling in fermentors and in complex media. *Analytica Chimica Acta*, **163**, 135.

[63] Miro, M. and Frenzel, W. (2004) Implantable flow-through capillary-type microdialysers for continuous in-situ monitoring of environmentally relevant parameters. *Analytical Chemistry*, **76**, 5974.

[64] Miro, M. and Frenzel, W. (2005) The potential of microdialysis as an automatic sample processing technique in environmental research. *Trends in Analytical Chemistry*, **24**, 324.

[65] Torto, N. (2009) A review of microdialysis sampling systems. *Chromatographia*, **70**, 1305.

[66] Davies, M.I. and Lunte, C.E. (1997) Microdialysis sampling coupled online to microseparation techniques. *Chemical Society Reviews*, **26**, 215.

[67] Torto, N., Hofte, A., van der Hoeven, R. *et al.* (1989) Microdialysis introduction high-performance anion-exchange chromatography/ionspray mass spectrometry for monitoring of on-line desalted carbohydrate hydrolysates. *Journal of Mass Spectrometry*, **33**, 334.

[68] Goodman, J.C., Valadka, A.B., Gopinath, S.P. *et al.* (1999) Simultaneous measurement of cortical potassium, calcium, and magnesium levels measured in head injured patients using microdialysis with ion chromatography. *Acta Neurochirurgica Supplementum*, **75**, 35.

[69] Waelchli, R.O., Jaworski, T., Ruddock, W.D., and Betteridge, K.J. (2000) Estimation of sodium and potassium concentrations in the uterine fluid of mares by microdialysis and ion chromatography. *Journal of Reproduction and Fertility Supplement*, **56**, 327.

[70] Nilsson, C., Nilsson, F., Turner, P. *et al.* (2006) Characterisation of two novel cyclodextrinases using on-line microdialysis sampling with high-performance anion exchange chromatography. *Analytical and Bioanalytical Chemistry*, **385**, 1421.

[71] Tanaka, Y. (2007) Donnan dialysis, ion exchange membranes – fundamentals and applications, Chapter 7. *Membrane Science and Technology*, **12**, 495.

[72] Cox, J.A., Gray, T., Yoon, K.S. *et al.* (1984) Selection of receiver electrolyte for the Donnan dialysis enrichment of cations. *Analyst*, **109**, 1603.

[73] Pyrzynska, K. (2006) Preconcentration and recovery of metal ions by Donnan dialysis. *Microchimica Acta*, **153**, 117.

[74] Cox, J.A. and Dabek-Zlotorzynska, E. (1987) Determination of anions in polyelectrolyte solutions by ion chromatography after Donnan dialysis sampling. *Analytical Chemistry*, **59**, 534.

[75] Jackson, P.E. and Jones, W.R. (1991) Membrane-based sample preparation device for the pretreatment of acidic samples prior to cation analysis by ion chromatography. *Journal of Chromatography A*, **586**, 283.

[76] Di Nunzio, J.E. and Jubara, M. (1983) Donnan dialysis preconcentration for ion chromatography. *Analytical Chemistry*, **55**, 1013.

[77] Liu, Y., Srinivasan, K., Pohl, C., and Avdalovic, N. (2004) Recent developments in electrolytic devices for ion chromatography. *Journal of Biochemical and Biophysical Methods*, **60**, 205.

[78] Martens, D.A. and Loeffelmann, K.L. (2004) Automatic generation of ultra-pure hydroxide eluent for carbohydrate analysis of environmental samples. *Journal of Chromatography A*, **1039**, 33.

[79] Haddad, P.R., Laksana, S., and Simons, R.G. (1993) Electrodialysis for clean-up of strongly alkaline samples in ion chromatography. *Journal of Chromatography A*, **640**, 135.

[80] Haddad, P.R. and Laksana, S. (1994) On-line analysis of alkaline samples with flow-through electrodialysis device coupled to an ion chromatograph. *Journal of Chromatography A*, **671**, 131.

[81] Okamoto, Y., Sakamoto, N., Yamamoto, M., and Kumamaru, T. (1991) Electrodialysis pretreatment system for ion chromatography of strongly acidic samples and its application to the determination of magnesium and calcium. *Journal of Chromatography A*, **539**, 221.

[82] Jönsson, J.A. (2009) *Membrane extraction in preconcentration, sampling and trace analysis*, in: Handbook of Membrane Separations (eds A.K. Pabby, S.S.H. Rizvi, and A.M. Sastre), CRC Press: Taylor & Francis, p. 345.

[83] Pawliszn, J. (2002) Membrane extraction. *Comprehensive Analytical Chemistry*, **37**, 479.

[84] Jakubowska, N., Polkowska, Z., Namiesnik, J., and Przyjazny, A. (2005) Analytical applications of membrane extraction for biomedical and environmental liquid sample preparation. *Critical Reviews in Analytical Chemistry*, **35**, 217.

[85] Barri, T. and Jönsson, J.A. (2008) Advances and developments in membrane extraction for gas chromatography: techniques and applications. *Journal of Chromatography A*, **1186**, 16.

[86] Araujo, C.S.T., Lira de Carvalho, J., Mota, D.R. *et al.* (2005) Determination of sulphite and acetic acid in foods by gas permeation flow injection analysis. *Food Chemistry*, **92**, 765.

[87] Goodwin, L.R., Francom, D., Urso, A., and Dieken, F.P. (1988) Determination of trace sulfides in turbid waters by gas dialysis/ion chromatography. *Analytical Chemistry*, **60**, 216.

[88] Liu, Y., Rocklin, R.D., Joyce, R.J., and Doyle, M.J. (1990) Photodissociation/gas diffusion/ion chromatography system for determination of total and labile cyanide in waters. *Analytical Chemistry*, **62**, 766.

[89] Gibbs, S.W., Wood, J.W., Fauzi, R., and Mantoura, C. (1995) Automation of flow injection gas diffusion/ion chromatography for the nanomolar determination of methylamines and ammonia in seawater and atmospheric samples. *Journal of Automatic Chemistry*, **17**, 205.

[90] Fäldt, S., Karlberg, B., and Frenzel, W. (2001) Hyphenation of gas-diffusion separation and ion chromatography. Part 1: Determination of free sulfite in wines. *Fresenius Journal of Analytical Chemistry*, **371**, 425.

[91] Kiesow, T. and Frenzel, W. (2004) *Membrane based sample preparation for chemical analysis in flow systems*, Handbuch Umweltwissenschaften, Alpha Informations-GmbH, Lampertheim, p. 223.

[92] Santoyo, E., Santoyo-Gutiérrez, S., and Verma, S.P. (2000) Trace analysis of heavy metals in groundwater samples by ion chromatography with post-column reaction and ultraviolet–visible detection. *Journal of Chromatography A*, **884**, 229.

[93] Kailu, L., Jin, L., and Siming, Y. (1993) Ion chromatographic analysis of anions based on solid-phase extraction. *Journal of Liquid Chromatography*, **16**, 3083–3092.

[94] Toofan, M., Pohl, C., Stillian, J.R., and Jackson, P.E. (1997) Factors affecting the ion chromatographic preconcentration behaviour of inorganic anions and organic acids. *Journal of Chromatography A*, **775**, 109–115.

[95] Kaykhaii, M., Dicinoski, G.W., and Haddad, P.R. (2010) Solid-phase microextraction for the determination of inorganic ions: applications and possibilities. *Analytical Letters*, **43**, 1546.

[96] Krynitsky, A.J., Niemann, R.A., Williams, A.D., and Hopper, M.L. (2006) Streamlined sample preparation procedure for determination of perchlorate anion in foods by ion chromatography–tandem mass spectrometry. *Analytica Chimica Acta*, **567**, 94.

[97] Nicolai, M., Rosin, C., Tousset, N., and Nicolai, Y. (1999) Trace metals analysis in estuarine and seawater by ICP-MS using on line preconcentration and matrix elimination with chelating resin. *Talanta*, **50**, 433.

[98] Verette, E., Qian, F., and Mangini, F. (1995) Online dialysis with high-performance liquid chromatography for the automated preparation and analysis of sugars and organic acids in food and beverages. *Journal of Chromatography A*, **705**, 195.

[99] Chiap, P., Hubert, P., and Crommen, J. (2001) Strategy for the development of automated methods involving dialysis and trace enrichment as on-line sample preparation for the determination of basic drugs in plasma by liquid chromatography. *Journal of Chromatography A*, **948**, 151.

[100] Debets, A.J.J., Kok, W.T., Hupe, K.P., and Brinkman, U.A.T. (1990) Electrodialytic sample treatment coupled on-line with high-performance liquid chromatography. *Chromatographia*, **30**, 361.

[101] Schminke, G. and Seubert, A. (2000) Comparison of ion chromatography methods based on conductivity detection, post-column reaction and on-line coupling ICP-MS for the determination of bromate. *Fresenius Journal of Analytical Chemistry*, **366**, 387.

[102] Wagner, H.P. (2007) Selective method for the analysis of perchlorate in drinking waters at nanogram per liter levels, using two-dimensional ion chromatography with suppressed conductivity detection. *Journal of Chromatography A*, **1155**, 15–21.

[103] US-EPA method 302.0 (2009) Determination of bromate in drinking water using two dimensional ion chromatography with suppressed conductivity detection.

[104] Colina, M. and Gardiner, P.H.E. (1999) Simultaneous determination of total nitrogen, phosphorus and sulphur by means of microwave digestion and ion chromatography. *Journal of Chromatography A*, **847**, 285.

[105] Rey, M.A., Riviello, J.M., and Pohl, C.A. (1997) Column switching for difficult cation separations. *Journal of Chromatography A*, **789**, 149.

[106] Vermeiren, K. (2005) Trace anion determination in concentrated hydrofluoric acid solutions by two-dimensional ion chromatography: I. Matrix elimination by ion-exclusion chromatography. *Journal of Chromatography A*, **1085**, 60.

[107] Glombitza, C., Pedersen, J., Roy, H., and Jorgensen, B.B. (2014) Direct analysis of volatile fatty acids in marine sediment porewater by two-dimensional ion chromatography–mass spectrometry. *Limnology and Oceanography: Methods*, **12**, 455.

[108] Michalski, R. (2010) Sample preparation for ion chromatography, in *Encyclopedia of Chromatography*, 3rd edn, vol. **III** (ed J. Cazes), Taylor & Francis, CRC Press, pp. 2106–2110, ISBN 978-1-4200-8459-7.

[109] Wang, Q.Y., Mkishima, A., and Nakamura, E. (2010) Determination of fluorine and chlorine by pyrohydrolysis and ion chromatography: comparison with alkaline fusion digestion and ion chromatography. *Geostandards and Geoanalytical Research*, **34**, 175.

[110] Oleksy-Frenzel, J., Wischnack, S., and Jekel, M. (2000) Application of ion-chromatography for the determination of the organic-group parameters AOCl, AOBr and AOI in water. *Fresenius Journal of Analytical Chemistry*, **366**, 89.

[111] Kwon, D.H., Kim, S.H., Lee, S.H., and Min, B. (2007) Determination of halogen elements in volatile organohalogen compounds by the Wickbold combustion pretreatment method and ion chromatography. *Bulletin of the Korean Chemical Society*, **28**, 59.

[112] Wagner, A., Raue, B., Brauch, H.-J. *et al.* (2013) Determination of adsorbable organic fluorine from aqueous environmental samples by adsorption to polystyrene–divinylbenzene based activated carbon and combustion ion chromatography. *Journal of Chromatography A*, **1295**, 82–89.

[113] Fung, Y.S. and Dao, K.L. (1995) Oxygen bomb combustion ion chromatography for elemental analysis of heteroatoms in fuel and wastes development. *Analytica Chimica Acta*, **315**, 347–355.

[114] ASTM D7359-14a. (2014) Standard test method for total fluorine, chlorine and sulfur in aromatic hydrocarbons and their mixtures by oxidative pyrohydrolytic combustion followed by ion chromatography detection (combustion ion chromatography—CIC).

[115] Brown, R.J.C. and Edwards, P.R. (2009) Measurement of anions in ambient particulate matter by ion chromatography: a novel sample preparation technique and development of a generic uncertainty budget. *Talanta*, **80**, 1020.

[116] Weber, R.J., Orsini, D., Daun, Y. *et al.* (2001) A particle-into-liquid collector for rapid measurement of aerosol bulk chemical composition. *Aerosol Science and Technology*, **35**, 718.

[117] Dabek-Zlotorzynska, E. and Dlouhy, J. (1993) Automatic simultaneous determination of anions and cations in atmospheric aerosols by ion chromatography. *Journal of Chromatography A*, **640**, 217.

[118] Berlin, A., Brown, R.H., and Saunders, K.J. (eds) (1987, CEC Pub. No. 10555EN) *Diffusive Sampling: An Alternative Approach to Workplace Air Monitoring*, European Communities, Brussels-Luxembourg.

[119] Krochmal, D. and Kalina, A. (1997) Measurements of nitrogen dioxide and sulphur dioxide concentrations in urban and rural areas of Poland using a passive sampling method. *Environmental Pollution*, **96**, 401.

[120] Salem, A.A., Soliman, A.A., and El-Haty, I.A. (2009) Determination of nitrogen dioxide, sulfur dioxide, ozone, and ammonia in ambient air using the passive sampling method associated with ion chromatographic and potentiometric analyses. *Air Quality, Atmosphere and Health*, **2**, 133.

[121] Kitto, A.N. and Colbeck, I. (1999) Filtration and denuder sampling techniques, in *Analytical Chemistry of Aerosols* (ed K.R. Spurny), Lewis, Boca Raton.

[122] Amati, B., di Palo, V., and Possanzini, M. (1999) Simultaneous determination of inorganic and organic acids in air by use of annular denuders and ion chromatography. *Chromatographia*, **50**, 150.

[123] VDI 3869 (2010). Measurement of ammonia in ambient air. Sampling with coated diffusion separators (denuders). Photometric or ion chromatographic analysis, Part. 3.

[124] Simon, P.K., Dasgupta, P.K., and Vecera, Z. (1991) Wet effluent denuder coupled liquid/ion chromatography systems. *Analytical Chemistry*, **63**, 1237–1242.

[125] Dasgupta, P.K. (1993) *Automated measurement of atmospheric trace gases: diffusion-based collection and analysis*, Advances in Chemistry Series, vol. **232**, Royal Society of Chemistry, p. 41.

[126] Dasgupta, P.K., Ni, L.Z., Poruthoor, S.K., and Hindes, D.C. (1997) A multiple parallel plate wetted screen diffusion denuder for high-flow air sampling applications. *Analytical Chemistry*, **69**, 5018.

[127] http://www.urgcorp.com/index.php/systems/realtime-aim-systems/anion-and-cation-particle-and-gas-system-9000d.

[128] http://www.metrohm.com/en-us/products-overview/process-analyzers/applikon-marga/ and AIM-9000 URG.

[129] Small, H., Stevens, T.S., and Bauman, W.C. (1975) Novel ion exchange chromatographic method using conductometric detection. *Analytical Chemistry*, **47**, 1801.

[130] Buchberger, W.W. (2001) Detection techniques in ion chromatography of inorganic ions. *Trends in Analytical Chemistry*, **20**, 296.

[131] Michalski, R. (2010) Detection in ion chromatography, in *Encyclopedia of Chromatography*, 3rd edn, vol. **I** (ed J. Cazes), Taylor & Francis, CRC Press, pp. 576–580.

[132] Seubert, A., Schminke, G., Nowak, M. *et al.* (2000) Comparison of on-line coupling of ion-chromatography with atmospheric pressure ionization mass spectrometry and with inductively coupled plasma mass spectrometry as tools for the ultra-trace analysis of bromate in surface water samples. *Journal of Chromatography A*, **884**, 191.

[133] Wang, Z., Lau, B.P.Y., Tague, B. *et al.* (2011) Determination of perchlorate in infant formula by isotope dilution ion chromatography/tandem mass spectrometry. *Food Additives and Contaminants, Part A*, **28**, 799.

[134] Barron, L. and Gilchrist, E. (2014) Ion chromatography–mass spectrometry: a review of recent technologies and applications in forensic and environmental explosive analysis. *Analytica Chimica Acta*, **806**, 27.

[135] Soukup-Hein, R.J., Remsburg, J.W., Dasgupta, P.K., and Armstrong, D.W. (2007) A general, positive ion mode ESI-MS approach for the analysis of singly charged inorganic and organic anions using a dicationic reagent. *Analytical Chemistry*, **79**, 7346.

INDEX

Application of IC-MS and IC-ICP-MS in Environmental Research, First Edition.
Edited by Rajmund Michalski.
© 2016 John Wiley & Sons, Inc. Published 2016 by John Wiley & Sons, Inc.